T0092481

Lecture Notes in Physics

The Lecture Notes in Physics

The series Lecture Notes in Physics (LNP), founded in 1969, reports new developments in physics research and teaching – quickly and informally, but with a high quality and the explicit aim to summarize and communicate current knowledge in an accessible way. Books published in this series are conceived as bridging material between advanced graduate textbooks and the forefront of research and to serve three purposes:

- to be a compact and modern up-to-date source of reference on a well-defined topic

- to serve as an accessible introduction to the field to postgraduate students and nonspecialist researchers from related areas

- to be a source of advanced teaching material for specialized seminars, courses and schools

Both monographs and multi-author volumes will be considered for publication. Edited volumes should, however, consist of a very limited number of contributions only. Proceedings will not be considered for LNP.

Volumes published in LNP are disseminated both in print and in electronic formats, the electronic archive being available at springerlink.com. The series content is indexed, abstracted and referenced by many abstracting and information services, bibliographic networks, subscription agencies, library networks, and consortia.

Proposals should be sent to a member of the Editorial Board, or directly to the managing editor at Springer:

Christian Caron
Springer Heidelberg
Physics Editorial Department I
Tiergartenstrasse 17
69121 Heidelberg / Germany
christian.caron@springer.com

A. Buchleitner
C. Viviescas
M. Tiersch (Eds.)

Entanglement and Decoherence

Foundations and Modern Trends

 Springer

Andreas Buchleitner
Univ. Freiburg
Inst. Physik
Hermann-Herder-Str. 3
79104 Freiburg
Germany
abu@uni-freiburg.de

Carlos Viviescas
Universidad Nacional Colombia
Depto. Física
Carrera 30 No. 45-03
Bogota D. C.
Colombia
clviviescas@unal.edu.co

Markus Tiersch
Univ. Freiburg
Inst. Physik
Hermann-Herder-Str. 3
79104 Freiburg
Germany
markus.tiersch@physik.uni-freiburg.de

Buchleitner, A. et al. (Eds.), *Entanglement and Decoherence: Foundations and Modern Trends*, Lect. Notes Phys. 768 (Springer, Berlin Heidelberg 2009), DOI 10.1007/978-3-540-88169-8

ISBN: 978-3-540-88168-1   e-ISBN: 978-3-540-88169-8

DOI 10.1007/978-3-540-88169-8

Lecture Notes in Physics ISSN: 0075-8450   e-ISSN: 1616-6361

Library of Congress Control Number: 2008936814

Cover design: Integra Software Services Pvt. Ltd.

Printed on acid-free paper

9 8 7 6 5 4 3 2 1

springer.com

Preface

Entanglement and (de-)coherence arguably define the central issues of concern in present day quantum information theory. In state-of-the-art experiments, ever larger numbers of quantum particles are entangled in a controlled way, and ever heavier particles are brought to interfere. Some sub-fields of quantum information science, in particular quantum cryptography, already find commercial applications, and communal communication networks that rely on quantum information technology are in preparation, as well as satellite-based quantum communication. Moreover, entanglement is no more considered as just an important resource for quantum information processing, but it allows for a better characterization of "complex" quantum systems, realized, e.g., in engineered, interacting many-particle systems, as well as in the solid state. Thus, there is a permanent and in many respects enhanced need for a deeper understanding of – and fresh approaches to – quantum entanglement, notably in high-dimensional quantum systems. Equally so, entanglement being a consequence of the quantum mechanical superposition principle for composite systems, we need a better understanding of the environment-induced destruction of coherent superposition states and of those interference phenomena that may survive the action of a noisy environment. Such research will allow us to identify realistic scales and possibly novel strategies for harvesting quantum interference phenomena.

The present book collects a series of advanced lectures on the theoretical foundations of this active research field and illustrates the breadth of present day theoretical efforts – from mathematics to mesoscopic transport theory. Uhlmann and Crell start out with a mathematical introduction to the geometry of state space, followed by an elementary introduction to entanglement theory by Mintert et al. Back again in the mathematical realm, Kauffman and Lomonaco discuss topological aspects of quantum computation, with some close relation to the theory of braids and knots. Ozorio de Almeida sheds new light on entanglement, in phase space, and touches some issues related to decoherence theory, which are then systematically expanded by Hornberger. Müller is subsequently concerned with dephasing and decoherence in the context of spintronics and disordered systems, thus establishing the bridge to real-life quantum transport, and the solid state.

All lecture notes start out from an elementary level and proceed along a steep learning curve, what makes the material equally suitable for student seminars on the more fundamental theoretical aspects of quantum information, as well as to supplement advanced lectures on this topic.

The material assembled here was first taught by the authors during an international summer school on "Quantum Information" at the Max Planck Institute for the Physics of Complex Systems in Dresden, in September 2005, thus inspiring the idea to compile the present book. The editors' special thanks therefore go to the authors, as well to Markus Grassl, Martin Rötteler, Christian Roos, Hartmut Häffner, Herbert Wagner, Per Delsing, Daniel Estève, Steffen Glaser, Gilles Nogues, Mauro d'Ariano, Robin Hudson, Reinhard Werner, Maciej Lewenstein, Andrzej Kossakowski, Karol Życzkowski, Mark Fannes, Richard Gill, Rainer Blatt, Marita Schneider, Christian Caron, Gabriele Hakuba, Andreas Erdmann, Helmut Deggelmann, Torsten Goerke, Heidi Naether, Andreas Schneider, Hubert Scherrer, Andreas Wagner, Karsten Batzke, and Jan-Michael Rost, who all have their share in getting the present volume into press.

Freiburg im Breisgau and Bogotá,   *Andreas Buchleitner*
August 2008   *Carlos Viviescas*
                                                                        Markus Tiersch

Contents

1 Geometry of State Spaces

A. Uhlmann and B. Crell

Institut für Theoretische Physik, Universität Leipzig, Augustusplatz 10/11,
D-04109 Leipzig, Germany

1.1 Introduction

In the Hilbert space description of quantum theory, one has two major inputs: First its linearity, expressing the superposition principle, and, second, the scalar product, allowing to compute transition probabilities. The scalar product defines an Euclidean geometry. In quantum physics, one may ask for the physical meaning of geometric constructs in this setting. As an important example, we consider the length of a curve in Hilbert space and the *velocity*, i.e., the length of the tangents, with which the vector runs through the Hilbert space.

The Hilbert spaces are generically complex in quantum physics: There is a multiplication with complex numbers; two linear-dependent vectors represent the same state. By restricting to unit vectors, one can diminish this arbitrariness to phase factors.

As a consequence, two curves of unit vectors represent the same curve of states, if they differ only in phase. They are physically equivalent. Thus, considering a given curve – for instance, a piece of a solution of a Schrödinger equation – one can ask for an equivalent curve of minimal length. This minimal length is the *Fubini–Study length*. The geometry, induced by the minimal length requirement in the set of vector states, is the *Fubini–Study metric*.

There is a simple condition from which one can read off whether all pieces of a curve in Hilbert space fulfill the minimal length condition, so that their Euclidean and their Fubini–Study length coincide piecewise: It is the parallel transport condition, defining the geometric (or Berry) phase of closed curves by the following mechanism. We replace the closed curve by changing its phases to a minimal length curve. Generically, the latter will not close. Its initial and its final point will differ by a phase factor, called the geometric phase (factor). We only touch these aspects in our essay and advise the reading of [1] instead. We discuss, as quite another application, the Mandelstam–Tamm inequalities.

The set of vector states associated to a Hilbert space can also be described as the set of its 1-dimensional subspaces or, equivalently, as the set of all rank one projection operators. Geometrically, it is the *projective space* given by the

Uhlmann, A., Crell, B.: *Geometry of State Spaces.* Lect. Notes Phys. **768**, 1–60 (2009)
DOI 10.1007/978-3-540-88169-8_1   © Springer-Verlag Berlin Heidelberg 2009

Hilbert space in question. In finite-dimension, it is a well studied manifold.[1] Again, we refer the reader to a more comprehensive monograph, say [2], to become acquainted with projective spaces in quantum theory. We just like to point to one aspect: Projective spaces are rigid. A map, transforming our projective space one-to-one onto itself and preserving its Fubini–Study distances, is a Wigner symmetry. On the Hilbert space level, this is a theorem of Wigner asserting that such a map can be induced by a unitary or by an anti-unitary transformation.

After these lengthy introduction to our first section, we have not much to comment on our third one. It is just devoted to the (partial) extension of what has been said above to general state spaces. It will be done mainly, but not only, by purification methods.

The central objects are the generalized transition probability (*fidelity*), the Bures distance, and its Riemann metric. These concepts can be defined, and show similar features, in all quantum state spaces. They are "universal" in quantum physics.

However, at the beginning of quantum theory, people were not sure whether density operators describe states of a quantum system or not. In our days, we think, the question is completely settled. There are genuine quantum states described by density operators. But not only that, the affirmative answer opened new insight into the basic structure of quantum theory. The second section is dedicated to these structural questions.

To a Hilbert spaces $\mathcal{H}$, one associates the algebra $\mathcal{B}(\mathcal{H})$ of all bounded operators that map $\mathcal{H}$ into itself. With a density operator ω and any operator $A \in \mathcal{B}(\mathcal{H})$, the number

$$\underline{\omega}(A) = \operatorname{tr} A\omega$$

is the expectation value of A, provided the *system is in the state given by* ω. Now, $\underline{\omega}$ is linear in A, it attains positive values or zero for positive operators, and it returns 1, if we compute the expectation value of the identity operator **1**. These properties are subsumed by saying $\underline{\omega}$ *is a normalized positive linear functional on the algebra* $\mathcal{B}(\mathcal{H})$. Exactly such functionals are also called *states of* $\mathcal{B}(\mathcal{H})$, asserting that every one of these objects can possibly be a state of our quantum system.

If the Hilbert space is of finite-dimension, then every state of $\mathcal{B}(\mathcal{H})$ can be characterized by a density operator. But in the infinite-dimensional cases, there are in addition states of quite different nature, the so-called singular ones. They can be ignored in theories with finitely many degrees of freedom, for instance, in Schrödinger theory. But in treating unbounded many degrees of freedom, we have to live with them.

One goes an essential step further in considering more general algebras than $\mathcal{B}(\mathcal{H})$. The idea is that a quantum system is defined, ignoring other demands, by its observables. States should be identified by their expectation

---

[1] It is certainly the most important algebraic variety.

values. However, not any set of observables should be considered as a quantum system. There should be an algebra, say $\mathcal{A}$, associated to a quantum system containing the relevant observables. Besides being an algebra (over the complex numbers), an Hermitian adjoint $A^\dagger$ should be defined for every $A \in \mathcal{A}$ and, finally, there should be "enough" states of $\mathcal{A}$.

As a matter of fact, these requirements are sufficient, if $\mathcal{A}$ is of finite-dimension as a linear space. Then, the algebra can be embedded as a so-called *-subalgebra in any algebra $\mathcal{B}(\mathcal{H})$ with dim $\mathcal{H}$ sufficiently large or infinite. Relying on Wedderburn's theorem, we describe all these algebras and their state spaces; they all can be gained by performing direct products and direct sums of some algebras $\mathcal{B}(\mathcal{H})$. Intrinsically, they are enumerated by their *type*, a finite set of positive numbers. We abbreviate this set by $\mathbf{d}$, to shorten the more precise notation $I_{\mathbf{d}}$, for the so-called type-one algebras.

If the algebras are not finite, things are much more involved. There are von Neumann (i.e., concrete W*-) algebras, C*-algebras, and more general classes of algebras. About them we say (almost) nothing but refer, for a physical motivated introduction, to [3].

Let us stress, however, a further point of view. In a bipartite system, which is the direct product of two other ones – say Alice's and Bob's, both systems are embedded in the larger one as subsystems. Their algebras become subalgebras of another, larger algebra.

There is a more general point of view: It is a good idea to imagine the quantum world as a hierarchy of quantum systems. An "algebra of observables" is attached to each one, together with its state space. The way an algebra is a subalgebra of another one describes how the former one should be understood as a quantum subsystem of a "larger" system.

Let us imagine such a hierarchical structure. A state in one of these systems determines a state in every of its subsystems: We just have to look at the state by using the operators of the subsystem in question only, i.e., we recognize what possibly can be observed by the subsystems observables.

Restricting a state of a quantum system (of an algebra) to a subsystem (to a subalgebra) is equivalent to the *partial trace* in direct products. It extends the notion to more general settings.

On the other hand, starting with a system, every operator remains an operator in all larger systems containing the original one as a subsystem. To a large amount, the physical meaning of a quantum system, its operators and its states, is determined by its relations to other quantum systems.

The last section is an appendix devoted to the geometric mean, perhaps the most important operator mean. It provides a method to handle two positive operators in general position. Only one subsection of the appendix is directly connected with the third section: How parallel lifts of Alice's states are seen by Bob.

1.2 Geometry of Pure States

1.2.1 Norm and Distance in Hilbert Space

Let us consider a Hilbert space[2] $\mathcal{H}$. Its elements are called *vectors*. If not explicitly stated otherwise, we consider complex Hilbert spaces, i.e., the multiplication of vectors with complex numbers is allowed. Vectors will be denoted either by their mathematical notation, say ψ, or by Dirac's, say $|\psi\rangle$.

For the scalar product, we write accordingly[3]

$$\langle \varphi, \psi \rangle \ \text{ or } \ \langle \varphi | \psi \rangle \,.$$

The *norm* or *vector norm* of $\psi \in \mathcal{H}$ reads

$$\| \psi \| := \sqrt{\langle \psi, \psi \rangle} = \text{ vector norm of } \psi \,.$$

It is the Euclidian length of the vector ψ. One defines

$$\| \psi - \psi' \| = \text{ distance between } \psi \text{ and } \psi',$$

which is an *Euclidean distance*: If in any basis, $|j\rangle$, $j = 1, 2, \ldots$, one gets

$$|\psi\rangle = \sum z_j |j\rangle, \quad z_j = x_j + i y_j \,,$$

and, with coefficients z'_j, the similar expansion for ψ', then

$$\| \psi - \psi' \| = \left(\sum [x_j - x'_j]^2 + [y_j - y'_j]^2 \right)^{1/2} ,$$

justifying the name *Euclidean distance*.

The scalar product defines the norm and the norm the Euclidean geometry of $\mathcal{H}$. In turn, one can obtain the scalar product from the vector norm:

$$4\langle \psi, \psi' \rangle = \| \psi + \psi' \|^2 - \| \psi - \psi' \|^2 - i \| \psi + i\psi' \|^2 + i \| \psi - i\psi' \|^2 \,.$$

The scalar product allows for the calculation of quantum probabilities. Now we see that, due to the complex structure of $\mathcal{H}$, these probabilities are also encoded in its Euclidean geometry.

1.2.2 Length of Curves in $\mathcal{H}$

We ask for the *length* of a curve in Hilbert space. The curve may be given by

$$t \to \psi_t, \quad 0 \leq t \leq 1, \tag{1.1}$$

---

[2] We only consider Hilbert spaces with countable bases.

[3] We use the "physicist's convention" that the scalar product is anti-linear in φ and linear in ψ.

where t is a parameter, not necessarily the time. We assume that for all $\varphi \in \mathcal{H}$, the function $t \to \langle \varphi, \psi_t \rangle$ of t is continuous.

To get the length of (1.1), we have to take all subdivisions

$$0 \leq t_0 < t_1 < \cdots < t_n \leq 1$$

in performing the sup,

$$\text{length of the curve} = \sup \sum_{j=1}^{n} \| \psi_{t_{j-1}} - \psi_{t_j} \| . \tag{1.2}$$

The length is independent of the parameter choice.

If we can guaranty the existence of

$$\dot{\psi}_t = \frac{\mathrm{d}}{\mathrm{d}t} \psi_t \in \mathcal{H}, \tag{1.3}$$

then one knows

$$\text{length of the curve} = \int_0^1 \sqrt{\langle \dot{\psi}_t, \dot{\psi}_t \rangle} \, \mathrm{d}t . \tag{1.4}$$

The vector $\dot{\psi}_t$ is the (contra-variant) tangent along (1.1). Its length is the *velocity* with which[4] $\psi = \psi_t$ travels through $\mathcal{H}$, i.e.,

$$\frac{\mathrm{d}s}{\mathrm{d}t} = \sqrt{\langle \dot{\psi}, \dot{\psi} \rangle} . \tag{1.5}$$

Interesting examples are solutions $t \to \psi_t$ of a Schrödinger equation,

$$H\psi = \mathrm{i}\hbar\dot{\psi} . \tag{1.6}$$

In this case, the tangent vector is time independent and we get

$$\frac{\mathrm{d}s}{\mathrm{d}t} = \hbar^{-1} \sqrt{\langle \psi, H^2 \psi \rangle} . \tag{1.7}$$

The length of the solution between the times t_0 and t_1 is

$$\text{length} = \hbar^{-1}(t_1 - t_0) \sqrt{\langle \psi, H^2 \psi \rangle} . \tag{1.8}$$

Anandan [4, 5] has put forward the idea to consider the Euclidean length (1.5) as an intrinsic and universal parameter in Hilbert space. For example, consider

$$\frac{\mathrm{d}t}{\mathrm{d}s} = \langle \dot{\psi}, \dot{\psi} \rangle^{-1/2} = \hbar \left(\langle \psi, H^2 \psi \rangle \right)^{-1/2}$$

instead of $\mathrm{d}s/\mathrm{d}t$ and interpret it as the *quantum velocity with which time is elapsing during a Schrödinger evolution*. Also other metrical structures, to which we come later on, allow for similar interpretations.

---

[4] We often write just ψ instead of ψ_t.

Remark 1. Though we shall be interested mostly in finite-dimensional Hilbert spaces or, in the infinite-dimensional case, in bounded operators, let us have a short look at the general case.

In the Schrödinger theory H, the Hamilton operator, is usually *unbounded* and there are vectors not in the domain of definition of H. However, there is always an integrated version: A unitary group

$$t \to U(t) = \exp\left(\frac{t}{i\hbar}H\right)$$

which can be defined rigorously for self-adjoint H. Then ψ_0 belongs to the domain of definition of H exactly, if the tangents (1.3) of the curve $\psi_t = U(t)\psi_0$ exist. If the tangents exist, then the Hamiltonian can be gained by

$$i\hbar \lim_{\epsilon \to 0} \frac{U(t+\epsilon) - U(t)}{\epsilon}\psi = H\psi$$

and (1.7) and (1.8) apply. If, however, ψ_0 does not belong to the domain of definition of H, then (1.2) returns ∞ for the length of every piece of the curve $t \to U(t)\psi_0$. In this case, the vector runs, during a finite time interval, through an infinitely long piece of $t \to \psi_t$. The velocity ds/dt must be infinite.

1.2.3 Distance and Length

Generally, a distance *dist* in a space attaches a real and not negative number to any pair of points satisfying

(a) $\text{dist}(\xi_1, \xi_2) = \text{dist}(\xi_2, \xi_1)$,
(b) $\text{dist}(\xi_1, \xi_2) + \text{dist}(\xi_2, \xi_3) \geq \text{dist}(\xi_1, \xi_3)$,
(c) $\text{dist}(\xi_1, \xi_2) = 0 \Leftrightarrow \xi_1 = \xi_2$.

A set with a distance is a *metric space*.

Given the distance, $\text{dist}(.,.)$, of a metric space and two different points, say ξ_0 and ξ_1, one may ask for the length of a continuous curve connecting these two points.[5] The inf of the lengths over all these curves is again a distance, the *inner distance*. The inner distance, $\text{dist}_{\text{inner}}(\xi_0, \xi_1)$, is never smaller than the original one,

$$\text{dist}_{\text{inner}}(\xi_0, \xi_1) \geq \text{dist}(\xi_0, \xi_1).$$

If equality holds, then the distance (and the metric space) is called *inner*. A curve connecting ξ_0 and ξ_1, the length of which equals the distance between the two points, is called a *short geodesic* arc. A curve, which is a short geodesic between all sufficiently near points, is a *geodesic*.

The Euclidian distance is an inner distance. It is easy to present the shortest curves between to vectors, ψ_1 and ψ_0, in Hilbert space:

---

[5] We assume that for every pair of points, such curves exist.

$$t \to \psi_t = (1 - t)\psi_0 + t\psi_1 , \quad \dot{\psi} = \psi_1 - \psi_0 . \tag{1.9}$$

It is a short geodesic arc between both vectors.

In Euclidean spaces, the shortest connection between two points is a piece of a straight line, and this *geodesic arc* is unique. Indeed, from (1.9) we conclude

$$\parallel \psi_t - \psi_r \parallel = |t - r| \parallel \psi_1 - \psi_0 \parallel . \tag{1.10}$$

With this relation, we can immediately compute (1.2) and we see that the length of (1.9) is equal to the distance between the starting and ending points.

We have seen something more general: *If in a linear space the distance is defined by a norm, the metric is inner and the geodesics are of the form (1.9).*

1.2.4 Curves on the Unit Sphere

Restricting the geometry of $\mathcal{H}$ to the unit sphere $\{\psi \in \mathcal{H}, \parallel \psi \parallel = 1\}$ can be compared with the change from Euclidean geometry to spherical geometry in an Euclidean 3-space. In computing a length by (1.2), only curves on the sphere are allowed.

The geodesics on a unit sphere are *great circles*. These are sections of the sphere with a plane that contains the center of the sphere. The spherical distance of two points, say ψ_0 and ψ_1, is the angle, α, between the rays from the center of the sphere to the two points in question:

$$\text{dist}_{\text{spherical}}(\psi_1, \psi_0) = \text{angle between the radii pointing to } \psi_0 \text{ and } \psi_1, \tag{1.11}$$

with the restriction $0 \leq \alpha \leq \pi$. By the additivity modulo 2π of the angle, one can compute (1.2) along a great circle to see that (1.11) is an inner metric.

If the two points are not antipodes, $\psi_0 + \psi_1 \neq 0$, then the great circle crossing through them and short geodesic arc between the two vectors is unique. For antipodes, the great circle crossing through them is not unique, and there are many short geodesic arcs of length π connecting them.

By elementary geometry,

$$\parallel \psi_1 - \psi_0 \parallel = \sqrt{2 - 2\cos\alpha} = 2\sin\frac{\alpha}{2}, \tag{1.12}$$

and $\cos\alpha$ can be computed by

$$\cos\alpha = \frac{\langle \psi_0, \psi_1 \rangle + \langle \psi_1, \psi_0 \rangle}{2} . \tag{1.13}$$

We see from (1.13) that

$$\cos\alpha \leq |\langle \psi_0, \psi_1 \rangle| . \tag{1.14}$$

Therefore, we have the following statement: *The length of a curve on the unit sphere connecting ψ_0 with ψ_1 is at least*

$$\arccos |\langle \psi_0, \psi_1 \rangle| \, .$$

Applying this observation to the solution of a Schrödinger equation, one gets the *Mandelstam–Tamm inequalities* [6, 7]. To get them, one simply combines (1.14) with (1.8): If a solution ψ_t of the Schrödinger equation (1.6) goes at time $t = t_0$ through the unit vector ψ_0 and at time $t = t_1$ through ψ_1, then

$$(t_1 - t_0) \sqrt{\langle \psi, H^2 \psi \rangle} \geq \hbar \arccos |\langle \psi_0, \psi_1 \rangle| \qquad (1.15)$$

must be valid. (Remember that H is conserved along solutions of (1.6) and we can use any $\psi = \psi_t$ from the assumed solution.)

However, a sharper inequality holds

$$(t_1 - t_0) \sqrt{\langle \psi, H^2 \psi \rangle - \langle \psi, H \psi \rangle^2} \geq \hbar \arccos |\langle \psi_0, \psi_1 \rangle| \, . \qquad (1.16)$$

Namely, the right-hand side is invariant against *gauge transformations*

$$\psi_t \mapsto \psi_t' = \exp(i\gamma t) \, \psi_t \, .$$

The left side of (1.16) does not change in substituting H by $H' = H - \gamma 1$, and ψ_t' is a solution of

$$H'\psi' = i\hbar\dot{\psi}' \, .$$

Hence, we can "gauge away" the extra term in (1.16) to get the inequality (1.15).

Remark 2. The reader will certainly identify the square-root expression in (1.16) as the *uncertainty* $\triangle_\psi(H)$ of H in the state given by the unit vector ψ. More specific, (1.16) provides the strict lower bound $T\triangle_\psi(H) \geq h/4$ for the time T to convert ψ to a vector orthogonal to ψ by a Schrödinger evolution.

Remark 3. If $U(r)$ is a one-parameter unitary group with generator A, then

$$|r|\triangle_\psi(A) \geq \arccos |\langle \psi, U(r)\psi \rangle| \, .$$

Interesting candidates are the position and the momentum operators, the angular momentum along an axis, occupation number operators, and so on.

Remark 4. The tangent space consists of pairs $\{\psi, \dot{\psi}\}$ with a tangent or velocity vector $\dot{\psi}$, reminiscent from a curve crossing through ψ. The fiber of all tangents based at ψ carries the positive quadratic form

$$\psi_1, \psi_2 \rightarrow \langle \psi, \psi \rangle \langle \psi_1, \psi_2 \rangle - \langle \psi_1, \psi \rangle \langle \psi, \psi_2 \rangle$$

gained by polarization.

Remark 5. More general than proposed in (1.16), one can say something about time-dependent Hamiltonians, $H(t)$, and the Schrödinger equation

$$H(t)\psi = i\hbar\dot\psi.$$
(1.17)

If a solution of (1.17) is crossing the unit vectors ψ_j at times t_j, then

$$\int_{t_0}^{t_1} \sqrt{\langle\psi, H^2\psi\rangle - \langle\psi, H\psi\rangle^2}\,dt \geq \hbar \arccos|\langle\psi_0, \psi_1\rangle|.$$
(1.18)

For further application to the speed of quantum evolutions, see [8–10].

1.2.5 Phases

If the vectors ψ and ψ' are linearly dependent, they describe the same state.

$$\psi \mapsto \pi_\psi = \frac{|\psi\rangle\langle\psi|}{\langle\psi, \psi\rangle}$$
(1.19)

maps the vectors of $\mathcal{H}$ onto the pure states, with the exception of the zero vector. Multiplying a vector by a complex number different from zero is the *natural gauge transformation* offered by $\mathcal{H}$. From this freedom in choosing a state vector for a pure state,

$$\psi \to \epsilon\psi, \quad |\epsilon| = 1,$$
(1.20)

the *phase change* is of primary physical interest.

In the following, we consider parameterized curve as in (1.1) on the unit sphere of $\mathcal{H}$. First we observe that

$$\langle\psi_t, \dot\psi_t\rangle$$
(1.21)

is purely imaginary. To see this, one differentiates

$$0 = \frac{d}{dt}\langle\psi, \psi\rangle = \langle\dot\psi, \psi\rangle + \langle\psi, \dot\psi\rangle,$$

and this is equivalent with the assertion. The curves

$$t \to \psi_t \text{ and } t \to \psi'_t := \epsilon_t\psi_t, \quad \epsilon_t = \exp(i\gamma_t),$$
(1.22)

are *gauge equivalent.* The states themselves,

$$t \to \pi_t = |\psi_t\rangle\langle\psi_t|,$$
(1.23)

are gauge invariant.

From the transformation (1.22), we deduce for the tangents

$$\dot\psi' = \dot\epsilon\psi + \epsilon\dot\psi, \quad \epsilon^{-1}\dot\epsilon = i\dot\gamma$$
(1.24)

with real γ. Thus, by an appropriate choice of the gauge, one gets

$$\langle \psi', \dot{\psi}' \rangle = 0 \,, \tag{1.25}$$

the geometric phase transport condition [11]. Indeed, (1.25) is the equation

$$\langle \psi', \dot{\psi}' \rangle = \mathrm{i}\dot{\gamma}\langle \psi, \psi \rangle + \langle \psi, \dot{\psi} \rangle = 0 \,.$$

Because of $\langle \psi, \psi \rangle = 1$ and $\langle \psi, \dot{\psi} \rangle = -\langle \dot{\psi}, \psi \rangle$, we get

$$\epsilon_t = \exp \int_{t_0}^{t} \langle \dot{\psi}, \psi \rangle \, \mathrm{d}t \,. \tag{1.26}$$

For a curve $t \to \psi_t$, $0 \le t \le 1$, with $\psi_1 = \psi_0$, the integral is the *geometric* or *Berry phase* [12]. For more about phases, see [1].

Remark 6. This is true on the unit sphere. If the vectors are not normalized, one has to replace (1.25) by the vanishing of the "gauge potential"

$$\frac{\langle \psi, \dot{\psi} \rangle - \langle \dot{\psi}, \psi \rangle}{2\mathrm{i}} \quad \text{or} \quad \frac{\langle \psi, \dot{\psi} \rangle - \langle \dot{\psi}, \psi \rangle}{2\mathrm{i}\langle \psi, \psi \rangle} \,. \tag{1.27}$$

In doing so, we conclude the following: The phase transport condition and the Berry phase do not depend on the normalization.

1.2.6 Fubini–Study Distance

With the Fubini–Study distance [13, 14], the set of pure states becomes an inner metric space. In our approach, we introduce a slight deviation from its original form, which is defined on the positive operators of rank one. To this end, we look at (1.19) in two steps. First, we skip normalization and replace (1.19) by

$$\psi \mapsto |\psi\rangle\langle\psi| \,,$$

and only after that we shall normalize.

Let ψ_0 and ψ_1 be two vectors from $\mathcal{H}$. We start with the first form of the Fubini–Study distance:

$$\mathrm{dist}_{\mathrm{FS}}(|\psi_1\rangle\langle\psi_1|, |\psi_0\rangle\langle\psi_0|) = \min_{\epsilon} \| \psi_1 - \epsilon\psi_0 \| \,, \tag{1.28}$$

where the minimum is over the complex numbers ϵ, $|\epsilon| = 1$. One easily finds

$$\mathrm{dist}_{\mathrm{FS}}(|\psi_1\rangle\langle\psi_1|, |\psi_0\rangle\langle\psi_0|) = \sqrt{\langle\psi_0, \psi_0\rangle + \langle\psi_1, \psi_1\rangle - 2|\langle\psi_1, \psi_0\rangle|} \,. \tag{1.29}$$

Therefore, (1.28) coincides with $\| \psi_1 - \psi_0 \|$, after choosing the relative phase appropriately, i.e., after choosing $\langle\psi_1, \psi_0\rangle$ real and not negative.

Equation (1.28) is a distance in the set of positive rank one operators: Choosing the phases between ψ_2, ψ_1 and between ψ_1, ψ_0 appropriately,

$$\mathrm{dist}_{\mathrm{FS}}(|\psi_2\rangle\langle\psi_2|, |\psi_1\rangle\langle\psi_1|) + \mathrm{dist}_{\mathrm{FS}}(|\psi_1\rangle\langle\psi_1|, |\psi_0\rangle\langle\psi_0|)$$

becomes equal to

$$\| \psi_2 - \psi_1 \| + \| \psi_1 - \psi_0 \| \geq \| \psi_2 - \psi_0 \|$$

and, therefore,

$$\| \psi_2 - \psi_0 \| \geq \mathrm{dist}_{\mathrm{FS}}(|\psi_2\rangle\langle\psi_2|, |\psi_0\rangle\langle\psi_0|).$$

Now we can describe the geodesics belonging to the distance $\mathrm{dist}_{\mathrm{FS}}$ and see that (1.28) is an inner distance: If the scalar product between ψ_0 and ψ_1 is real and not negative, then this is true for the scalar products between any pair of the vectors

$$t \to \psi_t := (1 - t)\psi_0 + t\psi_1, \quad \langle\psi_1, \psi_0\rangle \geq 0. \qquad (1.30)$$

Then we can conclude

$$\mathrm{dist}_{\mathrm{FS}}(|\psi_r\rangle\langle\psi_r|, |\psi_t\rangle\langle\psi_t|) = \| \psi_r - \psi_t \|, \qquad (1.31)$$

and (1.30) is geodesic in $\mathcal{H}$. Furthermore,

$$t \to |\psi_t\rangle\langle\psi_t|, \quad 0 \leq t \leq 1, \qquad (1.32)$$

is the shortest arc between $|\psi_0\rangle\langle\psi_0|$ and $|\psi_1\rangle\langle\psi_1|$. Explicitly,

$$t \to (1 - t)^2|\psi_0\rangle\langle\psi_0| + t^2|\psi_1\rangle\langle\psi_1| + t(1 - t)\left(|\psi_0\rangle\langle\psi_1| + |\psi_1\rangle\langle\psi_0|\right). \qquad (1.33)$$

If ψ_0 and ψ_1 are unit vectors, $\pi_j = |\psi_j\rangle\langle\psi_j|$ are (density operators of) pure states. Then (1.31) simplifies to

$$\mathrm{dist}_{\mathrm{FS}}(\pi_1, \pi_0) = \sqrt{2 - 2|\langle\psi_1, \psi_0\rangle|} = \sqrt{2 - 2\sqrt{\mathrm{Pr}(\pi_0, \pi_1)}}, \qquad (1.34)$$

where we have used the notation $\mathrm{Pr}(\pi_1, \pi_0)$ for the *transition probability*

$$\mathrm{Pr}(\pi_1, \pi_0) = \mathrm{tr}\,\pi_0\pi_1. \qquad (1.35)$$

The transition probability is the probability to get an affirmative answer in testing whether the system is in the state π_1, if it was actually in state π_0.

However, the geodesic arc (1.33) cuts the set of pure states only at π_0 and at π_1. Therefore, the distance (1.28) is not an inner one for the space of pure states. To obtain the appropriate distance, which we call $\mathrm{Dist}_{\mathrm{FS}}$, we have to minimize the length with respect to curves consisting of pure states only. This problem is quite similar to the change from Euclidean to spherical geometry in $\mathcal{H}$ (and, of course, in ordinary 3-space). We can use a great circle on the unit sphere of our Hilbert space, which obeys the condition (1.25), $\langle\dot{\psi}, \psi\rangle = 0$. Then the map $\psi \to |\psi\rangle\langle\psi| = \pi$ is one to one within small intervals of the parameter: The map identifies antipodes in the unit sphere of the Hilbert space. Thus "in the small," the map is one to one. Using this normalization, we get

$$\text{Dist}_{\text{FS}}(\pi_0, \pi_1) = \arccos \sqrt{\Pr(\pi_0, \pi_1)}\,. \tag{1.36}$$

The distance of two pure states become maximal if π_0 and π_1 are orthogonal. This occurs at the angle $\pi/2$. As in the unit sphere, the geodesics are closed, but now have length π.

Remark 7. If Dist is multiplied by a positive real number, we get again a distance. (This is obviously so for any distance.) Therefore, another normalization is possible. Fubini and Study, who considered these geodesics at first, "stretched" them to become metrical isomorph to the unit circle [14]:

$$\text{Dist}_{\text{Study}}(\pi_0, \pi_1) = 2\arccos \sqrt{\Pr(\pi_0, \pi_1)}\,.$$

1.2.7 Fubini–Study Metric

As we have seen, with dist$_{\text{FS}}$, the set of positive operators of rank one becomes an inner metric space. We now convince ourselves that it is a Riemannian manifold. Its Riemannian metric, called *Fubini–Study metric,* reads

$$ds_{\text{FS}} = \sqrt{\langle\psi,\psi\rangle\langle\dot\psi,\dot\psi\rangle - \langle\dot\psi,\psi\rangle\langle\psi,\dot\psi\rangle}\,dt \tag{1.37}$$

for curves

$$t \mapsto \psi_t \mapsto |\psi_t\rangle\langle\psi_t|\,, \tag{1.38}$$

where in (1.37), the index t in ψ_t is suppressed.

To prove it, we consider firstly normalized curves ψ_t remaining on the unit sphere of $\mathcal{H}$. Imposing the geometric phase transport condition (1.25), the map (1.38) becomes an isometry for small-parameter intervals. Simultaneously, (1.37) reduces to the Euclidean line element along curves fulfilling (1.25). Hence, for curves on the unit sphere, (1.37) has been proved. To handle arbitrary normalization, we scale by

$$\psi'_t = z_t\psi_t\,, \quad z_t \neq 0\,, \tag{1.39}$$

and obtain

$$\langle\psi',\psi'\rangle\langle\dot\psi',\dot\psi'\rangle - \langle\dot\psi',\psi'\rangle\langle\psi',\dot\psi'\rangle = (z^*z)^2\left[\langle\psi,\psi\rangle\langle\dot\psi,\dot\psi\rangle - \langle\dot\psi,\psi\rangle\langle\psi,\dot\psi\rangle\right]. \tag{1.40}$$

Therefore, (1.37) shows the correct scaling as required by dist$_{\text{FS}}$, and it is valid on the unit sphere. Thus, (1.37) is valid generally, i.e., for curves of positive operators of rank one.

We now express (1.37) in terms of states. If

$$\pi_t = |\psi_t\rangle\langle\psi_t|\,, \quad \langle\psi_t,\psi_t\rangle = 1\,, \quad \langle\dot\psi_t,\psi_t\rangle = 0\,, \tag{1.41}$$

then one easily sees that $\operatorname{tr} \dot{\pi}\dot{\pi} = 2\langle \dot{\psi}, \dot{\psi} \rangle$ or

$$ds_{FS} = \sqrt{\frac{1}{2}\operatorname{tr} \dot{\pi}^2}\, dt \tag{1.42}$$

for all (regular enough) curves $t \to \pi_t$ of pure states. These curves satisfy

$$\operatorname{tr} \pi = 1, \quad \operatorname{tr} \dot{\pi} = 0, \quad \operatorname{tr} \dot{\pi}\pi = 0. \tag{1.43}$$

The latter assertion follows from $\pi^2 = \pi$ by differentiation, $\dot{\pi} = \dot{\pi}\pi + \pi\dot{\pi}$, and after taking the trace on both sides.

Let now $\rho = |\psi\rangle\langle\psi|$ with $\psi = \psi_t$ being a curve in the Hilbert space. We lost normalization of ψ, but we are allowed to require the vanishing of the gauge potential (1.27). Then

$$\operatorname{tr} \dot{\rho} = 2\langle \psi, \dot{\psi} \rangle, \quad \operatorname{tr} \dot{\rho}^2 = 2\langle \psi, \psi \rangle \langle \dot{\psi}, \dot{\psi} \rangle + \frac{1}{2}(\operatorname{tr} \dot{\rho})^2,$$

and we conclude

$$ds_{FS}^2 = \frac{1}{2}[\operatorname{tr} \dot{\rho}^2 - (\operatorname{tr} \dot{\rho})^2]\, dt^2 \tag{1.44}$$

for curves $t \to \rho_t$ of positive operators of rank one.

There is a further expression for the Fubini–Study metric since $\rho^2 = (\operatorname{tr}\rho)\rho$ for a positive operator of rank one. By differentiating and after some algebraic manipulations, one arrives at

$$ds_{FS}^2 = \left[(\operatorname{tr}\rho)^{-1}\operatorname{tr} \rho\dot{\rho}^2 - (\operatorname{tr} \dot{\rho})^2\right] dt^2. \tag{1.45}$$

1.2.8 Symmetries

It was Wigner's famous idea [15] to use the transition probability to define the concept of symmetry in the set of pure states. If $\pi \to T(\pi)$ maps the set of pure states onto itself, T is a *symmetry*, if it satisfies

$$\operatorname{Pr}(T(\pi_1), T(\pi_2)) = \operatorname{Pr}(\pi_1, \pi_2). \tag{1.46}$$

Looking at (1.34) or (1.30), it becomes evident that (1.46) is valid if and only if T is an isometry with respect to dist$_{FS}$ and also to Dist$_{FS}$.

Before stating the main results of this section, we discuss the case dim $\mathcal{H} = 2$, i.e., the *1-qubit space*. Here, the pure states are uniquely parameterized by the points of a 2-sphere, the Bloch sphere. Indeed, π is a pure state if

$$\pi = \frac{1}{2}\left(1 + \sum_{j=1}^{3} x_j\sigma_j\right), \quad x_1^2 + x_2^2 + x_3^2 = 1, \tag{1.47}$$

with a *Bloch vector* $\{x_1, x_2, x_3\}$. Clearly,

$$\operatorname{tr} \dot{\pi}^2 = \frac{1}{2} \sum \dot{x}_j^2$$

and, by (1.42),

$$ds_{FS} = \frac{1}{2} \sqrt{\sum \dot{x}_j^2} \, dt \, . \tag{1.48}$$

It follows already that a symmetry T, in the sense of Wigner, is a map of the 2-sphere into itself conserving the metric induced on the sphere by the Euclidean one. Hence, there is an orthogonal matrix with entries O_{jk} changing the Bloch vector as

$$\pi \to T(\pi) \Leftrightarrow x_j \to \sum_k O_{jk} x_k \, . \tag{1.49}$$

As it is well known, proper orthogonal transformations can be implemented by a unitary transformation, i.e., with a suitable unitary U,

$$U \left(\sum_j x_j \sigma_j \right) U^{-1} = \sum_j x_j' \sigma_j \, .$$

An anti-unitary, say V, can be written as $V = U \theta_f$ with a spin-flip

$$\theta_f \big(c_0 |0\rangle + c_1 |1\rangle \big) = \big(c_1^* |0\rangle - c_0^* |1\rangle \big),$$

producing the inversion $x_j \to -x_j$ of the Bloch sphere. This says, in short, that

$$T(\pi) = V \pi V^{-1} \, , \tag{1.50}$$

V is either unitary or anti-unitary. The validity of (1.50) for all Hilbert spaces was proposed by Wigner.

There is a stronger result[6] for $\dim \mathcal{H} > 2$, saying that it suffices that T preserves orthogonality:

Theorem 1. *In order that (1.50) holds for all pure states π, it is necessary and sufficient that one of the following conditions take place:*

(a) It is a symmetry in the sense (1.46) of Wigner.
(b) It is an isometry of the Fubini–Study distance.
(c) It is $\dim \mathcal{H} \geq 3$ and

$$\Pr(\pi_1, \pi_2) = 0 \Leftrightarrow \Pr(T(\pi_1), T(\pi_2)) = 0 \, . \tag{1.51}$$

If $\dim \mathcal{H} > 2$, the condition (c) is obviously a more advanced statement than (a) or (b). An elementary proof is due to Uhlhorn [16]. Indeed, the theorem is also a corollary of a deeper rooted result by Dye [17].

---

[6] The 1-qubit case is too poor in structure compared with higher dimensional ones.

1.2.9 Comparison with Other Norms

While for the vectors of a Hilbert space one has naturally only one norm, i.e., the vector norm; there are many norms to estimate on an operator, say A. For instance, one defines

$$\| A \|_2 = \sqrt{\operatorname{tr} A^\dagger A} \quad \text{and} \quad \| A \|_1 = \operatorname{tr} \sqrt{A^\dagger A}. \tag{1.52}$$

The first one is called *Frobenius* or *von Neumann norm*. The second is the *functional-* or *1-norm*. If $\mathcal{H} = \infty$, these norms can be easily infinite, and their finiteness is a strong restriction to the operator. If A is of finite rank, then

$$\| A \|_2 \le \| A \|_1 \le \sqrt{\operatorname{rank}(A)} \| A \|_2. \tag{1.53}$$

The rank of A is at most as large as the dimension of the Hilbert space.

For $r \ge 1$, one also defines the *Schatten norms*

$$\| A \|_r = \left[\operatorname{tr} \left(A^\dagger A \right)^{r/2} \right]^{1/r}. \tag{1.54}$$

If π is a pure state's density operator, then $\| \pi \|_r = 1$. The Schatten norms of the difference $\nu = \pi_2 - \pi_1$ is also easily computed. One may assume $\dim \mathcal{H} = 2$ as all calculations are done in the space spanned by the vectors ψ_j with $\pi_j = |\psi_j\rangle\langle\psi_j|$. Now ν is Hermitian and with trace 0, its square is a multiple of $\mathbf{1}$. We get

$$\lambda^2 \mathbf{1} = \nu^2 = \pi_1 + \pi_2 - \pi_1\pi_2 - \pi_2\pi_1$$

and, taking the trace, $\lambda^2 = 1 - \operatorname{Pr}(\pi_1, \pi_2)$, by (1.35). Thus

$$\| \pi_2 - \pi_1 \|_r = 2^{1/r} \sqrt{1 - \operatorname{Pr}(\pi_1, \pi_2)}. \tag{1.55}$$

Comparing with

$$\operatorname{dist}_{FS}(\pi_1, \pi_2) = \sqrt{2}\sqrt{1 - |\langle \psi_1, \psi_2 \rangle|} = \sqrt{2}\sqrt{1 - \sqrt{\operatorname{Pr}(\pi_1, \pi_2)}}$$

results in

$$\| \pi_2 - \pi_1 \|_r = \frac{2^{1/r}}{\sqrt{2}} \operatorname{dist}_{FS}(\pi_1, \pi_2) \sqrt{1 + \sqrt{\operatorname{Pr}(\pi_1, \pi_2)}}. \tag{1.56}$$

As the value of transition probability is between 0 and 1, the identity provides tight inequalities between Schatten distances and the Fubini–Study distance for two pure states.

One important difference between the Schatten distances (1.55) and the Fubini–Study distance concerns the geodesics. We know that the geodesics with respect to a norm read

$$t \to \pi_t = (1 - t)\pi_0 + t\pi_1$$

and, therefore, they consist of mixed density operators for $0 < t < 1$. On the other hand, the Fubini-Study geodesics remain within the set of pure states.

1.3 Operators, Observables, and States

Let us fix some notions. We denote the algebra of all bounded linear operators acting on an Hilbert space $\mathcal{H}$ by $\mathcal{B}(\mathcal{H})$.

If $\dim \mathcal{H} < \infty$, every linear operator A is bounded. To control it, in general, one introduces the norm

$$\| A \|_\infty = \sup_\psi \| A\psi \|, \quad \| \psi \| = 1 \tag{1.57}$$

and calls A bounded, if this sup over all unit vectors is finite. To be bounded means that the operator cannot stretch unit vectors to arbitrary length. One has

$$\lim_{r \to \infty} \| A \|_r = \| A \|_\infty, \tag{1.58}$$

if the Schatten norms are finite for large enough r. The ∞-*norm* (1.57) of every unitary operator and of every projection operator (different from the operator $\mathbf{0}$) is one.

Equation (1.57) defines an *operator norm* because one has

$$\| AB \|_\infty \leq \| A \|_\infty \| B \|_\infty,$$

in addition to the usual norm properties. For $1 < r < \infty$, no Schatten norm is an operator norm. On the other hand, there are many operator norms. However, among them, the ∞-norm has a privileged position. It satisfies

$$\| A^\dagger A \| = \| A \|^2, \quad \| A^\dagger \| = \| A \| \ . \tag{1.59}$$

An operator norm satisfying (1.59) is called a C*-norm. *There is only one C*-norm in $\mathcal{B}(\mathcal{H})$, the ∞-norm.*

Remark 8. In mathematics and in mathematical physics, the operation $A \to A^\dagger$ is called *the star operation*. In these branches of science, the Hermitian adjoint of an operator A is called A^*. The notion $A^\dagger$ was used by Dirac in his famous book "The Principles of Quantum Mechanics" [18].

Let us come now to the density operators. Density operators describe states. We shall indicate them by using small Greek letters. Density operators are positive operators with trace one:

$$\omega \geq \mathbf{0}, \quad \mathrm{tr}\,\omega = 1 . \tag{1.60}$$

One can prove that

$$\| \rho \|_1 = \mathrm{tr}\,\rho = 1 \iff \rho \text{ is a density operator.} \tag{1.61}$$

A bounded operator on an infinite-dimensional Hilbert space is said to be of *trace class*, if its 1-norm is finite. The trace class operators constitute a tiny portion of $\mathcal{B}(\mathcal{H})$ in the infinite case.

1.3.1 States and Expectation Values

Let ω be a density operator and $A \in \mathcal{B}(\mathcal{H})$ an operator. The value $\operatorname{tr} A\omega$ is called the *expectation value of A in state ω*. There are always operators with different expectation values, for two different density operators. In this sense, one may say, *observables distinguish states*.

Remark 9. Not every operator in $\mathcal{B}(\mathcal{H})$ represents an observable in the strict sense: An observable should have a spectral decomposition. Therefore, observables are represented by normal operators, i.e., $A^{\dagger}A = AA^{\dagger}$ must be valid. (For historical but not physical reasons, often hermiticity or, if $\dim \mathcal{H} = \infty$, self-adjointness is required in textbooks. A critical overview can be found in [19].) On the other hand, to distinguish states, the expectation values of projection operators are sufficient.

As already said, observables (or operators) distinguish states, more observables allow for a finer description, i.e., they allow to discriminate between more states. To use less observables is like "coarse graining": Some states cannot be distinguished any more.

These rules will be condensed in a precise scheme later on. The first step in this direction is to describe a state in a different way, namely as the set of its expectation values. To do so, one considers a state as a function defined for all operators. In particular, if ω is a density operator, one considers the *function* (or *functional*, or *linear form*)

$$A \rightarrow \underline{\omega}(A) := \operatorname{tr} A\omega. \tag{1.62}$$

Let us stress the following properties of (1.62)

(1) Linearity: $\underline{\omega}(c_1 A_1 + c_2 A_2) = c_1 \underline{\omega}(A_1) + c_2 \underline{\omega}(A_2)$
(2) Positivity: $\underline{\omega}(A) \geq 0$ if $A \geq \mathbf{0}$
(3) It is normalized: $\underline{\omega}(\mathbf{1}) = 1$.

At this point, one inverts the reasoning. One considers (1)–(3) the essential conditions and calls *every* functional on $\mathcal{B}(\mathcal{H})$ which fulfills these three conditions, a *state of the algebra $\mathcal{B}(\mathcal{H})$*. In other words, (1)–(3) is the definition of the term *state of $\mathcal{B}(\mathcal{H})$!* Therefore, the definition does not discriminate between pure and mixed states from the beginning.

Let us see, how it works. If $\dim \mathcal{H} < \infty$, every functional obeying (1), (2), and (3) can be written

$$\underline{\omega}(A) = \operatorname{tr} A\omega, \quad \omega \geq \mathbf{0}, \quad \operatorname{tr} \omega = 1$$

as in (1.62). Here the definition just reproduces the density operators.

Indeed, every linear form can be written $\underline{\omega}(A) = \operatorname{tr} BA$ with an operator $B \in \mathcal{B}(\mathcal{H})$. However, if $\operatorname{tr} BA$ is a real and non-negative number for every $A \geq \mathbf{0}$, one infers $B \geq \mathbf{0}$ (take the trace with a basis of eigenvectors of B to see it). Finally, condition (3) forces B to have trace one. Now one identifies $\omega := B$.

The case $\dim \mathcal{H} = \infty$ is more intriguing. A measure in "classical" mathematical measure theory has to respect the condition of countable additivity. The translation to the non-commutative case needs the so-called *partitions of the unit element*, i.e., decompositions

$$\mathbf{1} = \sum_j P_j \tag{1.63}$$

with projection operators P_j. These decompositions are necessarily orthogonal, $P_k P_l = \mathbf{0}$ if $k \neq l$, and in one-to-one relation to decompositions of the Hilbert space into orthogonal sums,[7]

$$\mathcal{H} = \bigoplus_j \mathcal{H}_j, \quad \mathcal{H}_j = P_j \mathcal{H}. \tag{1.64}$$

A state $\underline{\omega}$ is called *normal* if for all partitions of $\mathbf{1}$,

$$\sum_j \underline{\omega}(P_j) = \underline{\omega}(\mathbf{1}) = 1 \tag{1.65}$$

is valid. $\underline{\omega}$ is normal exactly if its expectation values are given as in (1.62) with the help of a density operator ω.

There is a further class of states, the *singular states*. A state $\underline{\omega}$ of $\mathcal{B}(\mathcal{H})$ is called *singular*, if $\underline{\omega}(P) = 0$ for all projection operators of finite rank. Thus, if $\dim(P\mathcal{H}) < \infty$, one gets $\underline{\omega}(P) = 0$ for singular states.

A theorem asserts that every state $\underline{\omega}$ of $\mathcal{B}(\mathcal{H})$ has a unique decomposition

$$\underline{\omega} = (1 - p)\underline{\omega}_{\text{normal}} + p\underline{\omega}_{\text{singular}}, \quad 0 \leq p \leq 1. \tag{1.66}$$

In mathematical measure theory, a general $\underline{\omega}$ corresponds to an *additive measure*, in contrast to the genuine measures which are countably additive. Accordingly we are invited to consider a normal state of $\mathcal{B}(\mathcal{H})$ to be a *countably additive non-commutative probability measure*, and any other state to be an *additive non-commutative probability measure*.

We cannot but at this point of the lecture mention the 1957 contribution of Gleason [20]. He asked whether it will be possible to define states already by their expectation values at projections.

Assume $P \to f(P) \geq 0$, $f(\mathbf{1}) = 1$, is a function that is defined only on the projection operators $P \in \mathcal{B}(\mathcal{H})$ and which satisfies

---

[7] If a sum of projections is a projection, it must be an orthogonal sum. To see it, square the equation and take the trace. The trace of a product of two positive operators is not negative and can be zero only if the product of the operators is zero.

$$\sum_j f(P_j) = 1 \qquad\qquad (1.67)$$

for all orthogonal partitions (1.63) of the unity $\mathbf{1}$. Gleason has proved that if $\dim \mathcal{H} > 2$, there is a density operator ω with $\operatorname{tr} P\omega = f(P)$ for all $P \in \mathcal{B}(\mathcal{H})$,[8] i.e., $\underline{\omega}(P) = f(P)$.

The particular merit of Gleason's theorem consists in relating directly quantum probabilities to the concept of *state* as defined above: Suppose our quantum system is in state ω, and we test whether P is valid, the answer is YES with probability $\underline{\omega}(P) = \operatorname{tr} P\omega$.

It took about 30 years to find out an extension to general states. There is a lengthy proof by Maeda, Christensen, Yeadon, and others (see [21]), exhibiting a lot of steps (most of them not particularly difficult) and a rich architecture. Indeed, they examined the problem for general von Neumann algebras, but in the case at hand, they assert the following extension of Gleason's finding:

Theorem 2. *Assume* $\dim \mathcal{H} \geq 3$. *Given a function* $f \geq 0$ *on the projection operators satisfying (1.65) for all* finite *partitions of* $\mathbf{1}$. *Then there is a state* $\underline{\omega}$ *fulfilling* $\underline{\omega}(P) = f(P)$ *for all projection operators of* $\mathcal{B}(\mathcal{H})$.

1.3.2 Subalgebras and Subsystems

There is a consistent solution to the question: What is a subsystem of a quantum system with Hilbert space $\mathcal{H}$ and algebra $\mathcal{B}(\mathcal{H})$? The solution is unique in the finite-dimensional case. Below we list some necessary requirements that become sufficient if $\dim \mathcal{H} < \infty$. As already indicated, a subsystem of a quantum system should consist of less observables (operators) than the larger one. For the larger one, we start with $\mathcal{B}(\mathcal{H})$ to be on (more or less) known grounds.

Let $\mathcal{A} \subset \mathcal{B}(\mathcal{H})$ be a subset. $\mathcal{A}$ is called a *subalgebra* of $\mathcal{B}(\mathcal{H})$ or, equivalently, an *operator algebra* on $\mathcal{H}$

(a) if $\mathcal{A}$ is a linear space and
(b) if $A, B \in \mathcal{A}$ then $AB \in \mathcal{A}$.

Essential is also the condition:

(c) If $A \in \mathcal{A}$ then $A^\dagger \in \mathcal{A}$.

A subset $\mathcal{A}$ of $\mathcal{B}(\mathcal{H})$ satisfying (a), (b), and (c) is called an *operator *-algebra*. In an operator algebra, the scalar product of the Hilbert space is reflected by the star operation, $A \to A^\dagger$. A further point to mention concerns positivity of operators: $B \in \mathcal{B}(\mathcal{H})$ is positive if and only if it can be written $B = A^\dagger A$.

---

[8] As already mentioned, in two dimensions, the set of projections is too poor in relations.

Finally, an algebra $\mathcal{A}$ is called *unital* if it contains an *identity* or *unit element*, say $\mathbf{1}_\mathcal{A}$. The unit element, if it exists, is uniquely characterized by

$$\mathbf{1}_\mathcal{A}\, A = A\, \mathbf{1}_\mathcal{A} = A \text{ for all } A \in \mathcal{A}, \tag{1.68}$$

and we refer to its existence as

(d) $\mathcal{A}$ is unital.

Assume $\mathcal{A}$ fulfills all four conditions (a)–(d). Then one can introduce the concept of *state*. We just mimic what has been said to be a state of $\mathcal{B}(\mathcal{H})$ and obtain a core definition:

Definition 1. *A state of $\mathcal{A}$ is a function $A \to \underline{\omega}(A) \in \mathbb{C}$ of the elements of $\mathcal{A}$ satisfying for all elements of $\mathcal{A}*

(1') $\underline{\omega}(c_1 A_1 + c_2 A_2) = c_1 \underline{\omega}(A_1) + c_2 \underline{\omega}(A_2)$ *(linearity)*,
(2') $\underline{\omega}(A^\dagger A) \geq 0$ *(positivity)*,
(3') $\underline{\omega}(\mathbf{1}_\mathcal{A}) = 1$ *(normalization)*.

Let us stop for a moment to ask what has changed compared to the definition we gave in Sect. 1.3.1. The change is in (2) to (2'). In (2'), no reference is made to the Hilbert space. It is a purely algebraic definition, which only refers to operations defined in $\mathcal{A}$. It circumvents the way A is acting on $\mathcal{H}$. That implies the following: The concept of state does *not* depend how $\mathcal{A}$ is embedded in $\mathcal{B}(\mathcal{H})$, or "at what place $\mathcal{A}$ is sitting within a larger *-algebra". Indeed, to understand the abstract skeleton of the quantum world, one is confronted with (at least!) two questions:

–   What is a quantum system, what is its structure?
–   How is a quantum system embedded in other ones as a subsystem?

Now let us proceed more prosaically. $A, B \to \underline{\omega}(A^\dagger B)$ is a positive Hermitian form. Therefore,

$$\underline{\omega}(A^\dagger A)\, \underline{\omega}(B^\dagger B) \geq \underline{\omega}(A^\dagger B), \tag{1.69}$$

which is the important Schwarz inequality.

The set of all states of $\mathcal{A}$ is the *state space of $\mathcal{A}$*. It will be denoted by $\underline{\Omega}(\mathcal{A})$. The state space is naturally convex[9]:

$$\underline{\omega} := \sum p_j \underline{\omega}_j \in \underline{\Omega}(\mathcal{A}), \tag{1.70}$$

for any convex combination of the $\underline{\omega}_j$, i.e., for all these sums with

$$\sum p_j = 1 \text{ and } p_j > 0 \text{ for all } j. \tag{1.71}$$

A *face* of $\underline{\Omega}(\mathcal{A})$ is a subset with the following property: If $\underline{\omega}$ is contained in this subset, then for every convex decomposition (1.70), (1.71), of $\underline{\omega}$, also all states $\underline{\omega}_j$ belong to this subset.

---

[9] For more about convexity, see [22, 23].

Example 1. Let $P \in \mathcal{A}$ be a projection. Define $\underline{\Omega}(\mathcal{A})_P$ to be the set of all $\underline{\omega} \in \underline{\Omega}(\mathcal{A})$ such that $\underline{\omega}(P) = 1$. This set is a face of $\underline{\Omega}(\mathcal{A})$. To see it, one looks at the definition of state and concludes from (1.70) and (1.71) that $\underline{\omega}_j(P) = 1$ necessarily.

If $\mathcal{A}$ is a *-subalgebra of $\mathcal{B}(\mathcal{H})$ and $\dim \mathcal{H} < \infty$, then every face of $\underline{\Omega}(\mathcal{A})$ is of the form $\underline{\Omega}(\mathcal{A})_P$ with a projection $P \in \mathcal{A}$.

Definition 2. *If a face consists of just one state $\underline{\pi}$, then $\underline{\pi}$ is called* extremal *in $\underline{\Omega}(\mathcal{A})$. This is the mathematical definition. In quantum physics, a state $\underline{\pi}$ of $\mathcal{A}$ is called* pure *if and only if $\underline{\pi}$ is extremal in $\underline{\Omega}(\mathcal{A})$.*

These are rigorous and fundamental definitions. We do not assert that every $\mathcal{A}$ satisfying the requirements (a)–(d) above represents or "is" a quantum system. But we claim that every quantum system, which can be represented by bounded operators, can be based on such an algebra. Its structure gives simultaneously meaning to the concepts of *observable*, *state*, and *pure state*. It does so in a clear and mathematical clean way.

Subsystems

Now we consider some relations between operator algebras, in particular between quantum systems. We start by asking for the concept of *subsystems* of a given quantum system. Let $\mathcal{A}_j$ be *-subalgebras of $\mathcal{B}(\mathcal{H})$ with unit element 1_j respectively. From $\mathcal{A}_1 \subset \mathcal{A}_2$, it follows $1_1 1_2 = 1_1$, and 1_1 is a projection in $\mathcal{A}_2$. To be a *subsystem* of the quantum system $\mathcal{A}_2$, we require

$$\mathcal{A}_1 \subset \mathcal{A}_2, \quad 1_1 = 1_2. \tag{1.72}$$

In mathematical terms, $\mathcal{A}_1$ is a *unital subalgebra* of $\mathcal{A}_2$. Thus, if two quantum systems are represented by two unital *-algebras $\mathcal{A}_j$ satisfying (1.72), then $\mathcal{A}_1$ is said to be a *subsystem* of $\mathcal{A}_2$.

In particular, $\mathcal{A}$ is a subsystem of $\mathcal{B}(\mathcal{H})$ if it contains the identity operator, $1_{\mathcal{H}}$ or simply 1 of $\mathcal{H}$ because 1 is the unit element of $\mathcal{B}(\mathcal{H})$.

The case $1_1 \neq 1_2$ will be paraphrased by calling $\mathcal{A}_1$ an *incomplete subsystem* of $\mathcal{A}_2$.

Let $\mathcal{A}_1$ be a subsystem of $\mathcal{A}_2$ and let us ask for relations between their states. At first we see the following: A state $\underline{\omega}_2 \in \underline{\Omega}(\mathcal{A}_2)$ gives automatically a state $\underline{\omega}_1$ on $\mathcal{A}_1$ by just defining $\underline{\omega}_1(A) := \underline{\omega}_2(A)$ for all operators of $\mathcal{A}_1$. $\underline{\omega}_1$ is called the *restriction of $\underline{\omega}_2$ to $\mathcal{A}_1$*. Clearly, the conditions (1')–(3') remain valid in restricting a state to a subsystem.

Of course, it may be that there are many states in $\mathcal{A}_2$ with the same restriction to $\mathcal{A}_1$. Two (and more) different states of $\mathcal{A}_2$ may "fall down" to one and the same state of the subalgebra $\mathcal{A}_1$. From the point of view of a subsystem, two or more different states of a larger system can become identical.

Conversely, $\underline{\omega}_2$ is an *extension* or *lift* of $\underline{\omega}_1$. The task of extending $\underline{\omega}_1$ to a state of a larger system is not unique: Seen from the subsystem $\mathcal{A}_1$, (almost) nothing can be said about expectation values for operators that are in $\mathcal{A}_2$ but not in $\mathcal{A}_1$.

As a consequence, we associate to the words *a quantum system is a subsystem of another quantum system*, a precise meaning. Or, to be more cautious, we have a necessary condition for the validity of such a relation. Imagining that every system might be a subsystem of many other ones, one gets a faint impression how rich the architecture of that hierarchy may be.

Remark 10. The restriction of a state to an incomplete subsystem will conserve the linearity and the positivity conditions (1') and (2'). The normalization (3') cannot be guaranteed in general.

$\dim \mathcal{H} = \infty$. Some Comments

As a matter of fact, the conditions (a)–(c) for a *-subalgebra of $\mathcal{B}(\mathcal{H})$ are not strong enough for infinite-dimensional Hilbert spaces. There are two classes of algebras in the focus of numerous investigations, the C*- and the von Neumann algebras. We begin by defining[10] C*-algebras and then we turn to von Neumann ones. Much more can be found in [3].

Every subalgebra $\mathcal{A}$ of $\mathcal{B}(\mathcal{H})$ is equipped with the ∞-norm, $\| \cdot \|_\infty$. One requires the algebra to be closed[11] with respect to that norm: For every sequence $A_j \in \mathcal{A}$, which converges to $A \in \mathcal{B}(\mathcal{H})$ in norm, $\| A - A_j \|_\infty \to 0$, the operator A must be also in $\mathcal{A}$. In particular, a *-subalgebra is said to be a *C*-algebra*, if it is closed with respect to the operator norm. The ∞-norm is a C*-norm in these algebras, see (1.59). One can prove that in a C*-algebra, there exists just one operator norm that is a C*-norm.

In the same spirit there is an 1-norm (or *functional norm*) $\| \cdot \|_1$, estimating the linear functionals of $\mathcal{A}$. $\| \underline{\nu} \|_1$ is the smallest number λ for which $|\underline{\nu}(A)| \leq \lambda \| A \|_\infty$ is valid for all $A \in \mathcal{A}$.

With respect to a unital C*-algebra, we can speak of its states, and its normal operators are its observables. However, a C*-algebra does not necessarily provide sufficiently many projection operators: There are C*-algebras containing no projection different from the trivial ones, $\mathbf{0}$ and $\mathbf{1}_\mathcal{A}$.

In contrast, von Neumann algebras contain sufficiently many projections. $\mathcal{A}$ is called a *von Neumann algebra,* if it is closed with respect to the so-called weak topology.

To explain it, let $\mathcal{F}$ be a set of operators and $B \in \mathcal{B}(\mathcal{H})$. B is a *weak limit point* of $\mathcal{F}$ if for every n, every $\epsilon > 0$, and for every finite set $\psi_1, \ldots, \psi_n$ of vectors from $\mathcal{H}$, there is an operator $A \in \mathcal{F}$ fulfilling the inequality

---

[10] We define the so-called *concrete* C*-algebras.
[11] Then the algebra becomes a *Banach algebra*.

$$\sum_{j=1}^{n} |\langle \psi_j, (B - A)\psi_j \rangle| \leq \epsilon.$$

The set of all weak limit points of $\mathcal{F}$ is the *weak closure* of $\mathcal{F}$.

A von Neumann algebra $\mathcal{A}$ is a *-subalgebra of $\mathcal{B}(\mathcal{H})$, which contains all its weak limit points. In addition, one requires to every unit vector $\psi \in \mathcal{H}$ an operator $A \in \mathcal{A}$ with $A\psi \neq 0$. Because of the last requirement, the notion of a von Neumann algebra is defined relative to $\mathcal{H}$. (If $\mathcal{A}$ is just weakly closed, then there is a subspace, $\mathcal{H}_0 \subset \mathcal{H}$, relative to which $\mathcal{A}$ is von Neumann.)

J. von Neumann could give a purely algebraic definition of the algebras carrying his name. It is done with the help of commutants. For a subset $\mathcal{F} \subset \mathcal{B}(\mathcal{H})$, the *commutant*, $\mathcal{F}'$, of $\mathcal{F}$ is the set of all $B \in \mathcal{B}(\mathcal{H})$ commuting with all $A \in \mathcal{F}$. The commutant of a set of operators is always a unital and weakly closed subalgebra of $\mathcal{B}(\mathcal{H})$.

If $\mathcal{F}^\dagger = \mathcal{F}$, i.e., $\mathcal{F}$ contains with A always also $A^\dagger$, its commutant $\mathcal{F}'$ becomes a unital *-algebra that, indeed, is a von Neumann algebra.

But then also the double commutant $\mathcal{F}''$, the commutant of the commutant, is a von Neumann algebra. Even more, von Neumann could show $\mathcal{A}$ *is a von Neumann algebra if and only if* $\mathcal{A}'' = \mathcal{A}$.

We need one more definition.

Definition 3. *The* center *of an algebra consists of those of its elements, which commute with every element of the algebra.*

The center of $\mathcal{A}$ is in $\mathcal{A}'$ and vice versa. We conclude

$$\mathcal{A} \cap \mathcal{A}' = \text{ center of } \mathcal{A}. \tag{1.73}$$

If $\mathcal{A}$ is a von Neumann algebra, $\mathcal{A} \cap \mathcal{A}'$ is the center of both, $\mathcal{A}$ and $\mathcal{A}'$.

A von Neumann algebra is called a *factor* if its center consists of the multiples of $\mathbf{1}$ only. Thus, a factor may be characterized by

$$\mathcal{A} \cap \mathcal{A}' = \mathbb{C}\mathbf{1}. \tag{1.74}$$

1.3.3 Classification of Finite Quantum Systems

There are two major branches in group theory, the groups themselves and their representations. We have a similar situation with quantum systems, if they are seen as operator algebras: There is a certain *-algebra and its concrete realizations as operators on a Hilbert space. However, at least in finite-dimensions, our task is much easier than in group theory.

Wedderburn [24] has classified all finite-dimensional matrix algebras,[12] or, what is equivalent, all subalgebras of $\mathcal{B}(\mathcal{H})$ if $\dim \mathcal{H} < \infty$. Here we report and comment his results for *-subalgebras only. (These results could also be

---

[12] He extends the Jordan form from matrices to matrix algebras.

read from the classification of factors by Murray and von Neumann,[13] see [3], Sect. III.2.)

One calls two *-algebras,

$$\mathcal{A} \subset \mathcal{B}(\mathcal{H}) \text{ and } \tilde{\mathcal{A}} \subset \mathcal{B}(\tilde{\mathcal{H}}), \tag{1.75}$$

*-isomorph if there is a map Ψ from $\mathcal{A}$ onto $\tilde{\mathcal{A}}$,

$$A \mapsto \Psi(A) = \tilde{A} \in \tilde{\mathcal{A}}, \quad A \in \mathcal{A},$$

satisfying

(A) $\Psi(c_1 A_1 + c_2 A_2) = c_1 \Psi(A_1) + c_2 \Psi(A_2)$,
(B) $\Psi(AB) = \Psi(A)\Psi(B)$,
(C) $\Psi(A^\dagger) = \Psi(A)^\dagger$,
(D) $A \neq B \Rightarrow \Psi(A) \neq \Psi(B)$,
(E) $\Psi(\mathcal{A}) = \tilde{\mathcal{A}}$.

The first three conditions guarantee the conservation of all algebraic relations under the map Ψ. From them, it follows the positivity of the map Ψ because an element of the form $A^\dagger A$ is mapped to $\tilde{A}^\dagger \tilde{A}$.

Condition (E) says that $\mathcal{A}$ is mapped *onto* $\tilde{\mathcal{A}}$, i.e., every $\tilde{A}$ can be gained as $\Psi(A)$. It follows that the unit element of $\mathcal{A}$ is transformed into that of $\tilde{\mathcal{A}}$.

Condition (D) now shows that Ψ is invertible because to every $A \in \mathcal{A}$, there is exactly one $\tilde{A}$ with $\Psi(A) = \tilde{A}$.

If only (A)–(D) is valid, Ψ maps $\mathcal{A}$ *into* $\tilde{\mathcal{A}}$. Replacing (E) by

(E') $\Psi(\mathcal{A}) \subset \tilde{\mathcal{A}}$

and requiring (A)–(D) defines an *embedding of $\mathcal{A}$ into $\tilde{\mathcal{A}}$.*

If $\mathcal{A} \to \Psi(\mathcal{A}) \subseteq \mathcal{B}(\mathcal{H})$ is an embedding of $\mathcal{A}$, the embedding is also said to be a *-representation* of $\mathcal{A}$ as an operator algebra. A *unital *-representation* of $\mathcal{A}$ maps $\mathbf{1}_\mathcal{A}$ to the identity operator of $\mathcal{H}$.

Important examples of unital *-representations and *-isomorphisms of $\mathcal{B}(\mathcal{H})$ are given by *matrix representations*. Every orthonormal basis $\psi_1, \psi_2, \ldots \psi_n$ of the Hilbert space $\mathcal{H}$, $\dim \mathcal{H} = n$, induces via the map

$$A \to \text{matrix } \mathbf{A} = \{A_{ij}\} \text{ with matrix elements } A_{ij} = \langle \psi_i, A\psi_j \rangle, \ A \in \mathcal{B}(\mathcal{H}),$$

a unital *-isomorphism between $\mathcal{B}(\mathcal{H})$ and the algebra $M_n(\mathbb{C})$, of complex $n \times n$ matrices. If $\dim \mathcal{H} = \infty$, however, matrix representations are a difficult matter.

---

[13] Von Neumann and Murray introduced and investigated von Neumann algebras in a famous series of papers on *Rings of Operators* [25–30], for a general reference see e.g., [31].

Direct Product and the Direct Sum Constructions

Let us review some features of direct products. We start with

$$\mathcal{H} = \mathcal{H}^{\mathrm{A}} \otimes \mathcal{H}^{\mathrm{B}} . \tag{1.76}$$

The algebra $\mathcal{B}(\mathcal{H}^{\mathrm{A}})$ is not a subalgebra of $\mathcal{B}(\mathcal{H})$, but it becomes one by

$$\mathcal{B}(\mathcal{H}^{\mathrm{A}}) \mapsto \mathcal{B}(\mathcal{H}^{\mathrm{A}}) \otimes \mathbf{1}^{\mathrm{B}} \subset \mathcal{B}(\mathcal{H}^{\mathrm{A}} \otimes \mathcal{H}^{\mathrm{B}}) . \tag{1.77}$$

Here "$\mapsto$" points to the unital embedding

$$A \in \mathcal{B}(\mathcal{H}^{\mathrm{A}}) \mapsto A \otimes \mathbf{1}^{\mathrm{B}} \in \mathcal{B}(\mathcal{H}) \tag{1.78}$$

of $\mathcal{B}(\mathcal{H}^{\mathrm{A}})$ into $\mathcal{B}(\mathcal{H})$. It is an *-isomorphism from the algebra $\mathcal{B}(\mathcal{H}^{\mathrm{A}})$ onto $\mathcal{B}(\mathcal{H}^{\mathrm{A}}) \otimes \mathbf{1}^{\mathrm{B}}$. Similarly, $\mathcal{B}(\mathcal{H}^{\mathrm{B}})$ is *-isomorph to $\mathbf{1}^{\mathrm{A}} \otimes \mathcal{B}(\mathcal{H}^{\mathrm{B}})$ and embedded into $\mathcal{B}(\mathcal{H})$ as a *-subalgebra. $\mathbf{1}^{\mathrm{A}} \otimes \mathcal{B}(\mathcal{H}^{\mathrm{B}})$ is the commutant of $\mathcal{B}(\mathcal{H}^{\mathrm{A}}) \otimes \mathbf{1}^{\mathrm{B}}$ and vice versa. Based on $A \otimes B = (A \otimes \mathbf{1}^{\mathrm{B}}) (\mathbf{1}^{\mathrm{A}} \otimes B)$, there is the identity

$$\mathcal{B}(\mathcal{H}^{\mathrm{A}} \otimes \mathcal{H}^{\mathrm{B}}) = \mathcal{B}(\mathcal{H}^{\mathrm{A}}) \otimes \mathcal{B}(\mathcal{H}^{\mathrm{B}}) = \left(\mathcal{B}(\mathcal{H}^{\mathrm{A}}) \otimes \mathbf{1}^{\mathrm{B}} \right) \left(\mathbf{1}^{\mathrm{A}} \otimes \mathcal{B}(\mathcal{H}^{\mathrm{B}}) \right) . \tag{1.79}$$

The algebras of $\mathcal{B}(\mathcal{H}^{\mathrm{A}}) \otimes \mathbf{1}^{\mathrm{B}}$ and $\mathbf{1}^{\mathrm{A}} \otimes \mathcal{B}(\mathcal{H}^{\mathrm{B}})$ are not only subalgebras but also factors. In finite-dimensions, every von Neumann factor on $\mathcal{H}$ is of that structure: *If $\mathcal{A}$ is a sub-factor of $\mathcal{B}(\mathcal{H})$ and $\dim \mathcal{H} < \infty$, then there is a decomposition (1.76) such that $\mathcal{A} = \mathcal{B}(\mathcal{H}^{\mathrm{A}}) \otimes \mathbf{1}^{\mathrm{B}}$.*

It is worthwhile to notice the *information* contained in an embedding of $\mathcal{B}(\mathcal{H}^{\mathrm{A}})$ into $\mathcal{B}(\mathcal{H})$: We need a definite decomposition (1.76) of $\mathcal{H}$ into a direct product of Hilbert spaces with correct dimensions of the factors. Most unitary transformations of $\mathcal{H}$ would give another possible decomposition of the form (1.77) resulting in another embedding (1.77). Generally speaking, distinguishing a subsystem of a quantum system enhances our knowledge and can be well compared with the information gain by a measurement.

One knows how to perform direct sums of linear spaces. To apply it to algebras, one has to say how the multiplication between direct summands is working. Indeed, it works in the most simple way: $\mathcal{A}$ is the *direct sum* of its subalgebras $\mathcal{A}_1, \ldots, \mathcal{A}_m$ if every $A \in \mathcal{A}$ can be written as a sum

$$A = A_1 + \cdots + A_m , \quad A_j \in \mathcal{A}_j \tag{1.80}$$

and the multiplication obeys

$$A_j A_k = \mathbf{0} \text{ whenever } j \neq k . \tag{1.81}$$

One can rewrite the direct sum construction in block matrix notation. Let us illustrate it for the case $m = 3$:

$$A = A_1 + A_2 + A_3 = \begin{pmatrix} A_1 & \mathbf{0} & \mathbf{0} \\ \mathbf{0} & A_2 & \mathbf{0} \\ \mathbf{0} & \mathbf{0} & A_3 \end{pmatrix} , \quad A_j \in \mathcal{A}_j , \tag{1.82}$$

is the *block matrix representation* of the direct sum. If one considers, say $\mathcal{A}_2$, as an algebra in its own right, its embedding into $\mathcal{A}$ is given by

$$A_2 \leftrightarrow \begin{pmatrix} 0 & 0 & 0 \\ 0 & A_2 & 0 \\ 0 & 0 & 0 \end{pmatrix}. \tag{1.83}$$

In contrast to the direct product construction, the embedding (1.83) is not a unital one. Equation (1.82) illustrates the two ways to direct sums: Either an algebra $\mathcal{A}$ can be decomposed as in (1.80), (1.81), or there are algebras $\mathcal{A}_j$ and we build up $\mathcal{A}$ by a direct sum construction out of them. In the latter case, one can write

$$\mathcal{A} = \mathcal{A}_1 \oplus \cdots \oplus \mathcal{A}_m.$$

We shall use both possibilities below.

Types

Our aim is to characterize invariantly the set of *-isomorphic finite von Neumann algebras and to choose in it distinguished ones. The restriction to finite-dimensions makes the task quite simple: *Any *-subalgebra of $\mathcal{B}(\mathcal{H})$ is *-isomorph to a direct sum of factors.*
Let $\mathbf{d}$ be a set of natural numbers,

$$\mathbf{d} = \{d_1, \ldots, d_m\}, \quad |\mathbf{d}| = \sum d_j. \tag{1.84}$$

The number m is called the *length of* $\mathbf{d}$.
We say that $\mathbf{d}' = \{d_1', \ldots, d_m'\}$ is *equivalent* to $\mathbf{d}$, and we write $\mathbf{d} \sim \mathbf{d}'$ if the numbers d_j' are a permutation of the d_j. If this takes place, i.e., if $\mathbf{d} \sim \mathbf{d}'$, we say that $\mathbf{d}$ *is of the same type* as $\mathbf{d}'$.
Given $\mathbf{d}$ as in (1.84) and Hilbert spaces $\mathcal{H}_j$ of dimensions $\dim \mathcal{H}_j = d_j$, we consider

$$\mathcal{B}_{\mathbf{d}} = \mathcal{B}_{d_1, \ldots, d_m} := \mathcal{B}(\mathcal{H}_1) \oplus \mathcal{B}(\mathcal{H}_2) \oplus \cdots \oplus \mathcal{B}(\mathcal{H}_m). \tag{1.85}$$

Similarly we can proceed with $\mathbf{d}'$ and Hilbert spaces $\mathcal{H}_j'$ of dimensions d_j'. We assert

$$\mathbf{d} \sim \mathbf{d}' \iff \mathcal{B}_{\mathbf{d}} \text{ is *-isomorph to } \mathcal{B}_{\mathbf{d}'}. \tag{1.86}$$

To see the claim, we use the permutation $d_j \to d_{i_j}$. In $\mathcal{H}_j$, we choose a basis $|k\rangle_j$, $k = 1, \ldots, d_j$ and a basis $|k\rangle_{i_j}$ in $\mathcal{H}_{i_j}$. Obviously, there is a unitary U with $U|k\rangle_j = |k\rangle_{i_j}$ for all j, k. We see that both algebras become *-isomorphic by $A' = UAU^{-1}$ for any operator A out of (1.85). We are now allowed to state the following:

Definition 4. *A *-algebra is of type $\mathbf{d}$, if it is *-isomorph to the algebra (1.85).*

Remark 11. The number $|\mathbf{d}|$ is occasionally called the *algebraic dimension of* $\mathcal{A}$. Its logarithm (in bits or nats) is called *entropy of* $\mathcal{A}$.

It is often convenient to choose within a type a standard one. This can be done by convention. A usual way is to require $d_1 \geq \cdots \geq d_m$. One then calls $\mathbf{d}$ *standardly* or *decreasingly ordered.* It opens the possibility to visualize the types with Young tableaux (see [32]).

The following example is with $|\mathbf{d}| = 3$. The standard representations are

$$\{3\}, \quad \{2,1\}, \quad \{1,1,1\}.$$

The first one is the *full* algebra $\mathcal{B}(\mathcal{H})$, dim $\mathcal{H} = 3$, the last one is a maximally commutative subalgebra, while the middle one is $\mathcal{B}(\mathcal{H}_2) \oplus \mathbb{C}$. ($\mathbb{C}$ stands for the algebra over an 1-dimensional Hilbert space.) Their Young diagrams are as follows: One may put (part of) Wedderburn's theorem in the form:

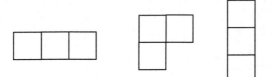

Theorem 3. *Every finite-dimensional* *-*subalgebra of an algebra* $\mathcal{B}(\mathcal{H})$ *is* *-*isomorph to an algebra (1.85), i.e., it is of a certain type* $\mathbf{d}$.

The algebra (1.85) can be identified with a subalgebra of $\mathcal{B}(\mathcal{H})$ where

$$\mathcal{H} = \mathcal{H}_1 \oplus \cdots \oplus \mathcal{H}_m, \quad \mathcal{H}_j = Q_j \mathcal{H}, \quad \mathbf{1} = \sum Q_j, \qquad (1.87)$$

with projections Q_j. The Q_j sum up to $\mathbf{1}$, the identity operator of $\mathcal{H}$. In the course of constructing $\mathcal{B}_{\mathbf{d}}$, the unit element $\mathbf{1}_j \in \mathcal{B}(\mathcal{H}_j)$ is mapped onto the projection $Q_j \in \mathcal{B}(\mathcal{H})$. (We may use alternatively both notations. $\mathbf{1}_j$ can indicate a use "inside" the algebra, while Q_j indicates a definite embedding in a larger algebra.)

Let us restrict the trace over $\mathcal{H}$ to operators $A = A_1 + \cdots + A_m$, $A_j \in \mathcal{B}(\mathcal{H}_j)$. We get

$$\operatorname{tr} A = \sum \operatorname{tr}_j A_j, \quad \operatorname{tr}_j \text{ is the trace over } \mathcal{H}_j.$$

Notice dim $\mathcal{H} = \operatorname{tr} \mathbf{1} = |\mathbf{d}|$. The restriction of the trace of $\mathcal{H}$ to $\mathcal{B}_{\mathbf{d}}$ is called *the canonical trace of* $\mathcal{B}_{\mathbf{d}}$.

Let us denote the canonical trace of $\mathcal{B}_{\mathbf{d}}$ by $\operatorname{tr}^{\text{can}}$, and let us try to explain the word *canonical*. The point of this extra notation is its "intrinsic" nature: Let us think of $\operatorname{tr}^{\text{can}}$ as a linear functional over $\mathcal{B}_{\mathbf{d}}$. It can be characterized by two properties: $\mathrm{tr}^{\mathrm{can}}$ is a positive integer valued at the projections $P \neq \mathbf{0}$ of $\mathcal{B}_\mathbf{d}$, and it is the smallest with that property. It means that *the canonical trace is a type invariant*. We can recover the canonical trace in every algebra *-isomorph to $\mathcal{B}_\mathbf{d}$.

There is yet another aspect to consider. The Lüders–von Neumann, or projective measurements [33–35], are in one-to-one correspondence with the partition of the identity (1.87) of $\mathcal{H}$. We can associate

$$\mathbf{d} = \{d_1, \ldots, d_m\}, \quad d_j = \mathrm{rank}(Q_j) \tag{1.88}$$

with the measurement. The average measurement result is given by a unital, trace preserving, and completely positive map,[14]

$$A \to \Phi(A) := \sum Q_j A Q_j, \quad A \in \mathcal{B}(\mathcal{H}). \tag{1.89}$$

In the direct sum (1.85), the term $\mathcal{B}(\mathcal{H}_j)$ can be identified with $Q_j \mathcal{B}(\mathcal{H}) Q_j$, the algebra of all operators that can be written as $Q_j A Q_j$. Hence,

$$\mathcal{B}_{d_1, \ldots, d_m} := \bigoplus Q_j \mathcal{B}(\mathcal{H}) Q_j, \quad d_j = \mathrm{rank}(Q_j). \tag{1.90}$$

Φ maps $\mathcal{B}(\mathcal{H})$ onto $\mathcal{B}_\mathbf{d}$.

Remark 12. Φ is a completely positive unital map, which maps the algebra onto a subalgebra, though it does not preserve multiplication: Generally $QABQ$ is not equal to $QAQBQ$ with a projection Q. Several interesting questions appear. For instance, which channels result after several applications of projective ones? The problem belongs to the theory of conditional expectations.

The State Space of $\mathcal{B}_\mathbf{d}$

To shorten the notation, we shall write $\Omega(\mathcal{H})$ instead of $\Omega(\mathcal{B}(\mathcal{H}))$. Let us now examine the state space $\Omega(\mathcal{B}_\mathbf{d})$, which is a subset of $\Omega(\mathcal{H})$. Indeed, a state $\underline{\omega}$ of $\mathcal{B}_\mathbf{d}$ can be written as $\underline{\omega}(A) = \mathrm{tr}\,\omega A$ and we conclude

$$\mathrm{tr}\,\omega A = \mathrm{tr}\,\omega \sum Q_j A Q_j = \mathrm{tr}\left(\sum Q_j \omega Q_j\right) A$$

by (1.87). Hence, we can choose $\omega \in \mathcal{B}_\mathbf{d}$ and, after doing so, ω becomes unique. In conclusion, $\Omega(\mathcal{B}_\mathbf{d}) \subset \Omega(\mathcal{H})$ and

$$\omega \in \Omega(\mathcal{B}_\mathbf{d}) \Leftrightarrow \sum Q_j \omega Q_j = \omega, \tag{1.91}$$

for density operators $\omega \in \Omega(\mathcal{H})$.

---

[14] Complete positive maps respect the superposition principle in tensor products [35, 36].

A density operator ω_j of $\mathcal{B}(\mathcal{H}_j)$ can be identified with a density operator on $\mathcal{H}$ supported by $\mathcal{H}_j = Q_j\mathcal{H}$. Equivalently, we have $\underline{\omega}_j(Q_j) = 1$ for the corresponding states. These states form a face of $\Omega(\mathcal{H})$, and these faces are orthogonal one to another. We get the convex combination

$$\omega \in \Omega(\mathcal{B}_\mathbf{d}) \Leftrightarrow \omega = \sum_{j=1}^{m} p_j\omega_j, \quad \operatorname{tr} Q_j\omega_j = \underline{\omega}_j(Q_j) = 1. \tag{1.92}$$

The convex combination (1.92) is uniquely determined by ω, a consequence of the orthogonality $\omega_j\omega_k = \mathbf{0}$ if $j \neq k$.

Theorem 4. *The state space of $\mathcal{B}_\mathbf{d}$, embedded in $\Omega(\mathcal{H})$, $\dim\mathcal{H} = |\mathbf{d}|$, is the direct convex sum of the state spaces $\Omega(\mathcal{H}_j)$ with $\dim\mathcal{H}_j = d_j$ and $d_j \in \mathbf{d}$.*

We see further: Φ defined in (1.89) maps $\Omega(\mathcal{H})$ onto $\Omega(\mathcal{B}_\mathbf{d})$.

A picturesque description is in saying we have a simplex with m corners and we "blow up", for all j, the jth corner to the convex set $\Omega(\mathcal{H}_j)$. Then we perform the convex hull.

From (1.87), (1.92), and the structure of $\mathcal{B}_\mathbf{d}$, we find the pure (i.e., extremal) density operators by selecting j and a unit vector $|\psi\rangle \in \mathcal{H}_j$ to be $P = |\psi\rangle\langle\psi|$. (We may also write $\pi = P$, but presently, we like to see the density operator of a pure state as a member of the projections. This double role of rank one projectors is a feature of discrete type I von Neumann algebras.) Let $\underline{\pi}$ be the state of $\mathcal{B}_\mathbf{d}$ with density operator P. Just by insertion, we see

$$PAP = \underline{\pi}(A)P \text{ for all } A \in \mathcal{B}_\mathbf{d}. \tag{1.93}$$

On the other hand, if for any projector P, there is a linear form $\underline{\pi}$ such that (1.93) is valid, $\underline{\pi}$ must be a state and P its density operator. (Inserting $A = P$, we find $\underline{\pi}(P) = 1$. Because $PA^\dagger AP$ is positive, $\underline{\pi}(A^\dagger A)$ must be positive. Hence, it follows from (1.93), if P is a projection P, $\underline{\pi}$ is a state.) It is also not difficult to see that (1.93) requires P to be of rank one and $\underline{\pi}$ is pure. We now have another criterion for pure states, which refers to the algebra only.

*Let $\mathcal{A}$ be *-isomorph to an algebra $\mathcal{B}_\mathbf{d}$. A state $\underline{\pi}$ of $\mathcal{A}$ is pure if and only if there is a projection P such that (1.93) is valid for all $A \in \mathcal{A}$. Then P is the density operator of the pure state, or, in other terms, $\pi = P$.*

The projections, which are density operators of pure states, enjoy a special property, they are *minimal*. A projection P is minimal in an algebra, if from $P = P_1 + P_2$ with P_j projections, it follows either $P_1 = P$ or $P_1 = \mathbf{0}$.

It is quite simple to see $P = |\psi\rangle\langle\psi|$ for a minimal projection operator of $\mathcal{B}_\mathbf{d}$ and, hence, it is a density operator of a pure state. Therefore, in algebras *-isomorph to an algebra $\mathcal{B}_\mathbf{d}$, we can assert the following: *A projection P of $\mathcal{A}$ is minimal if and only if it is the density operator of a pure state of $\mathcal{A}$.*

A further observation: Let $\mathcal{A}$ be of type $\mathbf{d}$. There is a linear functional over $\mathcal{A}$, which attains the value 1 for all minimal projections. This linear form is the canonical trace of $\mathcal{A}$.

By slightly reformulating some concepts from Hilbert space, we have obtained purely algebraic ones. This way of thinking will also dominate our next issue.

Transition Probabilities for Pure States

We start again with $\mathcal{B}_{\mathbf{d}}$ as a subalgebra of $\mathcal{B}(\mathcal{H})$ with $\dim \mathcal{H} = |\mathbf{d}|$. Let us consider some pure states $\underline{\pi}_j$ of $\mathcal{B}_{\mathbf{d}}$. They can be represented by unit vectors,

$$\underline{\pi}_j(A) = \langle \psi_j, A\psi_j \rangle, \quad \pi_j \equiv P_j = |\psi_j\rangle\langle\psi_j|. \tag{1.94}$$

Let us agree, *as usual*, that

$$\Pr(\underline{\pi}_1, \underline{\pi}_2) = \Pr(\pi_1, \pi_2) = |\langle \psi_1, \psi_2 \rangle|^2 \tag{1.95}$$

is the transition probability. To obtain the same value for two pure states of an algebra $\mathcal{A}$ *-isomorph to $\mathcal{B}_{\mathbf{d}}$, we reformulate (1.95) in an invariant way: The right-hand side of (1.95) is the trace of $\pi_1\pi_2$. In $\mathcal{B}_{\mathbf{d}}$, the canonical trace coincides with the trace over $\mathcal{H}$. Hence, for a general algebra $\mathcal{A}$, we have to use the canonical trace. We get

$$\Pr(\underline{\pi}_1, \underline{\pi}_2) = \Pr(\pi_1, \pi_2) = \mathrm{tr}^{\mathrm{can}} \pi_1\pi_2. \tag{1.96}$$

Switching, for convenience, to the notation $P_j = \pi_j$, we get $P_1 P_2 P_1 = \underline{\pi}_1(P_2)P_1$ by inserting $A = P_2$ in (1.93) for $\underline{\pi}_1$. By taking the trace, we get the expression (1.96) for the transition probability. Interchanging the indices, we finally get

$$\Pr(\underline{\pi}_1, \underline{\pi}_2) = \Pr(\pi_1, \pi_2) = \underline{\pi}_1(P_2) = \underline{\pi}_2(P_1). \tag{1.97}$$

This and (1.96) express the transition probability for any two pure states of an algebra $\mathcal{A}$, *-isomorph to a finite-dimensional von Neumann algebra.

Our next aim is to prove

$$\Pr(\underline{\pi}_1, \underline{\pi}_2) = \inf_{A>0} \underline{\pi}_1(A)\underline{\pi}_2(A^{-1}), \tag{1.98}$$

A is running through all invertible positive elements of $\mathcal{A}$.

It suffices to prove the assertion for $\mathcal{B}_{\mathbf{d}}$. Relying on (1.94), we observe

$$|\langle \psi_1, \psi_2 \rangle|^2 \leq \langle A^{1/2}\psi_1, A^{1/2}\psi_1 \rangle \langle A^{-1/2}\psi_2, A^{-1/2}\psi_2 \rangle.$$

Therefore, the left-hand side of (1.98) cannot be larger than the right one. It remains to ask, whether the asserted infimum can be reached. For this purpose, we set

$$A_s = s\mathbf{1} + P_2 \,, \quad A_s^{-1} = \frac{1}{s}\mathbf{1} - \frac{1}{s(1+s)}P_2 \,.$$

A_s is positive for $s > 0$. We find

$$\pi_1(A_s) = s + \Pr(\pi_1, \pi_2)\,, \quad \pi_2(A_s^{-1}) = \frac{1}{s} - \frac{1}{s(1+s)} = (1+s)^{-1}\,,$$

and it follows

$$\lim_{s \to +0} \pi_1(A_s)\pi_2(A_s^{-1}) = \Pr(\pi_1, \pi_2),$$

which proves (1.98).

In [37], a similar inequality is reported:

$$2|\langle \psi_1, \psi_2 \rangle| = \inf_{A>0} \langle \psi_1, A\psi_1 \rangle + \langle \psi_2, A^{-1}\psi_2 \rangle,$$

with A varying over all invertible positive operators on a Hilbert space. The equation remains valid for pairs of pure states in a finite *-subalgebra $\mathcal{A}$ of $\mathcal{B}(\mathcal{H})$. The slight extension of the inequality reads

$$2\sqrt{\Pr(\pi_1, \pi_2)} = \inf_{0 < A \in \mathcal{A}} \pi_1(A) + \pi_2(A^{-1})\,. \tag{1.99}$$

To prove it, we write down the inequality

$$0 \le \left[t\sqrt{\pi_1(A)} - t^{-1}\sqrt{\pi_2(A^{-1})} \right]^2 ,$$

t a positive number. We get

$$2\sqrt{\pi_1(A)\pi_2(A^{-1})} \le t^2 \pi_1(A) + t^{-2}\pi_2(A^{-1})$$

and, by (1.98), the right-hand side of (1.99) is not less than the left one. Adjusting the operators A_s above to $B_s = t_s^2 A_s$ in such a way that $\pi_1(B_s) = \pi_2(A^{-1})$, then

$$2\sqrt{\pi_1(B_s)\pi_2(B_s^{-1})} = \pi_1(B_s) + \pi_2(B_s^{-1})\,.$$

Performing the limit $s \to 0$ as in the proof of (1.98) shows that the asserted infimum can be approached arbitrarily well.

Last but not least, we convince ourselves that the transition probability between pure states *is already fixed by the convex structure* of $\Omega(\mathcal{A})$ respectively of $\underline{\Omega}(\mathcal{A})$.

We prove it for $\Omega(\mathcal{A})$. Let l be a real linear form over the Hermitian operators of $\mathcal{A}$, such that for all density operators ω, one has $0 \le l(\omega) \le 1$. Then $l(A) \ge 0$ for all positive operators A. Now assume $l(P) = 1$ for a minimal projection. Combining both assumptions, we find $l(\mathbf{1}_{\mathcal{A}}) = 1$. Hence, l is a pure state π of $\mathcal{A}$. If P' is another minimal projection, i.e., an extremal element of $\Omega(\mathcal{A})$, we can calculate the transition probability $l(P') = \pi(P')$.

The result implies the following: Our state spaces are *rigid:* If a linear map Φ,

$$\Phi : \mathcal{A} \mapsto \mathcal{A} ,$$

maps $\Omega(\mathcal{A})$ one-to-one onto itself, it must preserve the transition probabilities between pure density operators.

In the particular case $\mathcal{A} = \mathcal{B}(\mathcal{H})$, the map Φ must be a Wigner symmetry. A useful reformulation of this statement reads as follows: *Let Φ_1 and Φ_2 denote invertible linear maps from $\mathcal{B}(\mathcal{H})$ onto $\mathcal{B}(\mathcal{H})$. Assume $\Omega(\mathcal{H})$ is mapped by both maps onto the same set of operators. Then there is a unitary or an anti-unitary V such that*

$$\Phi_2(X) = \Phi_1(VXV^*) \text{ for all } X \in \mathcal{B}(\mathcal{H}) .$$

Indeed, $\Phi_1^{-1}\Phi_2$ must be a Wigner symmetry.

Remark 13. Mielnik has defined a "transition probability" between extremal states of a compact convex set K in this way. Let P and P' be two extremal points of K. The "probability" of the transition $P \rightarrow P'$ is defined to be $\inf l(P')$, where l runs through all real affine functionals on K with values between 0 and 1 and with $l(P) = 1$. Indeed, for $\Omega(\mathcal{A})$, the procedure gives the correct transition probability as shown above.

1.3.4 All Subsystems for $\dim \mathcal{H} < \infty$

Here, we are interested in Wedderburn's description, of the *-subalgebras of $\mathcal{B}(\mathcal{H})$, $\dim \mathcal{H} < \infty$ [24, 38]. In short, such a subalgebra is *-isomorph to a certain algebra $\mathcal{B}_{\mathbf{d}}$.

We change our notations toward its use in quantum information. We think of a quantum system with algebra $\mathcal{B}^{\mathrm{A}}$, owned by some person, say Alice. We may assume the algebra $\mathcal{B}^{\mathrm{A}}$ to be a unital *-subalgebra of a larger algebra $\mathcal{B}(\mathcal{H}^{\mathrm{AB}})$. The type of $\mathcal{B}^{\mathrm{A}}$ is the not ordered list $\mathbf{d}^{\mathrm{A}} = \{d_1^{\mathrm{A}}, \ldots, d_m^{\mathrm{A}}\}$. Alice is allowed to operate freely within her subsystem, which is also called *the A-system.*

Theorem 5. *Let $\mathcal{B}^{\mathrm{A}}$ be a unital *-subalgebra of $\mathcal{B}(\mathcal{H}^{\mathrm{AB}})$ of type $\mathbf{d}^{\mathrm{A}}$. Then, there is a decomposition*

$$\mathcal{H}^{\mathrm{AB}} = \mathcal{H}_1 \oplus \cdots \oplus \mathcal{H}_m , \quad \mathcal{H}_j = \mathcal{H}_j^{\mathrm{A}} \otimes \mathcal{H}_j^{\mathrm{B}}, \tag{1.100}$$

$$d_j^{\mathrm{A}} = \dim \mathcal{H}_j^{\mathrm{A}} , \quad d_j^{\mathrm{B}} = \dim \mathcal{H}_j^{\mathrm{B}} ,$$

such that

$$\mathcal{B}^{\mathrm{A}} = \left(\mathcal{B}(\mathcal{H}_1^{\mathrm{A}}) \otimes \mathbf{1}_1^{\mathrm{B}}\right) \oplus \cdots \oplus \left(\mathcal{B}(\mathcal{H}_m^{\mathrm{A}}) \otimes \mathbf{1}_m^{\mathrm{B}}\right) . \tag{1.101}$$

Equally well, we may represent $\mathcal{B}^{\mathrm{A}}$ as a diagonal block matrix with diagonal blocks $\mathcal{B}(\mathcal{H}_j^{\mathrm{A}}) \otimes \mathbf{1}_j^{\mathrm{B}}$.

In the theorem, we denote by $\mathbf{1}_j^A$ the identity operator of $\mathcal{H}_j^A$ and by $\mathbf{1}_j^B$ the one of $\mathcal{H}_j^B$. Therefore, $\mathbf{1}_j^A \otimes \mathbf{1}_j^B$ is equal to $\mathbf{1}_j$, the identity operator of $\mathcal{H}_j$. The latter can be identified with the projection Q_j projecting $\mathcal{H}$ onto $\mathcal{H}_j$, i.e., $\mathbf{1}_j = Q_j$. Equations (1.100) and (1.101) describe how $\mathcal{B}_{\mathbf{d}^A}$ is embedded into $\mathcal{B}(\mathcal{H}^{AB})$ to become $\mathcal{B}^A$ by the embedding *-isomorphism

$$A_1 + \cdots + A_m \leftrightarrow A_1 \otimes \mathbf{1}_1 + \cdots + A_m \otimes \mathbf{1}_m , \quad A_j \in \mathcal{B}(\mathcal{H}_j^A) . \qquad (1.102)$$

Now we can see, why, by identifying $\mathcal{B}^A$ as a subsystem of $\mathcal{B}(\mathcal{H}^{AB})$, a second subsystem, called "Bob's system", appears quite naturally. It consists of those operators of $\mathcal{B}(\mathcal{H}^{AB})$, which can be executed independently of Alice's actions. These operators must commute with those of the A-system. Hence, all of them[15] constitute Bob's algebra $\mathcal{B}^B$. Therefore, Bob's algebra is the commutant of $\mathcal{B}^A$ in $\mathcal{B}(\mathcal{H}^{AB})$. By (1.101), we see

$$\mathcal{B}^B := (\mathcal{B}^A)' = \left(\mathbf{1}_1^A \otimes \mathcal{B}(\mathcal{H}_1^B)\right) \oplus \cdots \oplus \left(\mathbf{1}_m^A \otimes \mathcal{B}(\mathcal{H}_m^B)\right) . \qquad (1.103)$$

Further, we can find the center of $\mathcal{B}^A$, respectively, of $\mathcal{B}^B$. The center describes the actions that are allowed to both, Alice and Bob. These operators behave *classical* for them. We get

$$\mathcal{B}^A \cap \mathcal{B}^B = \mathbb{C}Q_1 + \cdots + \mathbb{C}Q_m , \quad Q_j = \mathbf{1}_j^A \otimes \mathbf{1}_j^B = \mathbf{1}_j . \qquad (1.104)$$

The type of the commutant consists of m-times the number one.

The types of $\mathcal{B}^A$ and of $\mathcal{B}^B$ are $\mathbf{d}^A = \{d_1^A, \ldots, d_m^A\}$ and $\mathbf{d}^B = \{d_1^B, \ldots, d_m^B\}$, respectively. In general, neither one can be assumed decreasingly ordered. Notice

$$\dim \mathcal{H}^{AB} = \sum d_j^A d_j^B .$$

Let us denote by $\mathcal{B}^{AB}$ the subalgebra generated by $\mathcal{B}^A$ and $\mathcal{B}^B$. Equivalently, $\mathcal{B}^{AB}$ is the smallest subalgebra of $\mathcal{B}(\mathcal{H})$ containing $\mathcal{B}^A$ and $\mathcal{B}^B$,

$$\mathcal{B}^{AB} = \mathcal{B}(\mathcal{H}_1) \oplus \cdots \oplus \mathcal{B}(\mathcal{H}_m) = Q_1 \mathcal{B}(\mathcal{H}) Q_1 + \cdots + Q_m \mathcal{B}(\mathcal{H}) Q_m . \qquad (1.105)$$

The fact that $\mathcal{B}^{AB}$ is generated in a larger algebra by the algebras $\mathcal{B}^A$ and $\mathcal{B}^B$ can be expressed also by $\mathcal{B}^{AB} = \mathcal{B}^A \vee \mathcal{B}^B$. The type of $\mathcal{B}^{AB}$ is

$$\mathbf{d}^{AB} := \{d_1^A d_1^B, \ldots d_m^A d_m^B\} .$$

As long as $\mathcal{B}^{AB}$ is not considered itself as a subsystem of a larger one, we are allowed to write $\mathcal{B}^{AB} = \mathcal{B}_{\mathbf{d}^{AB}}$.

Embedding and Partial Trace

Let us stick to the just introduced subalgebras of $\mathcal{B}(\mathcal{H}^{AB})$, namely $\mathcal{B}^A$, $\mathcal{B}^B$, $\mathcal{B}^{AB}$, and $\mathcal{C} = \mathcal{B}^A \cap \mathcal{B}^B$.

---

[15] We ignore that there may be further restrictions to Bob.

If $\underline{\omega}^{AB}$ is a state of $\mathcal{B}^{AB}$, its restriction to $\mathcal{B}^A$ is a state $\underline{\omega}^A$ of $\mathcal{B}^A$. The restriction map lets fall down any functional of $\mathcal{B}^{AB}$ to $\mathcal{B}^A$. After its application, we have obtained $\underline{\omega}^A$ from $\underline{\omega}^{AB}$ and all what has changed is as follows: Only arguments from $\mathcal{B}^A$ will be allowed for $\underline{\omega}^A$.

The *partial trace*,[16] $\omega^{AB} \to \omega^A$, concerns the involved density operators. It is a map from $\mathcal{B}^{AB}$ to $\mathcal{B}^A$. For its definition and for later use, we need the canonical traces of $\mathcal{B}^A$ and $\mathcal{B}^B$, which we now denote by tr^A and tr^B, respectively. It is

$$\mathrm{tr}^A \omega^A X = \underline{\omega}^{AB}(X) \equiv \mathrm{tr}\,\omega^{AB} X , \quad X \in \mathcal{B}^A . \tag{1.106}$$

Remark 14. The algebra $\mathcal{B}^{AB}$ is of the form (1.90) and (1.85). Therefore, its canonical trace tr^{AB} is the canonical trace over $\mathcal{B}(\mathcal{H})$, i.e., it is just the trace over $\mathcal{H}$.

We read (1.106) as follows: The right-hand side becomes a linear form over $\mathcal{B}^A$. Every linear functional over $\mathcal{B}^A$ can be uniquely written by the help of the canonical trace as done at the left-hand side. This defines the partial trace

$$\omega^{AB} \to \omega^A := \mathrm{tr}_B \omega^{AB} \tag{1.107}$$

from $\mathcal{B}^{AB}$ to $\mathcal{B}^A$. The partial trace is *dual* to the restriction map.

The algebra $\mathcal{B}^{AB}$ consists of all operators

$$Z = \sum_{j=1}^m X_j Y_j = \sum_{j=1}^m (A_j \otimes \mathbf{1}_j^B)(\mathbf{1}_j^A \otimes B_j), \tag{1.108}$$

with

$$A_j \in \mathcal{B}(\mathcal{H}_j^A), \quad B_j \in \mathcal{B}(\mathcal{H}_j^B) .$$

This follows from (1.100) and (1.101). Now

$$\mathrm{tr}\,Y_j = \mathrm{tr}\,(\mathbf{1}_j^A \otimes B_j) = d_j^A \,\mathrm{tr}\,B_j = d_j^A \,\mathrm{tr}^B Y_j . \tag{1.109}$$

The dimensional factors point to the main difference between the canonical trace of $\mathcal{B}^A$ and of the *induced trace*, which is the trace of $\mathcal{H}$ applied to the operators of the subalgebra $\mathcal{B}^A$. All together, we get the partial trace of the operator (1.108),

$$\mathrm{tr}_B Z = \sum (d_j^A)^{-1} (\mathrm{tr}\,Y_j) X_j = \sum X_j \,\mathrm{tr}^B Y_j . \tag{1.110}$$

An important conclusion is

$$\mathrm{tr}_B XZ = X\mathrm{tr}_B Z , \quad \mathrm{tr}_B ZX = (\mathrm{tr}_B Z)X , \quad X \in \mathcal{B}^A . \tag{1.111}$$

Similar to tr_B, one treats the partial trace tr_A. One can check

---

[16] The partial trace is a particular *conditional expectation*.

$$\mathrm{tr_B}\,\mathrm{tr_A} = \mathrm{tr_A}\,\mathrm{tr_B} = \mathrm{tr_{AB}}\,. \tag{1.112}$$

Because $\mathrm{tr_{AB}}$ projects an operator of $\mathcal{B}^{AB}$ into both $\mathcal{B}^A$ and $\mathcal{B}^B$, it projects onto the center, $\mathcal{C} = \mathcal{B}^A \cap \mathcal{B}^B$, of $\mathcal{B}^{AB}$. By inspection, we identify (1.112) with the partial trace of $\mathcal{B}^{AB}$ onto its center.

The ansatz (1.106) applies also to the partial trace from $\mathcal{B}(\mathcal{H}^{AB})$ to $\mathcal{B}^{AB}$. Because the latter is the commutant of the center $\mathcal{C} = \mathcal{B}^A \cap \mathcal{B}^B$, we have

$$\mathrm{tr_{A\cap B}}(Z) = \sum Q_j Z Q_j, \quad \mathcal{B}^A \cap \mathcal{B}^B = \sum Q_j \mathbb{C}, \tag{1.113}$$

see (1.87) and (1.89), where the map has been called Φ because at this occasion, the partial trace was not yet defined.

1.4 Transition Probability, Fidelity, and Bures Distance

The aim is to define transition probabilities [39–43] between two states of a quantum system, say $\mathcal{A}$, by operating in larger quantum systems. We call it $\mathrm{Pr}(\rho, \underline{\omega})$ or, with density operators, $\mathrm{Pr}(\rho, \omega)$.

The notation for the fidelity, $\mathrm{F}(\rho, \omega)$, used here is that of Nielsen and Chuang[17] [36], i.e., it is the square root of the transition probability,

$$\mathrm{F}(\rho, \omega) := \sqrt{\mathrm{Pr}(\rho, \omega)}\,. \tag{1.114}$$

This quantity is also denoted by *square-root fidelity* or by *overlap*. An analogous quantity between two probability measures is known as *Kakutani mean* [44], and, for probability vectors, as *Bhattacharyya coefficient*. Occasionally, the latter name is also used in the quantum case.

There is a related extension of the Fubini–Study distance to the Bures one [45]. The Bures distance $\mathrm{dist}_B(\rho, \underline{\omega})$ is an inner distance in the set of positive linear functionals, or, in finite-dimensions equivalently, in the set of positive operators. The Bures distance is a quantum version of the Fisher distance [46].

There is a Riemannian metric, the Bures metric, belonging to the Bures distance [47]. It extends the Fubini–Study metric to general (i.e., mixed) states. It also extends the Fisher metric, originally defined for spaces of probability measures, to quantum theory. (However, there is a large class of reasonable quantum versions of the Fisher metric, discovered by Petz [48, 49].)

Below we shall define transition probability and related quantities "operationally". Later we shall discuss several possibilities to get them "intrinsically", without leaving a given quantum system [39–43].

From the mathematical point of view, there are some quite useful tricks in handling two positive operators in general position.

---

[17] There are also quite different expressions called *fidelity*.

1.4.1 Purification

Purification is a tool to extend properties of pure states to general ones. It lives from the fact that given a state of a quantum system A, say $\underline{\omega}^A$, there are pure states in sufficiently larger systems, the restriction of which to the A-system coincides with $\underline{\omega}^A$. The same terminology is used for the corresponding density operators. Of special interest is the case of a larger system that purifies all states of the A-system.

We can lift any state of a quantum system to every larger system. We can require that a pure state is lifted to a pure state: Let $\mathcal{A}_1 \subset \mathcal{A}_2 \subset \mathcal{B}(\mathcal{H})$ and $\underline{\pi}_1$ a pure state of $\mathcal{A}_1$ with density operator P_1. Being a minimal projection in $\mathcal{A}_1$, P_1 may not be minimal in $\mathcal{A}_2$. But then, we can write P_1 as a sum of minimal projections of $\mathcal{A}_2$. If P_2 is one of them and $\underline{\pi}_2$ the corresponding pure state of $\mathcal{A}_2$, then $\underline{\pi}_2$ is a pure lift of $\underline{\pi}_1$.

As a matter of fact, every state $\underline{\omega}_2$ satisfying $\underline{\omega}_2(P_1) = 1$ is a lift of $\underline{\pi}_1$ to $\mathcal{A}_2$. These states exhaust all lifts of $\underline{\pi}_1$ to $\mathcal{A}_2$. They constitute a face of the state space of $\mathcal{A}_2$.

Assume the state $\underline{\omega}_1$ of $\mathcal{A}_1$ is written as a convex combination of pure states. After lifting them to pure states of $\mathcal{A}_2$, we get a convex combination, which extends $\underline{\omega}_1$ to $\mathcal{A}_2$.

Generally, there is a great freedom in extending states of a quantum system to a larger quantum system.

The most important case is the purification of the states of $\mathcal{B}(\mathcal{H})$ or, equivalently, of $\Omega(\mathcal{H})$, well described in [36, 50–52], and in other text books on quantum information theory. It works by embedding $\mathcal{B}(\mathcal{H})$ as the subalgebra $\mathcal{B}(\mathcal{H}) \otimes \mathbf{1}'$ into a bipartite system $\mathcal{B}(\mathcal{H} \otimes \mathcal{H}')$, provided $d = \dim \mathcal{H} \leq \dim \mathcal{H}'$. Given $\omega \in \Omega(\mathcal{H})$, a unit vector $\psi \in \mathcal{H} \otimes \mathcal{H}'$ *purifys* ω, and $\pi = |\psi\rangle\langle\psi|$ is a purification of ω, if

$$\langle\psi, (X \otimes \mathbf{1}')\psi\rangle = \operatorname{tr} X\omega \text{ for all } X \in \mathcal{B}(\mathcal{H}) \tag{1.115}$$

or, equivalently,

$$\underline{\omega}(X) = \underline{\pi}(X \otimes \mathbf{1}') \equiv \operatorname{tr}\pi(X \otimes \mathbf{1}') . \tag{1.116}$$

To get a suitable ψ, one chooses d ortho-normal vectors $|j\rangle'$ in $\mathcal{H}'$ and a basis $|j\rangle$ of eigenvectors of ω. Now

$$|\psi\rangle = \sum \lambda^{1/2}|j\rangle \otimes |j\rangle' \text{ with } \omega|j\rangle = \lambda_j|j\rangle \tag{1.117}$$

purifies ω. Indeed,

$$\langle\psi, (X \otimes \mathbf{1}')\psi\rangle = \sum \lambda_j\langle j|X|j\rangle = \operatorname{tr} X\omega .$$

Now let $\mathcal{A}$ be a unital *-subalgebra of $\mathcal{B}(\mathcal{H})$ and $\underline{\omega}^A$ one of its states. We have already seen that we can lift $\underline{\omega}^A$ to a state $\underline{\omega}$ of $\mathcal{B}(\mathcal{H})$. With the density operator ω of $\underline{\omega}$, we now proceed as above.

1.4.2 Transition Probability and Fidelity

Let $\mathcal{A}$ be a unital *-subalgebra of an algebra $\mathcal{B}(\mathcal{H})$ with finite-dimensional Hilbert space $\mathcal{H}$. Let $\underline{\omega}_1^A$ and $\underline{\omega}_2^A$ denote the two states of $\mathcal{A}$ and ω_1^A and ω_2^A denote their density operators.

The task is to prepare $\underline{\omega}_2$, if the state of our system is $\underline{\omega}_1$. To do so, one thinks of purifications $\underline{\pi}_j$ of our $\underline{\omega}_j^A$ in a larger quantum system in which $\mathcal{A}$ is embedded. One then tests, in the larger system, whether π_2 is true. If the answer of the test is "yes", then π_2 and, hence, ω_2^A is prepared. The probability of success is $\mathrm{Pr}(\underline{\pi}_1, \underline{\pi}_2)$ as defined in (1.95), (1.96), and (1.97).

One now asks for optimality of the described procedure, i.e., one looks for a projective measurement in a larger system that prepares a purification of ω_2^A with maximal probability.

This maximal possible probability for preparing ω_2^A given ω_1^A is called the *transition probability* from ω_1^A to ω_2^A or, as this quantity is symmetric in its entries, the transition probability of the pair $\{\omega_1^A, \omega_2^A\}$. The definition applies to any unital C*-algebra and, formally, to any unital *-algebra [39, 53, 54].

The definition can be rephrased

$$\mathrm{Pr}(\underline{\omega}_1^A, \underline{\omega}_2^A) := \sup \mathrm{Pr}(\underline{\pi}_1, \underline{\pi}_2), \qquad (1.118)$$

where $\underline{\pi}_1, \underline{\pi}_2$ is running through all simultaneous purifications of ω_1^A, ω_2^A. We also use the density operator notation

$$\mathrm{Pr}(\omega_1^A, \omega_2^A) \equiv \mathrm{Pr}(\underline{\omega}_1^A, \underline{\omega}_2^A).$$

In almost the same way, we define the fidelity by

$$\mathrm{F}(\omega_1, \omega_2) = \sup |\langle \psi_1, \psi_2 \rangle|, \qquad (1.119)$$

where ψ_1, ψ_2 run through all simultaneous purifications of ω_1, ω_2 in some $\mathcal{B}(\mathcal{H})$. Though we do not include all possible purifications (by using only "full" algebras), the relation (1.114) remains valid.

Remark 15. Let $\underline{\omega}_1$ and $\underline{\omega}_2$ be two states of a unital C*-algebra $\mathcal{A}$ and $\underline{\nu}$ one of its linear functionals. If and only if

$$|\underline{\nu}(A^\dagger B)|^2 \leq \underline{\omega}_1(A^\dagger A)\,\underline{\omega}_2(B^\dagger B), \qquad (1.120)$$

for all $A, B \in \mathcal{A}$, there is an embedding Ψ in an algebra $\mathcal{B}(\mathcal{H})$ such that there are purifying vectors ψ_1, ψ_2 satisfying

$$\underline{\nu}(A) = \langle \psi_1, \Psi(A)\psi_2 \rangle, \quad A \in \mathcal{A}. \qquad (1.121)$$

This relation implies

$$|\underline{\nu}(\mathbf{1})|^2 \leq \mathrm{Pr}(\underline{\omega}_1, \underline{\omega}_2). \qquad (1.122)$$

Now the definition above can be rephrased: The transition probability is the sup of $|\underline{\nu}(\mathbf{1})|^2$ with $\underline{\nu}$ running through all linear forms satisfying (1.121). There exist linear functionals $\underline{\nu}$ satisfying (1.120) with equality in (1.122). Their structure and eventual uniqueness has been investigated by Alberti [55–57].

The Bures Distance

For the next term, the Bures distance [45], it is necessary, not to insist in normalization of the vectors and not to require the trace one condition for the density operators in (1.119). Remembering (1.28) and (1.29), one defines the *Bures distance* by

$$\text{dist}_B(\omega_1, \omega_2) = \sup \text{dist}_{\text{FS}}(\pi_1, \pi_2) = \sup \| \psi_2 - \psi_1 \|, \qquad (1.123)$$

where the sup is running through all simultaneous purifications of ω_1 and ω_2. Because of (1.119), this comes down to

$$\text{dist}_B(\omega_1, \omega_2) = \sqrt{\text{tr}\,\omega_1 + \text{tr}\,\omega_2 - 2F(\omega_1, \omega_2)}\,. \qquad (1.124)$$

Rewritten for two density operators, it becomes

$$\text{dist}_B(\omega_1, \omega_2) = \sqrt{2 - 2\sqrt{\text{Pr}(\omega_1, \omega_2)}}, \quad \text{tr}\,\omega_j = 1\,.$$

If only curves entirely within the density operators are allowed in optimizing for the shortest path, we get a further variant of the Bures distance, namely

$$\text{Dist}_B(\omega_1, \omega_2) = \arccos\sqrt{\text{Pr}(\omega_1, \omega_2)}, \qquad (1.125)$$

in complete analogy to the discussion of the Fubini–Study case.

What remains is to express of (1.118) or (1.119) in a more explicit way. The dangerous thing in these definitions is the word "all". How to control all possible purifications of every embedding in suitable larger quantum systems? The answer is in a *saturation* property: One cannot do better in (1.118) than by the squared algebraic dimension of $\mathcal{A}$ for the purifying system.

1.4.3 Optimization

Let $\mathcal{A} = \mathcal{B}_{\mathbf{d}}$ with $d = |\mathbf{d}|$ as in (1.85) and (1.87). Hence, up to a slight change in notation, we have

$$\mathcal{A} = \mathcal{B}(\mathcal{H}_1^{\text{A}}) \oplus \mathcal{B}(\mathcal{H}_2^{\text{A}}) \oplus \cdots \oplus \mathcal{B}(\mathcal{H}_m^{\text{A}})\,, \qquad (1.126)$$

$$\mathcal{H}^{\text{A}} = \mathcal{H}_1^{\text{A}} \oplus \cdots \oplus \mathcal{H}_m^{\text{A}}\,.$$

$\mathcal{A} \subset \mathcal{B}(\mathcal{H}^{\text{A}})$ is an embedding with the least possible Hilbert space dimension. (In contrast to (1.101), the general case.) Our working space will be

$$\mathcal{H}^{\text{AB}} = \mathcal{H}^{\text{A}} \otimes \mathcal{H}^{\text{B}}\,, \quad \dim\mathcal{H}^{\text{B}} = \dim\mathcal{H}^{\text{A}} = d\,. \qquad (1.127)$$

The production of purifying vectors is simplified, first, by selecting a maximally entangled vector

$$|\varphi\rangle = \sum_{j=1}^{d} |jj\rangle \equiv \sum |j\rangle^{\mathrm{A}} \otimes |j\rangle^{\mathrm{B}} \tag{1.128}$$

of length d. $\{|j\rangle^{\mathrm{A}}\}$ and $\{|j\rangle^{\mathrm{B}}\}$ are bases of $\mathcal{H}^{\mathrm{A}}$ and $\mathcal{H}^{\mathrm{B}}$, respectively. For any $X \in \mathcal{B}(\mathcal{H}^{\mathrm{A}})$, we get

$$(X \otimes \mathbf{1}^{\mathrm{A}})|\varphi\rangle = \sum X|j\rangle^{\mathrm{A}} \otimes |j\rangle^{\mathrm{B}} \,.$$

Bases are linearly independent. Hence if the right-hand side is zero, then $X = \mathbf{0}$. Because the dimension of $\mathcal{B}(\mathcal{H}^{\mathrm{A}})$, as a linear space, is equal to the dimension of the Hilbert space (1.127), *every vector ψ in $\mathcal{H}^{\mathrm{AB}}$ has a unique representation* $(X \otimes 1)|\varphi\rangle$. One computes for $X_1, X_2 \in \mathcal{B}(\mathcal{H}^{\mathrm{A}})$ the partial trace

$$\psi_i = (X_i \otimes 1)|\varphi\rangle \;\Rightarrow\; \mathrm{tr_B}|\psi_1\rangle\langle\psi_2| = X_1 X_2^{\dagger} \tag{1.129}$$

because we have to trace out the B-system in

$$\sum X_1 (|j\rangle^{\mathrm{A}}\langle k|) X_2^{\dagger} \otimes (|j\rangle^{\mathrm{B}}\langle k|) \,.$$

Our choice of $\mathcal{A}$ implies $\Omega(\mathcal{A}) \subset \Omega(\mathcal{H}^{\mathrm{A}})$, see (1.91) and (1.92). Therefore, we can apply (1.129) above to the density operators of the A-system. Now let ω^{A} be a density or just a positive operator from $\mathcal{A}$. It is convenient to call an operator $W \in \mathcal{A}$ an *amplitude of ω^{A}* if $\omega^{\mathrm{A}} = WW^{\dagger}$. $(W \otimes 1)|\varphi\rangle$ is a purifying vector for ω^{A}, if W is an amplitude of ω^{A} and vice versa.

There are many amplitudes of ω^{A}, and the change from one to another one can be described[18] by *gauge transformations* $W \to W' = WU$ with unitary $U \in \mathcal{A}$. The gauge transformations respect ω^{A} as a gauge invariant.

Let us return to our problem with two density operators, ω_1^{A} and ω_2^{A}, and two purifying vectors, ψ_1 and ψ_2. There are two operators W_1, W_2 in our $\mathcal{A}$ satisfying

$$\psi_j = (W_j \otimes 1)\varphi, \quad \omega_j^{\mathrm{A}} = W_j W_j^{\dagger} \,. \tag{1.130}$$

With these amplitudes, we have

$$\langle \psi_1, \psi_2 \rangle = \langle (W_1 \otimes 1)\varphi, (W_2 \otimes 1)\varphi \rangle = \mathrm{tr}\, W_1^{\dagger} W_2 \,. \tag{1.131}$$

Gauging $\psi_2 \to \psi_2'$ by $W_2 \to W_2' = W_2 U$, we see

$$\langle \psi_1, \psi_2' \rangle = \mathrm{tr}\, W_1^{\dagger} W_2 U \,.$$

Let us stress that we fix W_1 and vary only W_2 in this relation. Hence

$$F(\omega_1^{\mathrm{A}}, \omega_2^{\mathrm{A}}) = \sup_{\psi'} |\langle \psi_1, \psi_2' \rangle| = \sup_{U \in \mathcal{A}} |\mathrm{tr}\, W_1^{\dagger} W_2 U| \,,$$

---

[18] Due to our finiteness assumptions.

provided one cannot get better results in higher dimensional purifications. But this is not the case, as one can prove. (Essentially, this is because the largest dimension of a cyclic representation of $\mathcal{B}(\mathcal{H}^A)$ is of dimension d^2.)

It is $|\operatorname{tr} BU| \le \operatorname{tr} B$ in case $B \ge 0$, and U is unitary. Hence, we are done if $W_1^\dagger W_2 \ge 0$ can be reached. This is possible because the polar decomposition theorem is valid in $\mathcal{A}$ (and, indeed, in every von Neumann algebra). In other words, we can choose a pair of amplitudes such that

$$F(\omega_1^A, \omega_2^B) = \operatorname{tr} W_1^\dagger W_2, \quad W_1^\dagger W_2 \ge 0. \tag{1.132}$$

Let us restate (1.132) to respect *-isomorphisms. If $\mathcal{A}$ is any (finite-dimensional) *-subalgebra of any $\mathcal{B}(\mathcal{H})$, we have to understand the trace in (1.132) as the canonical trace. (Remember: Only for the algebras $\mathcal{B}_\mathbf{d}$ with $\dim \mathcal{H} = \mathbf{d}$, the canonical trace coincides with the trace of $\mathcal{H}$.) Whenever for two density operators of $\mathcal{A}$

$$\omega_1^A = W_1 W_1^\dagger, \quad \omega_2^A = W_2 W_2^\dagger, \quad W_1^\dagger W_2 \ge 0,$$

we call the pair of amplitudes W_1, W_2 *parallel*. Parallelity implies

$$\mathrm{F}(\omega_1^A, \omega_2^A) = \operatorname{tr}^{\mathrm{can}} W_1^\dagger W_2, \quad \Pr(\omega_1^A, \omega_2^A) = (\operatorname{tr}^{\mathrm{can}} W_1^\dagger W_2)^2. \tag{1.133}$$

1.4.4 Why the Bures Distance Is a Distance

Before proceeding along the main line, the triangle inequality should be proved. Inserting (1.132) into (1.124) yields

$$\operatorname{dist}_B(\omega_1^A, \omega_2^A) = \sqrt{\operatorname{tr} W_1 W_1^\dagger + \operatorname{tr} W_2 W_2^\dagger - 2\operatorname{tr} W_1^\dagger W_2}.$$

Now observe that the traces of $WW^\dagger$ and $W^\dagger W$ are equal. Further, remember that $W_1^\dagger W_2$ is assumed to be positive and, therefore, hermitian:

$$W_1^\dagger W_2 = W_2^\dagger W_1. \tag{1.134}$$

Altogether we proved that if W_1, W_2 are parallel amplitudes, then

$$\operatorname{dist}_B(\omega_1^A, \omega_2^A) = \sqrt{\operatorname{tr}^{\mathrm{can}} (W_1 - W_2)^\dagger (W_1 - W_2)}, \tag{1.135}$$

and for two arbitrary amplitudes, the left-hand side cannot be larger than the right one. The latter can also be rewritten $\| W_2 - W_1 \|_2$.

Consider now three positive operators, ω_1^A, ω_2^A, and ω_3^A. Starting with W_2, we can choose W_1 and W_3, such that the pairs W_2, W_1 and W_2, W_3 are parallel amplitudes. This allows to convert the triangle inequality

$$\| W_1 - W_2 \|_2 + \| W_2 - W_3 \|_2 \ge \| W_1 - W_3 \|_2$$

into
$$\text{dist}_B(\omega_1^A, \omega_2^A) + \text{dist}_B(\omega_2^A, \omega_3^A) \geq \| W_1 - W_3 \|_2$$

and the last term cannot be smaller than the Bures distance. Hence,

$$\text{dist}_B(\omega_1^A, \omega_2^A) + \text{dist}_B(\omega_2^A, \omega_3^A) \geq \text{dist}_B(\omega_1^A, \omega_3^A). \tag{1.136}$$

It is instructive to rewrite our finding with purifying vectors. We extend our notation and call a pair of purifying vectors *parallel* if the amplitudes in (1.130), that is in $\psi_j = (W_j \otimes \mathbf{1})\varphi$, are parallel ones. We can express (1.135) by

$$\text{dist}_B(\omega_1^A, \omega_2^A) \leq \| \psi_2 - \psi_1 \|, \tag{1.137}$$

for all pairs of purifying vectors of ω_1^A, ω_2^A. Equality holds for pairs of parallel purifying vectors.

Some Geometric Properties of the Bures Distance

The Bures distance is an inner one: There are short geodesic arcs with length equal to the Bures distance of their end points. Given ω_0^A, ω_1^A, we chose the parallel amplitudes W_0, W_1. Then any pair of amplitudes belonging to the arc

$$t \mapsto W_t = (1 - t)W_0 + tW_1, \quad 0 \leq t \leq 1, \tag{1.138}$$

is a parallel pair. Exactly as in (1.10), we get

$$\text{dist}_B(\omega_s^A, \omega_t^A) = \| W_t - W_s \|_2, \quad \omega_s^A = W_s W_s^\dagger. \tag{1.139}$$

Bures did not ask wether his distance is based on a Riemannian metric. He was interested in cases with infinite tensor products of von Neumann algebras, and the theory of infinite-dimensional manifolds had not been developed. But for finite-dimension, the question is tempting.

There is, indeed, a Riemannian metric reproducing the Bures distance. Its line element is given by

$$\left(\frac{\text{d}s_B}{\text{d}t} \right)^2 = \text{tr}^{\text{can}} G^2 \omega^A = \frac{1}{2} \text{tr}^{\text{can}} \dot{\omega}^A G, \tag{1.140}$$

whenever there is a solution of

$$\dot{\omega}^A = \omega^A G + G \omega^A, \quad G = G^\dagger. \tag{1.141}$$

For invertible positive operators ω^A, there is a unique solution of (1.141). At the boundary, where the rank is smaller than the Hilbert space dimension, the existence of G depends on the direction of the tangent $\dot{\omega}^A$. For $\dim \mathcal{H} \geq 3$, there are directions for which the metric becomes singular (J. Dittmann, Private Communication).

However, for invertible ω^A, the metric behaves regularly. Let

$$t \rightarrow \omega_t^A = W_t W_t^\dagger, \quad |\psi_t\rangle = (W_t \otimes \mathbf{1}^B)|\varphi\rangle, \tag{1.142}$$

be an arc of invertible density operators. Then the curve $t \rightarrow W_t$ of the amplitudes is called *parallel*, if

$$\dot{W}_t^\dagger W_t = W_t^\dagger \dot{W}_t. \tag{1.143}$$

By straightforward computation, one proves the equivalence of (1.141) with the condition

$$\dot{W}_t = G_t W_t, \quad G_t^\dagger = G_t. \tag{1.144}$$

Now one easily gets

$$t \rightarrow W_t \text{ parallel}, \; \Rightarrow \; \left(\frac{ds_B}{dt}\right)^2 = \mathrm{tr}^{\mathrm{can}} \dot{W} \dot{W}^\dagger. \tag{1.145}$$

It is an easy, nice exercise to compute the Bures distance (1.139) by (1.145) to establish that the Bures distance can be gained from the metric (1.140).

After switching to the purifying arc $|\psi_t\rangle = (W_t \otimes \mathbf{1}^B)|\varphi\rangle$, another form of the results appears: The Hilbert space length of a purifying lift $t \rightarrow |\psi_t\rangle$ of $t \rightarrow \omega_t^A$ is never less than its Bures length. Equality is reached exactly with parallel amplitudes (1.143) in (1.142).

The Extended Mandelstam–Tamm Inequality

An application is the extended Mandelstam–Tamm inequality. Let

$$t \rightarrow \omega_t, \quad 0 \leq t \leq 1, \tag{1.146}$$

be a solution of time-dependent von Neumann–Schrödinger equation

$$i\hbar\dot{\omega} = [H, \omega], \quad H = H_t. \tag{1.147}$$

Then one can prove

$$\int_0^1 \sqrt{\mathrm{tr}(\omega H^2) - (\mathrm{tr}\,\omega H)^2} \geq \hbar \arccos \mathrm{F}(\omega_1, \omega_0), \tag{1.148}$$

see [58]. (One has to look for a lift $t \rightarrow W_t$ satisfying the differential parallel condition $\dot{W}^\dagger W = W^\dagger \dot{W}$ and a Schrödinger equation with a Hamiltonian $W \rightarrow HW + W\tilde{H}$, where $t \rightarrow \tilde{H}_t$ has to be chosen suitably.)

Using this, one can get a differential form of (1.148):

$$\mathrm{tr}(\omega H^2) - (\mathrm{tr}\,\omega H)^2 \geq \frac{\hbar}{2}\mathrm{tr}\,G\dot{\omega}. \tag{1.149}$$

One may compare this inequality with the *quantum Rao–Cramers inequality*, which, however, plays its role in a quite different context (hypothesis testing and other questions of mathematical statistics). A recent overview, discussing these relationships, can be found in I. Bengtsson's paper [59]. Another question has been discussed by A. Ericsson [60].

1.4.5 Expressions for Fidelity and Transition Probability

Now we return to (1.132) and (1.133) to benefit from the positivity of $W_1^\dagger W_2$ for parallel amplitudes. It holds

$$(W_1^\dagger W_2)^2 = W_1^\dagger W_2 W_2^\dagger W_1 = W_1^\dagger \omega_2^A W_1 \,.$$

There is a polar decomposition

$$W_1 W_1^\dagger = \omega_1^A \,, \quad W_1 = (\omega_1^A)^{1/2} U_1,$$

with a unitary U_1. Putting things together yields

$$(W_1^\dagger W_2)^2 = U_1^{-1}(\omega_1^A)^{1/2}\omega_2^A(\omega_1^A)^{1/2} U_1 \,. \tag{1.150}$$

We can take the positive root and obtain

$$W_1^\dagger W_2 = U_1^{-1}\sqrt{(\omega_1^A)^{1/2}\omega_2^A(\omega_1^A)^{1/2}}\, U_1 \,. \tag{1.151}$$

The canonical trace of (1.151) yields the fidelity,

$$\mathrm{F}(\omega_1^A, \omega_2^A) = \mathrm{tr}^{\mathrm{can}}\sqrt{(\omega_1^A)^{1/2}\omega_2^A(\omega_1^A)^{1/2}} \,. \tag{1.152}$$

Its square is the transition probability.

As an application, we consider direct products. With two pairs, ω_1, ω_2 and ρ_1, ρ_2 of density operators in two different Hilbert spaces, one can perform their direct products $\omega_j \otimes \rho_j$. The structure of the expression (1.152) allows to conclude

$$\mathrm{Pr}(\omega_1 \otimes \rho_1, \omega_2 \otimes \rho_2) = \mathrm{Pr}(\omega_1, \omega_2)\,\mathrm{Pr}(\rho_1, \rho_2) \,. \tag{1.153}$$

In what follows, we assume invertible positive operators, though the results do not depend on that assumption. As above, W_1, W_2 are parallel amplitudes of ω_1^A, ω_2^A. We define a positive gauge invariant, K,

$$W_1^\dagger W_2 > 0 \Leftrightarrow W_2 = K W_1 \,, \quad K > 0 \,. \tag{1.154}$$

Indeed, $W_1^\dagger W_2 = W_1^\dagger K W_1$ proves $K > 0$ equivalent to parallelity. Now

$$K = W_2 W_1^{-1} > 0 \,, \quad K^{-1} = W_1 W_2^{-1} > 0 \,, \tag{1.155}$$

and we conclude the existence of $K \in \mathcal{A}$ such that

$$\mathrm{tr}^{\mathrm{can}}\omega_1^A K = \mathrm{tr}^{\mathrm{can}}W_1^\dagger W_2 = \mathrm{tr}^{\mathrm{can}}\omega_2^A K^{-1} \,. \tag{1.156}$$

But $\mathrm{tr}^{\mathrm{can}}W_1^\dagger W_2$ is the fidelity, and with our K, we have

$$\mathrm{F}(\omega_1^A, \omega_2^A) = \mathrm{tr}^{\mathrm{can}}\omega_1^A K = \mathrm{tr}^{\mathrm{can}}\omega_2^A K^{-1} \,. \tag{1.157}$$

For a pair ψ_1, ψ_2 of parallel purifications and for every positive $C \in \mathcal{B}(\mathcal{H}^A \otimes \mathcal{H}^B)$, we know from (1.99)

$$\mathrm{F}(\omega_1^A, \omega_2^A) = |\langle \psi_1, \psi_2 \rangle| \leq (1/2)(\langle \psi_1, C\psi_1 \rangle + \langle \psi_2, C^{-1}\psi_2 \rangle).$$

Inserting $C = X \otimes \mathbf{1}^B$, it becomes clear that the right-hand side cannot become smaller than

$$\frac{1}{2} \inf_{X > 0} (\mathrm{tr}\,\omega_1 X + \mathrm{tr}\,\omega_2 X^{-1}), \quad X \in \mathcal{A}.$$

The particular case $X = K$ proves

$$\mathrm{F}(\omega_1^A, \omega_2^A) = \frac{1}{2} \inf_{X > 0} \left(\mathrm{tr}\,\omega_1^A X + \mathrm{tr}\,\omega_2^A X^{-1} \right). \tag{1.158}$$

Let us reformulate (1.158) to change from density operators to states. Finally, there is no reference on any bipartite structure. Let $\mathcal{A}$ be a *-subalgebra of $\mathcal{B}(\mathcal{H})$, and $\underline{\omega}$ and $\underline{\rho}$ two of its states or positive linear forms. Then

$$\mathrm{F}(\underline{\omega}, \underline{\rho}) = \frac{1}{2} \inf_{0 < X \in \mathcal{A}} \underline{\omega}(X) + \underline{\rho}(X^{-1}), \tag{1.159}$$

$$\mathrm{Pr}(\underline{\omega}, \underline{\rho}) = \inf_{0 < X \in \mathcal{A}} \underline{\omega}(X)\,\underline{\rho}(X^{-1}). \tag{1.160}$$

Thanks to the work of Araki and Raggio [61] and Alberti [62], the last two assertions are known to be true for any pair of states of any unital C*-algebra.

Super-Additivity

For all decompositions

$$\omega = \sum \omega_j, \quad \rho = \sum \rho_j, \tag{1.161}$$

of positive operators, the inequality

$$\mathrm{F}(\omega, \rho) \geq \sum_j \mathrm{F}(\omega_j, \rho_j) \tag{1.162}$$

is valid. The inequality expresses *super-additivity* of the fidelity.

For simplicity, we prove super-additivity assuming ω and ρ invertible and choose $K \in \mathcal{A}$ satisfying

$$\mathrm{F}(\omega, \rho) = \mathrm{tr}^{\mathrm{can}}\omega K = \mathrm{tr}^{\mathrm{can}}\rho K^{-1}$$

as in (1.156) and (1.157). We now have

$$2\mathrm{F}(\omega, \rho) = \sum \mathrm{tr}^{\mathrm{can}}\omega_j K + \sum \mathrm{tr}^{\mathrm{can}}\rho_j K^{-1}.$$

The proof terminates by estimating the right part by (1.158). This is the finite-dimensional case. (For von Neumann and C^*-algebras one returns to positive linear forms for which super-additivity of the fidelity can be proved equally well.)

Let us mention how (1.161) implies joint concavity. Because of

$$F(a\omega, b\rho) = \sqrt{ab}\,F(\omega, \rho)\,, \quad a, b \in \mathbb{R}^+, \tag{1.163}$$

it follows from (1.161) for convex sums of equal length

$$F\left(\sum_j p_j\omega_j, \sum_k q_k\rho_k\right) \geq \sum_j \sqrt{p_jq_j}\,F(\omega_j, \rho_j)\,. \tag{1.164}$$

From (1.152), one can conclude the following: *Equality holds in (1.164), if for $j \neq k$ it holds $\omega_j\rho_k = \mathbf{0}$.* Similar (indeed equivalent) statements are true for (1.159) and (1.160).

Monotonicity

Choi [63] proved for positive unital maps

$$\Psi(A^{-1}) \geq \Psi(A)^{-1} \text{ if } A \geq \mathbf{0}\,. \tag{1.165}$$

In the case of a 2-positive and unital Ψ, the conclusion

$$\begin{pmatrix} A & C \\ C^\dagger & B \end{pmatrix} \geq \mathbf{0} \Rightarrow \begin{pmatrix} \Psi(A) & \Psi(C) \\ \Psi(C^\dagger) & \Psi(B) \end{pmatrix} \geq \mathbf{0} \tag{1.166}$$

comes simply from the very definition of 2-positivity. Then, (1.165) follows with $B = A^{-1}$, $C = \mathbf{1}$, and unitality, $\Psi(\mathbf{1}) = \mathbf{1}$. However, according to Choi, in the particular case $C \geq \mathbf{0}$, just positivity and unitality are sufficient for the validity of (1.166). Therefore, (1.165) is valid for positive unital maps.

Let us apply (1.165) to the fidelity. To this end, we denote by Φ the map dual to Ψ,

$$\operatorname{tr} X\Psi(Y) = \operatorname{tr} \Phi(X)Y\,. \tag{1.167}$$

Ψ is positive if Φ is positive. Φ is trace preserving if Ψ is unital. Not every positive operator might be of the form $\Psi(X)$ with positive X. Therefore, by (1.158) or (1.159),

$$F(\omega, \rho) \leq \frac{1}{2} \inf_{\mathbf{0} < X \in \mathcal{A}} \operatorname{tr}^{\mathrm{can}}\Psi(X)\omega + \operatorname{tr}^{\mathrm{can}}\Psi(X)^{-1}\rho\,.$$

We can replace $\Psi(X)^{-1}$ by the larger $\Psi(X^{-1})$ in virtue of (1.165) to get an even larger right-hand side:

$$F(\omega, \rho) \leq \frac{1}{2} \inf_{\mathbf{0} < X \in \mathcal{A}} \operatorname{tr}^{\mathrm{can}}\Psi(X)\omega + \operatorname{tr}^{\mathrm{can}}\Psi(X^{-1})\rho\,.$$

Now we apply duality, (1.167), and obtain

$$F(\omega, \rho) \leq \frac{1}{2} \inf_{0 < X \in \mathcal{A}} \mathrm{tr}^{\mathrm{can}} X \Phi(\omega) + \mathrm{tr}^{\mathrm{can}} X^{-1} \Phi(\rho).$$

The right-hand side is an expression for the fidelity of the pair $\Phi(\omega), \Phi(\rho)$, and the proof of the *monotonicity property* is done:
Let Φ be positive and trace preserving. Then

$$F(\omega_1, \omega_2) \leq F(\Phi(\omega_1), \Phi(\omega_2)). \tag{1.168}$$

As a consequence, trace preserving positive maps are *Bures-contracting*,

$$\mathrm{dist}_B(\omega_1, \omega_2) \geq \mathrm{dist}_B(\Phi(\omega_1), \Phi(\omega_2)). \tag{1.169}$$

Density operators (and states) become closer one to another under the action of these maps.

Remark 16. It is well known that there are many Riemannian metrics in a state space $\Omega(\mathcal{H})$ which are monotone decreasing with respect to channels, i.e., with respect to completely positive and trace preserving maps.[19] Thanks to the work of Petz [48, 49] they can be constructed with the help of certain operator means. Kubo and Ando [64] enumerated all operator means by operator monotone functions. Another, but much related story is the question for functions $\mathrm{Pr}'(.,.)$, depending on two states, which are

(a) monotone increasing with respect to channels and which
(b) coincide with the transition probability for pure states.

Some of them are related to distances, i.e., inserting in dist_B the square root of Pr' for F returns a distance. Most of them, however, are not related to any distance. In any case, only the transition probability (1.118) is "operational" defined. Just by this very definition, one finds, for pairs of density operators,

$$\mathrm{Pr}(\omega, \rho) \geq \mathrm{Pr}'(\omega, \rho), \tag{1.170}$$

for all Pr' satisfying the two conditions (a) and (b). A nice example is

$$\mathrm{Pr}(\omega, \rho) \geq \mathrm{tr}\, \omega^{1-s} \rho^s, \quad 0 < s < 1. \tag{1.171}$$

Indeed, the right-hand side fulfills (a) and (b) (see [65], where one can also find a more "direct" proof of (1.171)).

---

[19] Though their geodesics and distances are mostly unknown.

1.4.6 Estimates and a "Hidden Symmetry"

We use the notation

$$\text{char}(A) = \text{ all roots of the characteristic equation of } A. \qquad (1.172)$$

Clearly, this is the eigenvalue, counted with the appropriate multiplicity, if A is diagonalizable. Because of

$$\omega_1^{1/2}\left(\omega_1^{1/2}\omega_2\omega_1^{1/2}\right)\omega_1^{-1/2} = \omega_1\omega_2,$$

one concludes

$$\text{char}\left(\omega_1^{1/2}\omega_2\omega_1^{1/2}\right) = \text{char}\left(\omega_1\omega_2\right). \qquad (1.173)$$

An Estimate

Denoting the characteristic values of (1.173) by $\lambda_1, \lambda_2, \ldots$, we get

$$\Pr(\omega_1, \omega_2) = \left(\sum \sqrt{\lambda_j}\right)^2. \qquad (1.174)$$

The sum of the λ_j is the trace of $\omega_1\omega_2$. Hence

$$\Pr(\omega_1, \omega_2) = \text{tr}\,\omega_1\omega_2 + 2\sum_{j<k} \sqrt{\lambda_j\lambda_k}.$$

We write $2r$ for the last term and use

$$\sqrt{r^2} = \left(\sum_{j<k}\lambda_j\lambda_k + \ldots\right)^{1/2}.$$

The dots abbreviate some non-negative terms. The other term in the sum is the second elementary symmetric function of the characteristic values λ_k of $\omega_1\omega_2$. Expressing the latter by traces yields

$$\Pr(\omega_1, \omega_2) \geq \text{tr}\,\omega_1\omega_2 + \sqrt{2}\sqrt{(\text{tr}\,\omega_1\omega_2)^2 - \text{tr}(\omega_1\omega_2)^2}, \qquad (1.175)$$

with equality for $\text{rank}(\omega_1\omega_2) \leq 2$. For $\dim\mathcal{H} = 3$, closer inspection produces

$$\Pr(\rho, \omega) = \text{tr}\,\rho\omega + \sqrt{2}\sqrt{(\text{tr}\,\rho\omega)^2 - \text{tr}(\rho\omega\rho\omega) + 4F(\rho, \omega)\sqrt{\det(\rho\omega)}}.$$

One Qubit, dim $\mathcal{H} = 2$

In the 1-qubit case, (1.175) becomes an equality. $\lambda_1 \lambda_2$ is the determinant of $\omega_1 \omega_2$. Thus,

$$\Pr(\omega_1, \omega_2) = \operatorname{tr} \omega_1 \omega_2 + 2\sqrt{\det \omega_1 \det \omega_2}. \tag{1.176}$$

Let us represent our density matrices by

$$\omega_1 = \frac{1}{2}\left(1 + \sum x_n \sigma_n\right), \quad \omega_2 = \frac{1}{2}\left(1 + \sum y_n \sigma_n\right), \tag{1.177}$$

and let us define a new coordinate by

$$x_4 := 2\sqrt{\det \omega_1}, \quad y_4 := 2\sqrt{\det \omega_2}. \tag{1.178}$$

We have now placed the density operators on the upper 3-hemisphere,

$$x_1^2 + \cdots + x_4^2 = y_1^2 + \cdots + y_4^2 = 1, \tag{1.179}$$

with $x_4 \geq 0$, $y_4 \geq 0$. The transition probability becomes

$$\Pr(\omega_1, \omega_2) = \frac{1}{2}\left(1 + \sum_{j=1}^{4} x_j y_j\right). \tag{1.180}$$

A "Hidden Symmetry"

Remember first the equality (1.173) for the characteristic numbers. Let Z be invertible and consider the change

$$\omega_1' = Z^{-1} \omega_1 (Z^{-1})^\dagger, \quad \omega_2' = Z^\dagger \omega_2 Z. \tag{1.181}$$

One immediately sees

$$\omega_1' \omega_2' = Z^{-1}(\omega_1 \omega_2) Z \tag{1.182}$$

and

$$\operatorname{char}(\omega_1' \omega_2') = \operatorname{char}(\omega_1 \omega_2). \tag{1.183}$$

Now (1.173) implies the following: *The eigenvalues of $\sqrt{\omega_1} \omega_2 \sqrt{\omega_1}$ do not change if ω_1, ω_2 are transformed according to (1.181).* In particular

$$F(\omega_1, \omega_2) = F\left(Z^{-1} \omega_1 (Z^{-1})^\dagger, Z^\dagger \omega_2 Z\right). \tag{1.184}$$

Indeed, the argument is valid for every symmetric function of the characteristic numbers in question.

We can even refrain from the invertibility of Z by substituting

$$\omega_1 \to Z \omega_1 Z^\dagger$$

in (1.184):

$$F(Z \omega_1 Z^\dagger, \omega_2) = F(\omega_1, Z^\dagger \omega_2 Z). \tag{1.185}$$

Relaying on continuity, we can state (1.185) for all operators Z.

1.4.7 "Operational Fidelity"

The question is whether the fidelity concept can be extended to pairs of quantum channels. Apparently, the first relevant studies were done by Raginski [66]. More recent developments can be seen from Belavkin et al. [67] and Kretschmann et al. [68]. Essentially, our aim is to define what is called *operational fidelity* and to arrive at it via the Bures distance. We restrict ourselves to maps Φ from $\mathcal{B}(\mathcal{H})$ into itself, and we assume $\dim \mathcal{H} = d$ finite.

We denote the identity map $I(X) = X$, $X \in \mathcal{B}(\mathcal{H})$ by I. Later on, we need the identity maps I_k of auxiliary algebras $\mathcal{B}(\mathcal{H}_k)$ with $\dim \mathcal{H}_k = k$.

With two positive maps, Φ_1 and Φ_2, and a density operator, $\omega \in \Omega(\mathcal{H})$, we observe that

$$\Phi_1, \Phi_2 \to \mathrm{dist}_B(\Phi_1(\omega), \Phi_2(\omega))$$

is symmetric in the maps and fulfills, for three positive maps, the triangle inequality. This is because the Bures distance does so. As $\Phi_1(\omega) = \Phi_2(\omega)$ may happen, we do not necessarily get a metrical distance, but only a *semi-distance*. As one can check, the sup of arbitrary many semi-distances is again a semi-distance. Therefore,

$$\mathrm{dist}_1(\Phi_1, \Phi_2) := \sup_{\omega \in \Omega} \mathrm{dist}_B(\Phi_1(\omega), \Phi_2(\omega)) \tag{1.186}$$

is a distance in the space of positive maps.

Indeed, as said above, it is a semi-distance. But if two maps are not equal one to another, there must be a density operator at which they take different values. The index "1" in (1.186) reflects our assumption that the maps are just positive, i.e., 1-positive. Obviously, the distance $\mathrm{dist}_1(\Phi, I)$ estimates how strongly Φ deviates from the identity map.

If a map Φ is k-positive, then the map $\Phi \otimes I_k$ is still positive. For pairs of k-positive maps, the expression

$$\mathrm{dist}_k(\Phi_1, \Phi_2) := \mathrm{dist}_1(\Phi_1 \otimes I_k, \Phi_2 \otimes I_k) \tag{1.187}$$

is well defined. More explicitly, (1.187) is a sup,

$$\sup_\rho \mathrm{dist}_B\left([\Phi_1 \otimes I_k](\rho), [\Phi_2 \otimes I_k](\rho)\right), \tag{1.188}$$

over all density operators $\rho \in \Omega(\mathcal{H} \otimes \mathcal{H}_k)$.

One can unitarily embed $\mathcal{H} \otimes \mathcal{H}_k$ into $\mathcal{H} \otimes \mathcal{H}_{k+1}$. This implies that the sup in (1.188) is running over less states as in the case dist_{k+1}, resulting in

$$\mathrm{dist}_k(\Phi_1, \Phi_2) \le \mathrm{dist}_{k+1}(\Phi_1, \Phi_2). \tag{1.189}$$

This enables the introduction of an *operational Bures distance*

$$\mathrm{dist}_\infty(\Phi_1, \Phi_2) := \lim_{k \to \infty} \mathrm{dist}_k(\Phi_1, \Phi_2), \tag{1.190}$$

which takes into account possible entanglement in the input. Possible entanglement based on $\mathcal{H} \otimes \mathcal{H}_k$ saturates with $\dim \mathcal{H} = k$,

$$\mathrm{dist}_\infty(\Phi_1, \Phi_2) = \mathrm{dist}_d(\Phi_1, \Phi_2), \quad d = \dim \mathcal{H}. \tag{1.191}$$

To obtain what has been called *operational fidelity* in the literature, we have to go back to the relation (1.124)

$$\mathrm{dist}_B(\omega_1, \omega_2) = \sqrt{\mathrm{tr}\,\omega_1 + \mathrm{tr}\,\omega_2 - 2\mathrm{F}(\omega_1, \omega_2)}$$

and try to replace accordingly states by maps. The task can be done quite naturally for completely positive, trace preserving maps: For all these maps, $\mathrm{tr}\,\Phi(\omega) = 1$ for density operators. It suggests

$$\mathrm{dist}_\infty(\Phi_1, \Phi_2) = \sqrt{2 - 2\mathrm{F}_{(\infty)}(\Phi_1, \Phi_2)}. \tag{1.192}$$

The quantity $\mathrm{F}_{(\infty)}(\Phi_1, \Phi_2)$ is called *operational fidelity*. The index (∞) is not standard and stands here only to respect the possibility of the same procedure with k-positive and trace preserving maps. For such maps, one can consider an *operational fidelity for k-positivity*, $\mathrm{F}_{(k)}$, as well.

As a matter of fact, one can transmit several properties of the fidelity and the Bures distance for states to completely or k-positive and trace preserving maps. The joint concavity for instance allows to perform the sup in (1.186) or in (1.187) over pure states only.

1.5 Appendix: The Geometrical Mean

Let A, B, and C be positive operators in a finite-dimensional Hilbert space. A remarkable observation by Pusz and Woronowicz [69] can be rephrased in the following form:

Given $A \geq \mathbf{0}$ and $B \geq \mathbf{0}$, there is a largest operator in the set of all C satisfying

$$\begin{pmatrix} A & C \\ C & B \end{pmatrix} \geq \mathbf{0}, \quad C \geq \mathbf{0}. \tag{1.193}$$

This unique element is called *the geometrical mean of A and B*, and it will be denoted, following Ando, by

$$A \# B. \tag{1.194}$$

In other words, (1.193) is valid if and only if

$$\mathbf{0} \leq C \leq A \# B. \tag{1.195}$$

From the definition, we get the relation

$$A \# B = B \# A, \quad A^{-1} \# B^{-1} = (A \# B)^{-1};\tag{1.196}$$

the latter is true for invertible positive operators. If just A is invertible, then the block matrix (1.193) is positive if and only if

$$B \geq C A^{-1} C,\tag{1.197}$$

and one concludes that $A \# B$ is the unique positive solution X of the equation

$$B = X A^{-1} X, \quad X \geq \mathbf{0}.\tag{1.198}$$

The equation can be solved, and one gets

$$A \# B = A^{1/2} \left(A^{-1/2} B A^{-1/2} \right)^{1/2} A^{1/2}.\tag{1.199}$$

To prove it, one rewrites (1.198) as

$$A^{-1/2} B A^{-1/2} = \left(A^{-1/2} X A^{-1/2} \right)^2$$

and takes the root.

In case B^{-1} exists too, we may rewrite (1.198) as

$$\mathbf{1} = B^{-1/2} X A^{-1} X B^{-1/2} = \left(B^{-1/2} X A^{-1/2} \right) \left(B^{-1/2} X A^{-1/2} \right)^{\dagger}.$$

Therefore, if A and B are strictly positive, the following three statements are equivalent:

(a) $X = A \# B$,
(b) $(B^{-1/2} X A^{-1/2})$ is unitary,
(c) $(A^{-1/2} X B^{-1/2})$ is unitary.

In turn, as shown in [70], an operator $X = (A^{1/2} U B^{1/2})$ with unitary U is positive exactly if $X = A \# B$.

If A and B commute, one can see from (1.193)

$$AB = BA \Rightarrow A \# B = (AB)^{1/2}.\tag{1.200}$$

To get it, one uses a common eigenbasis, which reduces (1.193) to

$$\begin{pmatrix} a & c \\ c & b \end{pmatrix} \geq \mathbf{0} \Leftrightarrow ab \geq c^2,$$

for three positive numbers a, b, and c.

1.5.1 Geometric Mean and Fidelity

Let us return to the operator K, defined in (1.154) by

$$W_2 = KW_1 \text{ if } W_1^\dagger W_2 > \mathbf{0},$$

for parallel and invertible amplitudes. One can calculate

$$K = \omega_1^{-1/2} \left(\omega_1^{1/2} \omega_2 \omega_1^{1/2} \right)^{1/2} \omega_1^{-1/2} \qquad (1.201)$$

and conclude the following: First, we can rewrite (1.201),

$$K = \omega_2 \# \omega_1^{-1}, \qquad (1.202)$$

and second, the expression remains meaningful for not invertible ω_2. In this sense, we understand the right-hand side of (1.201) and (1.202) to be valid for all pairs of positive operators in our finite-dimensional setting. One further concludes that for general pairs of positive operators, we have to substitute

$$K \to \omega_2 \# \omega_1^{-1} \text{ and } K^{-1} \to \omega_1 \# \omega_2^{-1},$$

in order that (1.156) and (1.157) can be applied to not necessarily invertible positive operators. With this convention, one arrives at

$$\mathrm{Pr}(\underline{\omega}_1, \underline{\omega}_2) = \underline{\omega}_1 \left(\omega_2 \# \omega_1^{-1} \right) = \underline{\omega}_2 \left(\omega_1 \# \omega_2^{-1} \right).$$

For parallel purifications in $\mathcal{H}^{AB} = \mathcal{H}^A \otimes \mathcal{H}^B$ of states ω_1^A and ω_2^A, one may ask how these purifications appear in Bob's system, i.e., after tracing out the A-system. To this end, let us first review the general situation, not requiring parallelity. We choose amplitudes, W_1, W_2, for ω_1^A and ω_2^A, so that, for $m = 1, 2$,

$$\omega_m^A = W_m W_m^\dagger = \mathrm{tr}_B |\psi_m\rangle\langle\psi_m|, \quad |\psi_m\rangle = \left(W_m \otimes \mathbf{1}^B \right) |\varphi\rangle, \qquad (1.203)$$

with $|\varphi\rangle$ chosen maximally entangled and with norm $d = \dim \mathcal{H}^A$ as in (1.128). For any $Y \in \mathcal{B}(\mathcal{H}^B)$, we get

$$\mathrm{tr} \left(|\psi_n\rangle\langle\psi_m| \right)^B Y = \langle\psi_m| \left(\mathbf{1}^A \otimes Y \right) |\psi_n\rangle = \langle\varphi|(W_m^\dagger W_n \otimes Y)|\varphi\rangle. \qquad (1.204)$$

The last expression is equal to

$$\sum \langle j|W_m^\dagger W_n|k\rangle \, \langle j|Y|k\rangle.$$

Let us define the transposition $X \to X^\top$ by

$$\langle j|X^\top|k\rangle_B = \langle k|X|j\rangle_A. \qquad (1.205)$$

In this equation, we used the two basis of the subsystems with which φ is represented in (1.128). It now follows

$$\text{tr}\left(|\psi_n\rangle\langle\psi_m|\right)^{\mathrm{B}} Y = \text{tr}\left(W_m^\dagger W_n\right)^\top Y$$

and, therefore,

$$\left(|\psi_n\rangle\langle\psi_m|\right)^{\mathrm{B}} = \left(W_m^\dagger W_n\right)^\top . \tag{1.206}$$

Next, we conclude, for $m = 1, 2$,

$$\omega_m^{\mathrm{B}} := \left(|\psi_m\rangle\langle\psi_m|\right)^{\mathrm{B}} = \left(W_m^\dagger W_m\right)^\top . \tag{1.207}$$

We are now prepared to use parallel amplitudes and proceed to show

$$W_m^\dagger W_n \geq \mathbf{0} \Rightarrow W_m^\dagger W_n = (W_m^\dagger W_m)\#(W_n^\dagger W_n) . \tag{1.208}$$

At first, we assume invertibility of W_m and use the identity

$$(W_n^\dagger W_n) = (W_n^\dagger W_m)(W_m^\dagger W_m)^{-1}(W_m^\dagger W_n),$$

which is true for parallel amplitudes. According to (1.198), our assertion must be true. In the general case, we mention that (1.193) forces the geometric mean $A\#B$ to give zero if applied to any null vector of either A or B. Therefore, the support of $A\#B$ is the intersection of the supports of A and of B. Thus, A and B become invertible if restricted onto the support of their geometric mean, and the reasoning above applies.

We can now return to (1.208) and state the following: *If W_m are parallel amplitudes for ω^{A} then*

$$W_1^\dagger W_2 = \left(\omega_1^{\mathrm{B}}\#\omega_2^{\mathrm{B}}\right)^\top \tag{1.209}$$

and, in particular,

$$F\left(\omega_1^{\mathrm{A}}, \omega_2^{\mathrm{A}}\right) = \text{tr}\left(\omega_1^{\mathrm{B}}\#\omega_2^{\mathrm{B}}\right) . \tag{1.210}$$

1.5.2 The Transformer Identity

To get further insight, one may use the fact that $A\#B$ is an *operator mean*. In particular, it satisfies the so-called *transformer identity*, i.e., for invertible Z it enjoys

$$Z(A\#B)Z^\dagger = (ZAZ^\dagger)\#(ZBZ^\dagger) . \tag{1.211}$$

For the proof, one relays on

$$\begin{pmatrix} A & C \\ C & B \end{pmatrix} \geq \mathbf{0} \Leftrightarrow \begin{pmatrix} ZAZ^\dagger & ZCZ^\dagger \\ ZCZ^\dagger & ZBZ^\dagger \end{pmatrix} \geq \mathbf{0},$$

for invertible Z.

Now we can combine (1.200) and (1.211) with $Z = A + B$. To check the positivity of (1.193), it is sufficient to do so on the support space of $A + B$. Thus, we may assume that this operator is invertible. Then

$$A' = (A + B)^{-1/2}A(A + B)^{-1/2} \text{ and } B' = (A + B)^{-1/2}B(A + B)^{-1/2}$$

commute. Indeed, it follows

$$A' + B' = 1, \quad A'\#B' = (A'B')^{1/2},$$

and we can apply (1.211). Therefore, we can express $A\#B$ by

$$(A+B)^{1/2}\left((A + B)^{-1/2}A(A + B)^{-1}B(A + B)^{-1/2}\right)^{1/2}(A+B)^{1/2}. \quad (1.212)$$

Resume. *An expression is equal to $A\#B$ if it does so for commuting positive operators and if it satisfies the transformer identity.*

A further application arises from the integral

$$1\#A = \sqrt{A} = \frac{1}{\pi}\int_0^1 \left(xA^{-1} + (1 - x)1\right)^{-1} \frac{dx}{\sqrt{x(1 - x)}}.$$

Substituting $A \to B^{-1/2}AB^{-1/2}$, the transformer identity

$$B\#A = B^{1/2}\left(1\#(B^{-1/2}AB^{-1/2})\right)B^{1/2}$$

allows to infer algebraically

$$A\#B = \frac{1}{\pi}\int_0^1 \left(xA^{-1} + (1 - x)B^{-1}\right)^{-1} \frac{dx}{\sqrt{x(1 - x)}}. \quad (1.213)$$

Super-Additivity

We prove super-additivity. Let

$$A = \sum A_j, \quad B = \sum B_j,$$

and

$$C_j = A_j\#B_j, \quad C = \sum C_j.$$

Then

$$\begin{pmatrix} A & C \\ C & B \end{pmatrix} = \sum \begin{pmatrix} A_j & C_j \\ C_j & B_j \end{pmatrix}$$

is a positive block matrix. Thus C is smaller than $A\#B$ and that proves

$$A\#B \geq \sum A_j \#B_j \,. \tag{1.214}$$

This nice inequality is the key to further estimates. We enumerate m positive operators, $A_1, \ldots A_m$, and modulo m by $A_{k+m} = A_m$. Then, we set $B_j = A_{j+1}$. Then the sum A of m consecutive A_j is equal to that of the B_j. Then, on the left side of (1.214), we have $A\#B = A$. Equation (1.214) yields

$$\sum_1^m A_j \geq \sum_j^m A_j \# A_{j+1} \,. \tag{1.215}$$

Take $m = 2$ as a particular case and respect (1.196). We get

$$(A + B)/2 \geq A\#B \,.$$

Replacing A and B by A^{-1} and B^{-1}, we get

$$\left(A^{-1} + B^{-1} \right) /2 \geq (A\#B)^{-1}$$

by (1.196). Taking the inverse of that inequality,

$$A\#B \geq 2 \left(A^{-1} + B^{-1} \right)^{-1} \,.$$

The right-hand side is the *harmonic mean* of A and B.

A hint to further developments: It is not obvious how to define a *geometrical mean* of more than two operators. One of the proposals is by Ando et al. [70]. It fits to the equality (1.215), and we describe it for just three positive operators, A, B, and C. We define recursively

$$A_{j+1} = B_j \# C_j \,, \quad B_{j+1} = C_j \# A_j \,, \quad C_{j+1} = A_j \# B_j \,, \tag{1.216}$$

starting with $A_0 = A, B_0 = B$, and $C_0 = C$. Equation (1.215) proves (1.216) to be a decreasing sequence of positive operators. Therefore, there is a limiting operator which is, up to a factor, the geometric mean $G(A, B, C)$ favored by Ando, Li, and Mathias,

$$G(A, B, C) = \frac{1}{3} \lim_{j \to \infty} A_j + B_j + C_j \,. \tag{1.217}$$

For three commuting operators, one gets $(ABC)^{1/3}$. However, for three positive operators, in general position, no explicit expression is known for (1.217) – even if the operators live on a 2-dimensional Hilbert space.

Monotonicity

Our next task is to prove a monotonicity theorem. Let Ψ be a positive super-operator and $\Psi(\mathbf{1}) > \mathbf{0}$. According to Choi [63], Ψ is *almost* 2-positive: A

2×2-positive block matrix with Hermitian off-diagonal remains positive by applying Ψ. That is

$$\begin{pmatrix} A & C \\ C & B \end{pmatrix} \geq \mathbf{0}, \; C = C^\dagger \Rightarrow \begin{pmatrix} \Psi(A) & \Psi(C) \\ \Psi(C) & \Psi(B) \end{pmatrix} \geq \mathbf{0}. \tag{1.218}$$

Therefore, applied to $C = A \# B$, with

$$\begin{pmatrix} A & A \# B \\ A \# B & B \end{pmatrix} \quad \text{also} \quad \begin{pmatrix} \Psi(A) & \Psi(A \# B) \\ \Psi(A \# B) & \Psi(B) \end{pmatrix}$$

must be a positive block operator with positive entries. Hence,

$$\Psi(A \# B) \leq \Psi(A) \# \Phi(B) \tag{1.219}$$

is valid by the very definition of the geometric mean.

A Rank Criterion

Here we like to prove the following: Let A, B, and C be positive operators in a Hilbert space of dimension d. We further assume that A^{-1} and B^{-1} exist. With these data, we consider the matrix

$$X = \begin{pmatrix} A & C \\ C & B \end{pmatrix}, \tag{1.220}$$

which is an operator in $\mathcal{B}(\mathcal{H} \oplus \mathcal{H})$. Then

$$\operatorname{rank} X \geq d = \dim \mathcal{H}, \tag{1.221}$$

and equality holds if and only if $C = A \# B$.

Proof. At first, we mention that the invertibility of A and B is essential. It allows to introduce the matrix

$$Y = \begin{pmatrix} 1 & D \\ D^\dagger & 1 \end{pmatrix} := \begin{pmatrix} A^{-1/2} & 0 \\ 0 & B^{-1/2} \end{pmatrix} X \begin{pmatrix} A^{-1/2} & 0 \\ 0 & B^{-1/2} \end{pmatrix}, \tag{1.222}$$

which is of the same rank as X. The set of eigenvectors of Y with eigenvalue 0 is a subspace $\mathcal{H}_0$ of $\mathcal{H} \oplus \mathcal{H}$. The unit vector $\psi \oplus \varphi$ belongs to $\mathcal{H}_0$ if

$$\psi + D\varphi = 0, \quad D^\dagger \psi + \varphi = 0.$$

At first, we see that neither ψ nor φ can be the zero vector of $\mathcal{H}$. (Otherwise both, ψ and φ, must be zero.) Hence, the dimension of $\mathcal{H}_0$ cannot exceed d, confirming (1.221). (This is so because, otherwise, one must have a non-zero vector in $\mathcal{H}_0$ with either ψ or φ equal to the zero vector. We had already excluded this.) Secondly, we deduce, by eliminating ψ, respectively, φ,

$$(1 - DD^\dagger)\psi = 0, \quad (1 - D^\dagger D)\varphi = 0.$$

If dim $\mathcal{H}_0 = d$, then these equations are valid for all $\psi \in \mathcal{H}$ and all $\varphi \in \mathcal{H}$. Therefore, D is unitary if the ranks of Y and X are equal to d. Then $D = A^{-1/2}CB^{-1/2}$ is unitary. However, we already know that this can be true if and only if $C = A\#B$. Finally, if D is unitary, we can easily find d linear-independent vectors of Y with eigenvalue zero.

Geometrical Mean in Two Dimensions

Let A and B be two positive operators acting on a 2-dimensional Hilbert space. Explicit expressions for the geometric mean are known:

$$A\#B = \frac{\sqrt{st}}{\sqrt{\det(A/s + B/t)}}(A/s + B/t)$$

with

$$s = \sqrt{\det A}, \quad t = \sqrt{\det B}.$$

1.5.3 #-Convexity

To handle more than two positive operators in general position is a very hard task. This is one of the problems we like to pose. We think it is important and, perhaps, not completely hopeless.

A set $\mathcal{K}$ of positive operators is called #-convex , if

(a) K contains all its limiting operators and
(b) K contains with A, B also $A\#B$.

Let us denote by $\#[A, B]$ the smallest #-convex set containing A and B. Assuming $AB = BA$ and both invertible, then

$$\#[A, B] = \left\{ A^s B^{1-s} \,|\, 0 \le s \le 1 \right\}. \tag{1.223}$$

If the operators A and B are not invertible, then A^0 and B^0 must be interpreted as the projections P_A and P_B onto the support of A and B, respectively. The general case of two non-commuting positive operators can be settled by the transformer inequality

$$Z \#[A, B] Z^\dagger = \# \left[ZAZ^\dagger, ZBZ^\dagger \right]. \tag{1.224}$$

Denote by $\#[A_1, A_2, \ldots A_m]$ the smallest #-convex set containing $A_1, \ldots A_m$. If the operators A_j are invertible and pairwise commuting, it is not hard to show that $\#[A_1, A_2, \ldots, A_m]$ consists of all operators

$$A_1^{s_1} A_2^{s_2} \cdots A_m^{s_m}, \quad \sum s_j = 1, \tag{1.225}$$

with all $s_j \ge 0$. What happens without commutativity is unknown.

References

1. D. Chruściński and A. Jamiotkowski: *Geometric Phases in Classical and Quantum Mechanics* (Birkhäuser, Boston 2004)
2. I. Bengtsson and K. Życzkowski: *Geometry of Quantum States: An Introduction to Quantum Entanglement* (Cambridge University Press, 2006)
3. R. Haag: *Local Quantum Physics: Fields, Particles, Algebras* (Springer, Berlin New York 1993)
4. J. Anandan: A geometric view of quantum mechanics. In: Quantum Coherence, ed by J. S. Anandan (World Scientific, Singapore 1990)
5. J. Anandan: A geometric approach to quantum mechanics. Found. Phys. 21, 1265–1284 (1991)
6. L. Mandelstam and I. Tamm: *The uncertainty relation between energy and time in non-relativistic quantum mechanics.* J. Phys. USSR **9**, 249–254 (1945).
7. I. Tamm: *Selected Papers* (Springer, Berlin New York 1991) pp. 116–123 (Reprinted)
8. V. Giovannetti, S. Lloyd, L. Maccone: *Quantum limits to dynamical evolution.* Phys. Rev. A **67**, 052109 (2003)
9. J. Batle, M. Casas, A. Plastino, A. R. Plastino: *On the connection between entanglement and the speed of quantum evolution.* Phys. Rev. A **72**, 032337 (2005)
10. A. Boras, M. Casas, A. R. Plastino, A. Plastino: *Entanglement and the lower bound on the speed of quantum evolution.* Phys. Rev. A **74**, 0222326 (2006)
11. V. Fock: *Über die Beziehung zwischen den Integralen der quantenmechanischen Bewegungsgleichungen und der Schrödingerschen Wellengleichung.* Zs. Physik **49**, 323–338 (1928)
12. M. Berry: *Quantal phase factors accompanying adiabatic changes.* Proc. Royal Soc. London A **392**, 45–57 (1984)
13. G. Fubini: *Sulle metriche definite da una forma Hermitiana.* Istituto Veneto **LXIII**, 2 (1904)
14. E. Study: *Kürzeste Wege im komplexen Gebiet.* Math. Annalen **LX**, 321–378 (1905)
15. E. Wigner: *Gruppentheorie und ihre Anwendung auf die Quantenmechanik der Atomspektren* (Vieweg, Braunschweig 1931)
16. U. Uhlhorn: *Representation of transformations in quantum mechanics.* Arkiv f. Fysik **23**, 307–340 (1962)
17. H. A. Dye: *On the geometry of projections in certain operator algebras.* Ann. Math. **61**, 73–89 (1955)
18. P. A. M. Dirac: *The Principles of Quantum Mechanics* (Clarendon Press, Oxford 1930)
19. J. M. Jauch: On bras and kets. In: *Aspects of Quantum Theory*, ed by A. Salam and E. Wigner (Cambridge University Press, Cambridge 1972) pp. 137–167
20. A. M. Gleason: *Measures on the closed subspaces of a Hilbert space.* J. Math. Mechanics **6**, 885–893 (1957)
21. S. Maeda: *Probability measures on projections in von Neumann algebras.* Rev. Math. Phys. 1, 235–290 (1990)
22. R. T. Rockafellar: *Convex Analysis* (Princeton University Press, Princeton 1970)
23. S. Boyd and L. Vandenberghe: *Convex Optimization* (Cambridge University Press, Cambridge 2004)

24. J. H. M. Wedderburn: *Lectures on Matrices* (American Mathematical Society, New York 1934). Online version: www.ams.org

25. J. v. Neumann: *On a certain topology for rings of operators*. Ann. Math. **37**, 111–115 (1936)

26. F. J. Murray and J. v. Neumann: *On rings of operators*. Ann. Math. **37**, 116–229 (1936)

27. F. J. Murray and J. v. Neumann: *On rings of operators II*. Trans. Amer. Math. Soc. **41**, 208–248 (1937)

28. J. v. Neumann: *On rings of operators III*. Ann. Math. **41**, 94–61 (1940)

29. F. J. Murray and J. v. Neumann: *On rings of operators IV*. Ann. Math. **44**, 716–808 (1943)

30. J. v. Neumann: *On rings of operators: Reduction theory*. Ann. Math. **50**, 401–485 (1949)

31. V. Kadison and J. R. Ringrose: *Fundamentals of the Theory of Operator Algebras –Vol. I and II*, Graduate Studies in Mathematics, vol 15 and 16, 2nd edn, (American Mathematical Society, Providence 1997) (A general reference is R)

32. W. Fulton: *Young Tableaux* (Cambridge University Press, Cambridge 1997)

33. J. v. Neumann: *Mathematische Grundlagen der Quantenmechanik* (Springer, Berlin 1932)

34. G. Lüders: *Über die Zustandsänderung durch den Meßprozeß*. Ann. Physik **8**, 322–328 (1951)

35. P. Busch, P. Lahti, and P. Mittelstaedt: *The Quantum Theory of Measurement*, 2nd edn (Springer, Berlin New York 1996)

36. M. Nielsen and I. Chuang: *Quantum Computation and Quantum Information* (Cambridge University Press, Cambridge 2000)

37. E. F. Beckenbach und R. Bellman: *Inequalities* (Springer, Berlin Heidelberg 1961)

38. R. Dubisch: *The Wedderburn structure theorems*. Am. Math. Monthly **54**, 253–259 (1947)

39. A. Uhlmann: *The "transition probability" in the state space of a $*$-algebra*. Rep. Math. Phys. **9**, 273–279 (1976)

40. R. Jozsa: *Fidelity for mixed quantum states*. J. Mod. Opt. **41**, 2315–2323 (1994)

41. Ch. A. Fuchs and C. M. Caves: *Mathematical techniques for quantum communication theory*. Open Sys. & Inf. Dyn. **3**, 345–356 (1995)

42. Ch. A. Fuchs: *Distinguishability and Accessible Information in Quantum Theory*. Dissertation, University of New Mexico (1995)

43. H. N. Barnum: *Quantum Information Theory*. Dissertation, University of New Mexico (1998)

44. S. Kakutani: *On equivalence of infinite product measures*. Ann. of Math. **49**, 214–224 (1948)

45. D. J. C. Bures: *An extension of Kakutani's theorem on infinite product measures to the tensor product of semifinite W^*-algebras*. Trans. Amer. Math. Soc. **135**, 199 (1969)

46. R. A. Fisher: *Theory of statistical estimation*. Proc. Camb. Philos. Soc. **22**, 700–725 (1925)

47. A. Uhlmann: The Metric of Bures and the Geometric Phase. In: *Quantum Groups and Related Topics*, ed by R. Gielerak et al. (Kluwer Academic Publishers, Dordrecht 1992) pp. 267–264

48. D. Petz: *Monotone metrics on matrix spaces*. Lin. Algebra and Appl. **244** 81–96 (1996)

49. D. Petz and Cs. Sudár: *Geometry of quantum states.* J. Math. Phys. **37**, 2662–2673 (1996)
50. J. Gruska: *Quantum Computing* (McGraw-Hill, London 1999)
51. J. L. Brylinski and G. Chen: *Mathematics of Quantum Computation* (Chapman & Hall, Boca Raton London New York 2002)
52. W.-H. Steeb and Y. Hardy: *Problems and Solutions in Quantum Computing and Quantum Information* (World Scientific, Singapore 2004)
53. P. M. Alberti and A. Uhlmann: *On Bures-distance and *-algebraic transition probability between inner derived positive linear forms over W^*-algebras.* Acta Applicandae Mathematicae **60**, 1–37 (2000)
54. P. M. Alberti and A. Uhlmann: *Transition probabilities on C^*- and W^*-algebras.* In: *Proceedings of the Second International Conference on Operator Algebras, Ideals, and Their Applications in Theoretical Physics, Leipzig 1983,* ed by H. Baumgärtel, G. Laßner, A. Pietsch, and A. Uhlmann, Teubner-Texte zur Mathematik 67 (B.G.Teubner, Leipzig 1984) pp. 5–11
55. P. M. Alberti: *On the geometry of pairs of positive linear forms (I).* Wiss. Z. KMU Leipzig, MNR **39**, 579–597 (1990)
56. P. M. Alberti: *A study on the geometry of pairs of positive linear forms, algebraic transition probablities and geometric phase over non-commutative operator algebras (I).* Z. Anal. Anwendungen **11**, 293–334 (1992)
57. V. Heinemann: *Geometrie und Transformation von Paaren positiver Linearformen über C^*-Algebren.* Dissertation, Universität Leipzig (1991)
58. A. Uhlmann: *An energy dispersion estimate.* Phys. Lett. A **161** 329–331 (1992)
59. I. Bengtsson: *Geometrical statistics – classical and quantum.* arXiv:quantph/0509017
60. A. Ericsson: *Geodesics and the best measurement for distinguishing quantum states.* J. Phys. A **38**, L725–L730 (2005)
61. H. Araki and G. Raggio: *A remark on transition probability.* Lett. Math. Phys. **6**, 237–240 (1982)
62. P. M. Alberti: *A note on the transition probability over C^*-algebras.* Lett.Math. Phys. **7**, 25–32 (1983)
63. M.-D. Choi: *Some assorted inequalities for positive linear maps on C^*-algebras.* J. Operator Theory **4**, 271–285 (1980)
64. F. Kubo and T. Ando: *Means of positive linear operators.* Math. Ann. **246**, 205–224 (1980)
65. K. M. R. Audenaert, J. Calsamiglia, R. Munoz-Tapia, E. Bagan, Ll. Masanes, A. Acin, and F. Verstraete: *The Quantum–Chernoff Bound,* Phys. Rev. Lett. **98**, 160501 (2007), arXiv:quant-ph/0610027
66. M. Raginski: *A fidelity measure for quantum channels.* Phys. Lett. A **290**, 11–18 (2001)
67. V. P. Belavkin, M. D'Ariano, and M. Raginski: *Operational distance and fidelity for quantum channels.* J. Math. Phys. **46**, 062106 (2005)
68. D. Kretschmann, D. Schlingemann, and R. F. Werner: *The information-disturbance tradeoff and the continuity of Stinespring's representation.* arXiv:quant-ph/0605009
69. W. Pusz and L. Woronowicz: *Functional calculus for sesquilinear forms and the purification map.* Rep. Math. Phys. **8**, 159–170 (1975)
70. T. Ando, Chi-Kwang Li, and R. Mathias: *Geometric means.* Lin. Algebra and Appl. **385**, 305–334 (2004)

2 Basic Concepts of Entangled States

F. Mintert[1,3], C. Viviescas[2,3], and A. Buchleitner[3]

[1] Department of Physics, Harvard University, 17 Oxford Street, Cambridge MA, USA
[2] Departamento de Física, Universidad Nacional de Colombia, Carrera 30 No. 45-03 Edif. 404, Bogotá D. C., Colombia
[3] Max-Planck-Institut für Physik komplexer Systeme, Nöthnitzer Str. 38, 01187 Dresden, Germany

2.1 Introduction

Quantum systems display properties that are unknown for classical ones, such as the superposition of quantum states, interference, or tunneling. These are all one-particle effects that can be observed in quantum systems, which are composed of a single particle. But these are not the only distinctions between classical and quantum objects – there are further differences that manifest themselves in composite quantum systems, that is, systems that are comprised of at least two subsystems. It is the correlations between these subsystems that give rise to an additional distinction from classical systems, whereas correlations in classical systems can always be described in terms of classical probabilities; this is not always true in quantum systems. Such non-classical correlations lead to apparent paradoxes like the famous Einstein Podolsky Rosen scenario [1] that might suggest, on the first glance, that there is remote action in quantum mechanics.

States that display such non-classical correlations are referred to as *entangled states*, and it is the aim of this chapter to introduce the basic tools that allow to understand the nature of such states, to distinguish them from those that are classically correlated, and to quantify non-classical correlations.

2.2 Entangled States

Composite quantum systems are systems that naturally decompose into two or more subsystems, where each subsystem itself is a proper quantum system. Referring to a decomposition as "natural" implies that it is given in an obvious fashion due to the physical situation. Most frequently, the individual susbsystems are characterized by their mutual distance that is larger than the size of a subsystem. A typical example is a string of ions, where each ion is a subsystem, and the entire string is the composite system. Formally, the Hilbert space $\mathcal{H}$ associated with a composite, or *multipartite system*, is given by the tensor product $\mathcal{H}_1 \otimes \cdots \otimes \mathcal{H}_N$ of the spaces corresponding to each of the subsystems.

Mintert, F. et al.: *Basic Concepts of Entangled States*. Lect. Notes Phys. **768**, 61–86 (2009)
DOI 10.1007/978-3-540-88169-8_2   © Springer-Verlag Berlin Heidelberg 2009

In the following, we shall focus on finite-dimensional *bipartite* quantum systems, i.e., systems composed of two distinct subsystems, described by the Hilbert space $\mathcal{H} = \mathcal{H}_1 \otimes \mathcal{H}_2$. Many of the concepts and ideas that we introduce can, nevertheless, be generalized to *multipartite systems*.

2.2.1 Pure States

We start out with a bipartite system with each subsystem prepared in a pure state $|\psi_i\rangle$ ($i = 1, 2$). The state of the composite system $|\Psi_s\rangle$ is the direct product thereof:

$$|\Psi_s\rangle = |\psi_1\rangle \otimes |\psi_2\rangle . \tag{2.1}$$

Suppose that one could perform only local measurements on the system, i.e., one had access to only one of the subsystems at a time. Then, after a measurement of any local observable $a \otimes \mathbb{1}$ on the first subsystem, where a is a hermitian operator acting on $\mathcal{H}_1$, and $\mathbb{1}$ is the identity acting on $\mathcal{H}_2$, the state of the first subsystem will be projected onto an eigenstate of a, but the state of the second subsystem remains unchanged. If later on, one performs a second local measurement, now on the second subsystem, it will yield a result that is independent of the result of the first measurement. Hence, the measurement outcomes on different subsystems are uncorrelated with each other and depend only on the states of each respective subsystem.

A general pure state in $\mathcal{H}$ can be given by a superposition of pure states of the form (2.1), for example,

$$|\Psi_e\rangle = \frac{1}{\sqrt{2}} \left(|\psi_1\rangle \otimes |\psi_2\rangle + |\phi_1\rangle \otimes |\phi_2\rangle \right) , \tag{2.2}$$

where $|\psi_i\rangle \neq |\phi_i\rangle$ ($i = 1, 2$). We may now ask what the state $|\Psi_e\rangle$ looks like if one has access to only one of the subsystems? For a local operator $a \otimes \mathbb{1}$ on the first subsystem, the expectation value observed in an experiment reads

$$\begin{aligned}
\langle a \rangle &= \langle \Psi_e | a \otimes \mathbb{1} | \Psi_e \rangle \\
&= \mathrm{tr}(a \otimes \mathbb{1} \, |\Psi_e\rangle\langle\Psi_e|) \\
&= \mathrm{tr}_1(a \, \mathrm{tr}_2 |\Psi_e\rangle\langle\Psi_e|) \\
&= \mathrm{tr}_1(a \varrho_1) ,
\end{aligned} \tag{2.3}$$

where $\mathrm{tr}_{1,2}$ denotes the partial trace over the first/second subsystem, and $\varrho_1 = \mathrm{tr}_2 |\Psi_e\rangle\langle\Psi_e|$ is the reduced density matrix of the first subsystem. Since (2.3) holds for any local operator a, we need to conclude that the state of the first subsystem alone is given by ϱ_1. An analogous reasoning leads to the conclusion that also the state of the second subsystem is described by its reduced density matrix $\varrho_2 = \mathrm{tr}_1 |\Psi_e\rangle\langle\Psi_e|$. The state of the composite system, however, is not equal to the product of both subsystem states, $\rho = |\Psi_e\rangle\langle\Psi_e| \neq \rho_1 \otimes \rho_2$. Moreover, if one performs a local measurement on one subsystem, this leads to a state reduction of the entire system state, not only of the subsystem on which the measurement had been performed. Therefore, the probabilities for an outcome of a measurement on one subsystem are influenced by prior measurements on the other subsystem. Thus, measurement results on – possibly distant and non-interacting – subsystems are correlated.

Based on these considerations, we can define that

states that can be written as a product of pure states, as in (2.1), are called *product* or *separable states*. If on the contrary, there are no local states $|\psi_1\rangle \in \mathcal{H}_1$ and $|\psi_2\rangle \in \mathcal{H}_2$, such that the state of the system $|\Psi\rangle$ can be written as a product thereof:

$$\nexists\, |\psi_1\rangle \in \mathcal{H}_1, |\psi_2\rangle \in \mathcal{H}_2 \quad \text{such that} \quad |\Psi\rangle = |\psi_1\rangle \otimes |\psi_2\rangle\ , \quad (2.4)$$

then $|\Psi\rangle$ is an *entangled state*.

2.2.2 Mixed States

So far we considered only pure states. More generally, however, the state of a quantum system can be mixed. Mixed states are in fact the most frequently encountered states in real experiments, since hardly any quantum system can be isolated completely from its surroundings. As elaborated in more detail in Sect. 5.3.1, it is in general not possible to keep track of the many environmental degrees of freedom, and the state of the system is given by the partial trace over the environment. This reduced state is then typically mixed.

Similarly to the case of pure states, *mixed product states*,

$$\varrho = \rho^{(1)} \otimes \rho^{(2)} \quad (2.5)$$

with $\rho^{(1)}$ and $\rho^{(2)}$ for the respective subsystems, do not exhibit correlations. A convex sum of different product states,

$$\varrho = \sum_i p_i\, \rho_i^{(1)} \otimes \rho_i^{(2)}\ , \quad (2.6)$$

with $p_i > 0$ and $\sum_i p_i = 1$, however, will in general yield correlated measurement results, i.e., there are local observables a and b such that $\mathrm{tr}(\varrho(a \otimes b)) \neq \mathrm{tr}(\varrho(a \otimes \mathbb{1}))\, \mathrm{tr}(\varrho(\mathbb{1} \otimes b)) = \mathrm{tr}_1 \varrho_1 a\ \mathrm{tr}_2 \varrho_2 b$. These correlations can be described in terms of the classical probabilities p_i, and are therefore considered classical. States of the form (2.6) thus are called *separable mixed states*.

Mixed entangled states, in turn, are defined by the non-existence of a decomposition into product states [2]:

A mixed state ϱ is entangled if there are no local states $\rho_i^{(1)}$, $\rho_i^{(2)}$, and non-negative weights p_i, such that ϱ can be expressed as a convex mixture thereof:

$$\nexists \varrho_i^{(1)}, \varrho_i^{(2)}, p_i \geq 0 \quad \text{such that} \quad \varrho = \sum_i p_i \, \rho_i^{(1)} \otimes \rho_i^{(2)} \, . \qquad (2.7)$$

Entangled states imply quantum correlations of measurements on different subsystems which, in contrast to classical correlations (see above), *cannot* be described in terms of only classical probabilities.

2.3 Separability Criteria

The above definitions of separable and entangled states appear simple on a first sight. But checking separability of a given state can turn out to be much more involved than one might expect. Separability is defined via the *existence* of a decomposition of a state into product states in the case of pure states, or into a convex sum of tensor products for mixed states. That is, in order to show that a given state is separable, one has to look for such decompositions. Once a decomposition is found one knows that a state is separable. But the failure to find one can have two different reasons: either the state is entangled and there is no decomposition into product states, or the state is actually separable, but the appropriate decomposition could not be identified.

For this reason, there is a need for potentially simple criteria to distinguish separable from entangled states that do not require an explicit search. For pure states, there are criteria that discriminate separable and entangled states unambiguously, but for mixed states similar tools are available only for low-dimensional system. For higher dimensional systems, these tools can provide only partial information, as we will see later on. But, before we discuss mixed states, we will start out with the comparatively simpler case of pure states.

2.3.1 Pure States

Let us consider the exemplary case

$$|\Psi\rangle = \frac{|0\rangle + |1\rangle}{\sqrt{2}} \otimes \frac{|0\rangle + 2|1\rangle}{\sqrt{5}} \, . \qquad (2.8)$$

One can see that $|\Psi\rangle$ factorizes into local states – it is separable, though could be rewritten also as

$$|\Psi\rangle = \frac{|00\rangle + 2|01\rangle + |10\rangle + 2|11\rangle}{\sqrt{10}} \, , \qquad (2.9)$$

where separability is less evident. It just turns out that separability is more easily identified if $|\Psi\rangle$ is expressed in the bases $\{(|0\rangle + |1\rangle)/\sqrt{2}, (|0\rangle - |1\rangle)/\sqrt{2}\}$ of $\mathcal{H}_1$ and $\{(|0\rangle + 2|1\rangle)/\sqrt{5}, (2|0\rangle - |1\rangle)/\sqrt{5}\}$ of $\mathcal{H}_2$ than in the basis $\{|0\rangle, |1\rangle\}$. As we shall see, the observation is generic, in the sense that there is always a basis that allows to reveal the entanglement properties. The representation of a state in this basis is called the *Schmidt decomposition* [3].

Schmidt Decomposition

Given two arbitrary local bases $\{|\varphi_i\rangle\}$ and $\{|\phi_i\rangle\}$ in the spaces $\mathcal{H}_1$ and $\mathcal{H}_2$, any pure state $|\Psi\rangle$ in $\mathcal{H} = \mathcal{H}_1 \otimes \mathcal{H}_2$ can be expressed in terms of the corresponding product basis

$$|\Psi\rangle = \sum_{ij} d_{ij}|\varphi_i\rangle \otimes |\phi_j\rangle . \tag{2.10}$$

The expansion coefficients d_{ij} are given by the overlap of the state with the basis vectors, $d_{ij} = \langle\varphi_i| \otimes \langle\phi_j|\Psi\rangle$. If one now makes a change of bases $|\tilde{\varphi}_i\rangle = \mathcal{U}|\varphi_i\rangle$ and $|\tilde{\phi}_i\rangle = \mathcal{V}|\phi_i\rangle$, with $\mathcal{U}$ and $\mathcal{V}$ arbitrary, local unitary transformations on $\mathcal{H}_1$ and $\mathcal{H}_2$, respectively, the d_{ij} change accordingly:

$$\begin{aligned}
\tilde{d}_{ij} &= \langle\tilde{\varphi}_i| \otimes \langle\tilde{\phi}_j|\Psi\rangle \\
&= \langle\varphi_i|\mathcal{U}^\dagger \otimes \langle\phi_j|\mathcal{V}^\dagger|\Psi\rangle \\
&= \sum_{pq}\langle\varphi_i|\mathcal{U}^\dagger|\varphi_p\rangle\langle\phi_j|\mathcal{V}^\dagger|\phi_q\rangle\langle\varphi_p| \otimes \langle\phi_q|\Psi\rangle \\
&= [udv]_{ij} ,
\end{aligned} \tag{2.11}$$

where in the third line we used the resolution of the identity on each subsystem, $\sum_i |\varphi_i\rangle\langle\varphi_i| = \mathbb{1}$ and $\sum_i |\phi_i\rangle\langle\phi_i| = \mathbb{1}$, and we defined the unitary matrices $u_{ip} = \langle\varphi_i|\mathcal{U}^\dagger|\varphi_p\rangle$, $v_{qj} = \langle\phi_j|\mathcal{V}^\dagger|\phi_q\rangle$. In the new basis, the state is given by

$$|\Psi\rangle = \sum_{ij} [udv]_{ij} |\tilde{\varphi}_i\rangle \otimes |\tilde{\phi}_j\rangle . \tag{2.12}$$

In order to obtain the Schmidt decomposition of $|\Psi\rangle$, we use the fact that for every complex matrix d, there always exist unitary transformations u and v such that udv is diagonal. This provides the *singular value decomposition* of d [4], with real, non-negative diagonal entries $\mathcal{S}_i$, called *singular values*. Therefore, for each state $|\Psi\rangle$, one can always find local bases $|\varphi_i^S\rangle$ and $|\phi_i^S\rangle$ in terms of which (2.12) reduces to

$$|\Psi\rangle = \sum_i \sqrt{\lambda_i} |\varphi_i^S\rangle \otimes |\phi_i^S\rangle , \tag{2.13}$$

where the $\lambda_i = \mathcal{S}_i^2$ are known as *Schmidt coefficients*, and the sum is limited by the dimension of the smaller subsystem. Like eigenvalues of a matrix, also the singular values are uniquely defined. Hence, for any state $|\Psi\rangle$ the Schmidt coefficients are unique. Furthermore, since the *Schmidt basis* $\{|\varphi_i^S\rangle \otimes |\phi_j^S\rangle\}$ is given by separable states, all information on the entanglement of a state is encoded in the Schmidt coefficients: If there is only one non-vanishing Schmidt coefficient, then $|\Psi\rangle$ is separable. Otherwise, when at least two Schmidt coefficients are different from zero, it is not possible to express $|\Psi\rangle$ in the form (2.1). Consequently, we can conclude that a pure state $|\Psi\rangle$ is separable if and only if it has only one non-vanishing Schmidt coefficient.

Reduced Density Matrix

Since the Schmidt coefficients are so useful for the distinction of separable and entangled states, we should focus on how to evaluate them. The reduced density matrices are particularly helpful in this context. The one of the first subsystem reads

$$
\begin{aligned}
\varrho_1 &= \mathrm{tr}_2 |\Psi\rangle\langle\Psi| \\
&= \mathrm{tr}_2 \sum_{ij} \sqrt{\lambda_i \lambda_j}\, |\varphi_i^S\rangle\langle\varphi_j^S| \otimes |\phi_i^S\rangle\langle\phi_j^S| \\
&= \sum_i \lambda_i |\varphi_i^S\rangle\langle\varphi_i^S|\,,
\end{aligned}
\tag{2.14}
$$

in terms of the Schmidt decomposition (2.13), where we used the orthonormality of the Schmidt basis while performing the trace over the second subsystem.

We see that the Schmidt coefficients are given by the eigenvalues of the reduced density matrix ϱ_1. An equivalent reasoning holds for the reduced density matrix of the second subsystem $\varrho_2 = \mathrm{tr}_1 |\Psi\rangle\langle\Psi|$; that is ϱ_1 and ϱ_2 have the same non-vanishing eigenvalues, and the basis vectors of the Schmidt basis are given by the eigenstates of ϱ_1 and ϱ_2.

We not only found a simple prescription to evaluate the Schmidt coefficients of any state $|\Psi\rangle$, but since separability requires that exactly one Schmidt coefficient is different from zero, we also have related the entanglement of a pure state $|\Psi\rangle$ to the degree of mixing of the reduced density matrices. That is, we can restate the separability criterion for pure states:

$$
\mathrm{tr}\varrho_r^2 = 1 \Rightarrow \varrho_r \text{ is pure} \quad \Rightarrow |\Psi\rangle \text{ is separable}
$$
$$
\mathrm{tr}\varrho_r^2 < 1 \Rightarrow \varrho_r \text{ is mixed} \Rightarrow |\Psi\rangle \text{ is entangled}
\tag{2.15}
$$

with r referring to either one of the two subsystems.

2.3.2 Mixed States

For pure states, the Schmidt decomposition provides a necessary and sufficient criterion for separability. Unfortunately, for mixed states such an elegant decomposition does not exist. In particular, if a state is mixed, the degree of mixing of its reduced density matrices is not an indicator of entanglement. Therefore, we need to find some new criteria to distinguish entangled from separable mixed states. The most prominent of such tools are *entanglement witnesses* and *positive maps*. As we shall see, both concepts are closely related.

Entanglement Witnesses

An entanglement witness W [5, 6] is a hermitian operator acting on $\mathcal{H}$ that is *not* positive definite, but that yields positive expectation values

$$\langle \Psi_{\rm s} | W | \Psi_{\rm s} \rangle \geq 0 \;, \tag{2.16}$$

for all separable pure states $|\Psi_{\rm s}\rangle$. Since any separable mixed state can be expressed as a convex sum of projectors onto pure separable states, $\varrho_{\rm s} = \sum_i p_i |\Psi_{\rm s}^{(i)}\rangle\langle\Psi_{\rm s}^{(i)}|$ with $p_i > 0$, and $\sum_i p_i = 1$, (2.18) implies that the expectation value of an entanglement witness with respect to any separable mixed state is also non-negative,

$$\mathrm{tr}(\varrho_{\rm s} W) = \sum_i p_i \langle \Psi_{\rm s} | W | \Psi_{\rm s} \rangle \geq 0 \;. \tag{2.17}$$

Thus, if a given density matrix ϱ leads to a negative expectation value

$$\mathrm{tr}(\varrho W) < 0 \;, \tag{2.18}$$

then ϱ is *entangled*, and one says that W *detects* ϱ.

The central benefit of witnesses is that there exists a witness for any entangled state that detects it [5]. Here we do not go into the details of the formal proof, but rather give some geometric, intuitive arguments that allow to understand why entanglement witnesses work.

Geometry of Quantum States

Let's try to understand quantum states in a geometrical setting. Density matrices can be conceived as vectors in a vector space that is referred to as Hilbert–Schmidt space [7]. For a geometric interpretation of this vector space, one needs a scalar product, and in the present context, this is defined as

$$\langle A | B \rangle = \mathrm{tr}\, A^\dagger B \;. \tag{2.19}$$

Now, separable states form a convex set. That means that, given two arbitrary separable states $\varrho_s^{(1)}$ and $\varrho_s^{(2)}$, any convex sum $\lambda \varrho_s^{(1)} + (1-\lambda)\varrho_s^{(2)}$ $(1 \geq \lambda \geq 0)$ is again separable. Geometrically, this means that the set of separable states has no trough, as illustrated in Fig. 2.1, where the shapes A, B, and C represent different convex sets, whereas D is not convex, since it has a trough on its right bottom part. Now, one can find several lines that separate the grey shaded areas from their white surrounding – the depicted lines W_i $(i = 1, \dots, 8)$ are only exemplary ones; one may find many more.

There is one crucial difference between cases A, B, and C on the one hand, and D on the other hand: For any point outside the convex sets A, B, and C, one can find a straight line that separates this point from the gray-shaded area. For D, this is not always possible. There is no straight line that separates D from point Z.

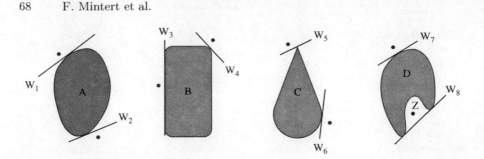

Fig. 2.1. Four different shapes, three of which (A, B, C) are convex, whereas shape D is not. To any point outside the convex shapes, there exists a line (like W_i, $i = 1, \ldots, 6$) that separates this point from the corresponding convex shape. For the non-convex shape D, the situation is different: for the point Z, there is no such line. The situation of entanglement witnesses is analogous: the set of separable states is convex; there exists a witness (the analogue of a line W_i) to any entangled state (the analogue of a point outside the grey shapes) that separates it from the set of separable states (the analogue of one of the convex shapes)

Although, the set of separable states is high dimensional and more complicated than the shapes in Fig. 2.1, the basic geometric picture of Fig. 2.1 still allows to understand the basic mechanism of entanglement witnesses.

Geometric Interpretation of Entanglement Witnesses

A separable state is characterized by the condition tr $\varrho W \geq 0$. The condition that trσW vanishes, requires σ to be a linear combination of operators $\mathcal{O}_i$ that are orthogonal to W:

$$\sigma = \sum_i \alpha_i \mathcal{O}_i , \quad \text{with} \quad \text{tr}(\mathcal{O}_i W) = 0 . \tag{2.20}$$

That is, the condition tr$\sigma W = 0$ defines a hyperplane in the space of operators – analogous to the lines W_i in Fig 2.1. The sign of tr ϱW then indicates on which side of the hyperplane ϱ is situated, and all separable states are situated on one side of this hyperplane (tr $\varrho W \geq 0$). Since the separable states form a convex set, there is a witness to any entangled state that detects it, just like there is a line to any point outside A, B, or C that separates it from the respective grey shaded areas.

Due to the complicated structure of the set of separable states that has curved borders, one needs infinitely many witnesses to characterize it completely. Given some specific entangled state ϱ, it can be rather complicated to find a witness that detects it, and the failure to find a suitable witness for a state ϱ does not necessarily allow to conclude that ϱ is separable. Therefore, a witness provides a *necessary separability criterion*: if a state is separable, it will yield a non-negative expectation value for any witness; but separability of a state *cannot* deduced from such a non-negative expectation value.

Positive Maps

An alternative tool to check on separability is the so-called positive linear maps Λ that map the set of operators $\mathcal{B}(\mathcal{H})$ acting on a Hilbert space $\mathcal{H}$ on the set $\mathcal{B}(\tilde{\mathcal{H}})$, where $\tilde{\mathcal{H}}$ can – though not necessarily needs to – be a different Hilbert space than $\mathcal{H}$. Such a map Λ is considered positive if $\tilde{\varrho} = \Lambda(\varrho)$ is a positive operator, for any positive operator ϱ. Now, let us consider the case of a bipartite system. One can extend this map to the product space $\mathcal{H} = \mathcal{H}_1 \otimes \mathcal{H}_2$, such that the extended map Λ_E acts on $\mathcal{B}(\mathcal{H}_1)$ like Λ, and Λ_E acts trivially on $\mathcal{B}(\mathcal{H}_2)$, i.e.,

$$\Lambda_E = \Lambda \otimes \mathbb{1} \ . \tag{2.21}$$

A very counterintuitive property of these positive maps is that the extended map Λ_E is *not* necessarily positive. That is, for some maps Λ, there are states ϱ on $\mathcal{H} = \mathcal{H}_1 \otimes \mathcal{H}_2$, such that $\Lambda_E(\varrho)$ is not a positive operator.

Now, let us take a separable state ϱ_s, i.e., one that has a convex decomposition into product states, and apply a positive linear map to it,

$$\Lambda_E(\varrho_s) = \sum_i p_i \Lambda(\rho_i^{(1)}) \otimes \rho_i^{(2)} \ . \tag{2.22}$$

Since Λ is positive, $\Lambda(\rho_i^{(1)})$ is a positive operator; and since also p_i and $\rho_i^{(2)}$ are positive, any expectation value of $\Lambda_E(\varrho_s)$ is positive, and therefore $\Lambda_E(\varrho_s)$ remains a positive operator. Thus, for any separable state, there is *no* positive map Λ, such that $\Lambda_E(\varrho_s)$ is *not* a positive operator. That is, if one can find a positive map Λ such that $\Lambda_E(\varrho)$ has at least one negative eigenvalue for a given state ϱ, then one knows for sure that ϱ is entangled.

The inverse statement is more involved. If one wants to prove separability of a state ϱ on $\mathcal{H}_1 \otimes \mathcal{H}_2$, then it is necessary to consider maps Λ that map $\mathcal{B}(\mathcal{H}_2)$ on $\mathcal{B}(\mathcal{H}_1)$ – that is $(\mathbb{1} \otimes \Lambda)(\varrho)$ is an operator acting on $\mathcal{H}_1 \otimes \mathcal{H}_1$. Now, a state is separable if and only if $(\mathbb{1} \otimes \Lambda)(\varrho)$ is positive for all positive linear maps of $\mathcal{B}(\mathcal{H}_2)$ on $\mathcal{B}(\mathcal{H}_1)$. But, since the characterization of positive maps is an open problem, such maps only provide a necessary separability criterion like above in the case of witnesses: if one has found a map Λ, such that $(\mathbb{1} \otimes \Lambda)(\varrho)$ is *not* a positive operator, then the state ϱ is entangled. But if one fails to find such a map, then one does not necessarily know whether this is due to separability of ϱ, or just due to the lack of success to find a suitable map.

Only in systems of small dimension the concept of positive maps allows to formulate a constructive criterion that is both necessary and sufficient: for a system of two qubits or a system of one qubit and one qutrit (three-level system), one can check separability by considering only a single positive map, and that is the transposition $T(\varrho) = \varrho^T$, i.e., the reflection of a matrix ϱ along the diagonal [5, 8]. The underlying reason for this is that *any* positive map from $\mathcal{B}(\mathbb{C}^2)$ on $\mathcal{B}(\mathbb{C}^2)$, or on $\mathcal{B}(\mathbb{C}^3)$, i.e., maps that take a qubit-operator to a qubit- or to a qutrit-operator can be written as

$$\Lambda = \Lambda_{CP}^1 + \Lambda_{CP}^2 \circ T , \qquad (2.23)$$

where Λ_{CP}^i ($i = 1, 2$) are completely positive maps, and T is the transposition [9, 10]. Therefore, the condition that $(\mathbb{1} \otimes \Lambda)(\varrho)$ be positive for any positive map Λ reduces to

$$
\begin{aligned}
(\mathbb{1} \otimes \Lambda)(\varrho) &= (\mathbb{1} \otimes \Lambda_{CP}^{(1)})(\varrho) + (\mathbb{1} \otimes \Lambda_{CP}^{(2)})(\mathbb{1} \otimes T)(\varrho) \\
&= (\mathbb{1} \otimes \Lambda_{CP}^{(1)})(\varrho) + (\mathbb{1} \otimes \Lambda_{CP}^{(2)})(\varrho^{\mathrm{pt}}) \geq 0 ,
\end{aligned}
\qquad (2.24)
$$

where $\varrho^{\mathrm{pt}} = (\mathbb{1} \otimes T)(\varrho)$ is called the *partial transpose* of ϱ. Since the $\Lambda_{CP}^{(i)}$ are completely positive, the extended maps $\mathbb{1} \otimes \Lambda_{CP}^{(i)}$ are positive maps. Therefore, $(\mathbb{1} \otimes \Lambda_{CP}^{(1)})(\varrho)$ is non-negative, i.e., it has no negative eigenvalue. The partial transpose ϱ^{pt}, however, is not necessarily a positive operator, since $\mathbb{1} \otimes T$ is *not* a positive map. But, if ϱ is such that its partial transpose is non-negative, then also $(\mathbb{1} \otimes \Lambda_{CP}^{(2)})(\varrho^{\mathrm{pt}})$ is non-negative. In that case, we can conclude that $(\mathbb{1} \otimes \Lambda)(\varrho)$ is non-negative for arbitrary positive maps Λ, and this implies that ϱ is separable. On the other hand, we already know that ϱ is entangled if its partial transpose has at least one negative eigenvalue. Therefore, the spectrum of ϱ^{pt} allows to unambiguously distinguish separable from entangled states in 2×2-dimensional and 2×3-dimensional systems.

In higher dimensional systems, however, (2.23) does not characterize all positive maps anymore, and there are entangled states with positive partial transpose (*ppt*). But also in high-dimensional systems, the so-called *ppt-criterion* is a frequently used separability criterion: despite being only a necessary separability criterion it still detects many entangled states, and it is rather straightforward to implement: a general state of a bipartite system can be expanded in some arbitrary product basis $\varrho = \sum_{ij,kl} \varrho_{ij,kl} |\varphi_i\rangle\langle\varphi_j| \otimes |\phi_k\rangle\langle\phi_l|$, and its partial transpose is obtained by a simple rearrangement of matrix elements. $\varrho^{\mathrm{pt}} = (\mathbb{1} \otimes T)(\varrho) = \sum_{ij,kl} \varrho_{ij,lk} |\varphi_i\rangle\langle\varphi_j| \otimes |\phi_k\rangle\langle\phi_l|$. One may check that ϱ^{pt} actually depends on the basis with the help of which it is constructed. However, it is only the spectrum of ϱ^{pt} that enters the present separability criterion, and the spectrum does *not* depend on this choice of basis.

Witnesses and Positive Maps

So far, we presented entanglement witnesses and positive maps as independent concepts. And indeed, they do not seem to have too much in common. Entanglement witnesses could be understood in a geometric setting, and positive maps have rather counterintuitive properties. However, these two concepts are more closely related than they seem to be on the first glance.

Let us consider a positive map Λ such that the extended map $\mathbb{1} \otimes \Lambda$ applied to some state ϱ yields a non-positive operator, i.e., Λ is not a completely positive map. Then $(\mathbb{1} \otimes \Lambda)(\varrho)$ has an eigenvector $|\chi\rangle$ with a negative eigenvalue λ,

$$(\mathbb{1} \otimes \Lambda)(\varrho)|\chi\rangle = \lambda|\chi\rangle \ . \tag{2.25}$$

We can now show that the observable $W = (\mathbb{1} \otimes \Lambda^\dagger)(|\chi\rangle\langle\chi|)$ is an entanglement witness. For an arbitrary separable state $|\Phi_s\rangle$, we have

$$
\begin{aligned}
\langle\Phi_s|W|\Phi_s\rangle &= \mathrm{tr}\Big[\big((\mathbb{1} \otimes \Lambda^\dagger)(|\chi\rangle\langle\chi|)\big)\ |\Phi_s\rangle\langle\Phi_s|\Big] \\
&= \mathrm{tr}\Big[|\chi\rangle\langle\chi|\ \big((\mathbb{1} \otimes \Lambda)(|\Phi_s\rangle\langle\Phi_s|)\big)\Big] \geq 0 \ ,
\end{aligned}
\tag{2.26}
$$

where the inequality is due to the positivity of Λ, such that $(\mathbb{1} \otimes \Lambda)(|\Phi_s\rangle\langle\Phi_s|)$ is a positive operator. And, indeed, this witness detects ϱ to be entangled:

$$
\begin{aligned}
\mathrm{tr}(\varrho W) &= \mathrm{tr}\Big[\varrho\ \big((\mathbb{1} \otimes \Lambda^\dagger)(|\chi\rangle\langle\chi|)\big)\Big] \\
&= \mathrm{tr}\Big[\big((\mathbb{1} \otimes \Lambda)(\varrho)\big)\ |\chi\rangle\langle\chi|\Big] \\
&= \langle\chi|(\mathbb{1} \otimes \Lambda)(\varrho)|\chi\rangle = \lambda < 0
\end{aligned}
\tag{2.27}
$$

because of the above eigenvector relation.

2.4 Entanglement Monotones and Measures

So far we contented ourselves with a qualitative distinction between separable and entangled states. This, however, does not allow to compare the amount of entanglement of two different states. For such purposes, one would need a quantitative description of entanglement. But the prior definition of entanglement in terms of the nonexistence of a decomposition of a state into product states (cf. (2.4),(2.7)) will not be helpful for finding such a quantification. Therefore, before we can introduce entanglement measures, we need to refine our concept of entanglement.

2.4.1 General Considerations

Let us forget for a while about the prior formal definition and focus more on the interpretation that entanglement is tantamount to correlations that cannot be described in terms of classical probabilities. This allows to arrive at a new concept that allows for a quantitative description of entangled states, and it will still be in agreement with the previous definitions of entanglement and separability.

The idea is to classify all operations that one could apply to a composite quantum system, and that can increase only classical correlations, that is those that are captured by probabilities p_i as in (2.6). Once this is done, one can make the decrease of correlations under all such operations a defining property of entanglement. Thus, before we can come to the promised quantification of entanglement, we first have to make a significant detour to end up with what is referred to as *local operations and classical communication.*

Quantum Operations

To do so, let us start out with the most general operations. The basic ones that are allowed by the laws of quantum mechanics comprise unitary evolutions

$$\varrho \mapsto \mathcal{U}\varrho\mathcal{U}^\dagger , \text{ with } \mathcal{U}\mathcal{U}^\dagger = \mathcal{U}^\dagger\mathcal{U} = \mathbb{1} , \qquad (2.28)$$

and v. Neumann measurements in which a quantum state ϱ is projected onto an eigenstate of the associated observable (see Sect. 5.1.3). Let us denote such a complete set of eigenstates $\{|\varphi_i\rangle\}$. Then, the corresponding measurement results in the collapse of ϱ on the state $|\varphi_i\rangle\langle\varphi_i|$, with probability $p_i = \langle\varphi_i|\varrho|\varphi_i\rangle$. That is, on average the state evolves as

$$\varrho \mapsto \sum_i p_i|\varphi_i\rangle\langle\varphi_i| = \sum_i |\varphi_i\rangle\langle\varphi_i|\varrho|\varphi_i\rangle\langle\varphi_i| . \qquad (2.29)$$

Thus, a v. Neumann measurement takes a state to a purely probabilistic mixture of the states $|\varphi_i\rangle$, and it destroys all coherences between them completely. Though, one might wonder if one could come up with a slightly less 'invasive' measurement with less dramatic effects. And, indeed, one can do so, if one uses an additional quantum system – often referred to as *ancilla* – lets this ancilla interact with the original system, and finally performs the measurement on the ancilla only. The original state ρ of the combined systems including the ancilla reads

$$\rho = \varrho \otimes |\Psi_a\rangle\langle\Psi_a| , \qquad (2.30)$$

where $|\Psi_a\rangle$ is an ancilla state. An interaction between the original system and the ancilla results in a global unitary evolution

$$U(\varrho \otimes |\Psi_a\rangle\langle\Psi_a|)U^\dagger , \qquad (2.31)$$

and a subsequent measurement in the basis $\{|\Psi_a^{(i)}\rangle\}$ of ancilla states projects this state on

$$\langle\Psi_a^{(i)}|U|\Psi_a\rangle\varrho\langle\Psi_a|U^\dagger|\Psi_a^{(i)}\rangle = A_i\varrho A_i^\dagger , \qquad (2.32)$$

with the operators $A_i = \langle\Psi_a^{(i)}|U|\Psi_a\rangle$ that act only on the original system. On average, the state evolves as

$$\varrho \mapsto \sum_i A_i\varrho A_i^\dagger . \qquad (2.33)$$

If one utilizes the completeness of the ancilla states $\sum_i |\Psi_a^{(i)}\rangle\langle\Psi_a^{(i)}| = \mathbb{1}$, and subsequently $U^\dagger U = \mathbb{1}$, one can convince oneself that the operators A_i satisfy the resolution of the identity

$$\sum_i A_i^\dagger A_i = \sum_i \langle\Psi_a|U^\dagger|\Psi_a^{(i)}\rangle\langle\Psi_a^{(i)}|U|\Psi_a\rangle = \mathbb{1} \ . \tag{2.34}$$

This property is crucial, since it guarantees the conservation of the trace

$$\mathrm{tr}\sum_i A_i \varrho A_i^\dagger = \mathrm{tr}\sum_i A_i^\dagger A_i \varrho = \mathrm{tr}\varrho \ , \tag{2.35}$$

and, therefore, of probability.

In Sect. 2.3.2, we were discussing positive maps and saw that a trivial extension of a map is not necessarily a positive map again. However, for any map that describes the evolution of a real quantum system, any such extension needs to be positive: if a map acts only on a subcomponent of a system, obviously the positivity of the state of the entire system has to be ensured; this is the case exactly if the extension of the map is positive, i.e., if the map is *completely positive*. Since any trace preserving, completely positive map can always be expressed in the form of (2.33), and, since any map of the form (2.33) is trace preserving and completely positive (see Sect. 5.3.1) [11–13], (2.33) is indeed the most general evolution a quantum state can undergo.

Some Examples

Let us look at a few exemplary cases of operations of the form (2.33) to see how they can affect entanglement properties. First, consider the specific unitary map

$$\mathcal{U} = \frac{1}{\sqrt{2}}\big(|00\rangle + |11\rangle\big)\langle00| + \frac{1}{\sqrt{2}}\big(|00\rangle - |11\rangle\big)\langle11| + |01\rangle\langle01| + |10\rangle\langle10| \ . \tag{2.36}$$

This is an example of a *global* operation, that is, it cannot be written as $\mathcal{U} = \mathcal{U}_1 \otimes \mathcal{U}_2$, and its implementation requires an interaction between the two individual subsystems. Applying the map to $|00\rangle$ takes this separable state to the entangled state $\mathcal{U}|00\rangle = (|00\rangle + |11\rangle)/\sqrt{2}$. Thus, such a global operation can indeed create entanglement.

A second example is given by a measurement in the Bell-basis

$$|\varphi_1\rangle = \frac{|00\rangle + |11\rangle}{\sqrt{2}} \ , \qquad\qquad |\varphi_2\rangle = \frac{|00\rangle - |11\rangle}{\sqrt{2}} \ , \tag{2.37}$$

$$|\varphi_3\rangle = \frac{|01\rangle + |10\rangle}{\sqrt{2}} \ , \qquad\qquad |\varphi_4\rangle = \frac{|01\rangle - |10\rangle}{\sqrt{2}} \ , \tag{2.38}$$

followed by a local unitary transformation that is conditioned on the measurement outcome. Let us start again with the separable state $|00\rangle$. Repeated measurements yield the two different outcomes $(|00\rangle + |11\rangle)/\sqrt{2}$, and

$(|00\rangle - |11\rangle)/\sqrt{2}$, with equal probability. A conditioned local unitary operation that is comprised of the identity operation in case of the first outcome, and of $u = |0\rangle\langle 0| - |1\rangle\langle 1|$ on the second subsystem in case of the second, yields the final state $(|00\rangle + |11\rangle)/\sqrt{2}$, which, once again, is entangled. This provides a second example of a global operation that can create entanglement.

We will see later, however, that the situation is different if we restrict ourselves to local operations, or, to *local operations and classical communication* that we introduce now.

Local Operations and Classical Communication

The most general local operation that acts non-trivially only on the first subsystem reads

$$\varrho \to \sum_i (a_i \otimes \mathbb{1})\varrho(a_i^\dagger \otimes \mathbb{1}) , \qquad \sum_i a_i^\dagger a_i = \mathbb{1} , \tag{2.39}$$

and analogously for operations on the second subsystem alone. Such operations do not induce any correlations: They map product states on product states,

$$\varrho = \rho^{(1)} \otimes \rho^{(2)} \mapsto \left(\sum_i a_i \rho^{(1)} a_i^\dagger \right) \otimes \rho^{(2)} , \tag{2.40}$$

and separable states on separable states

$$\varrho = \sum_i p_i \rho_i^{(1)} \otimes \rho_i^{(2)} \mapsto \sum_i p_i \left(\sum_j a_j \rho_i^{(1)} a_j^\dagger \right) \otimes \rho_i^{(2)} . \tag{2.41}$$

The situation changes if one allows for a correlated application of such local operations, where the operation that is applied at a certain instance depends on the outcomes of previous operations:

$$\varrho \mapsto \sum_i (a_i \otimes \mathbb{1})\varrho(a_i^\dagger \otimes \mathbb{1}) \tag{2.42a}$$

$$\mapsto \sum_{ij} (\mathbb{1} \otimes b_{ij})(a_i \otimes \mathbb{1})\varrho(a_i^\dagger \otimes \mathbb{1})(\mathbb{1} \otimes b_{ij}^\dagger) \tag{2.42b}$$

$$\mapsto \sum_{ijp} (c_{ijp} \otimes \mathbb{1})(\mathbb{1} \otimes b_{ij})(a_i \otimes \mathbb{1})\varrho(a_i^\dagger \otimes \mathbb{1})(\mathbb{1} \otimes b_{ij}^\dagger)(c_{ijp}^\dagger \otimes \mathbb{1}) \tag{2.42c}$$

$$\mapsto \sum_{ijp...q} (\mathbb{1} \otimes g_{ijp...q}) \dots (a_i \otimes \mathbb{1})\varrho(a_i^\dagger \otimes \mathbb{1}) \dots (\mathbb{1} \otimes g_{ijp...q}^\dagger) . \tag{2.42d}$$

In the first step, a local operation has been applied to the first subsystem. This can be understood as an interaction with an ancillary system and a subsequent measurement thereon, as discussed before (2.33). Conditioned on the measurement result that is associated with the collapse on the states $(a_i \otimes \mathbb{1})\varrho(a_i^\dagger \otimes \mathbb{1})$, the local operation associated with the operators b_{ij} is applied to the second subsystem in a consecutive step. And, conditioned on the outcome of this operation, another local operation is applied to the first subsystem, and so on.

Such operations are called *local operations and classical communication* (LOCC). The idea behind that terminology is that one could imagine two parties that have access to the individual subsystems, and those parties could apply their individual operations to their part of the composite system. But in order to arrive at the above operation, they would need to communicate with each other, i.e., tell the other party their measurement results. This communication, however, can be performed via a classical channel, does not require any quantum nature, and, therefore is referred to as 'classical'.

LOCC operations can take product states to states no more necessarily of product form. Thus, it is possible to create correlations with LOCC operations. Yet, since these correlations are based on the classical exchange of information, they remain correlations of classical nature. Therefore, we can refine our concept of entangled states by requiring [14, 15] that

> an *entanglement monotone* is a quantity that does not increase under *local operations and classical communication*.

Note that this requirement is perfectly compatible with the previous definition of separable and entangled states, since an entangled state cannot be created from a separable one by LOCC alone, but LOCC suffice to transform arbitrary separable states into each other.

Invariance of Entanglement Under Local Unitaries

Monotonicity under LOCC as the defining property of an entanglement monotone is in general difficult to verify. We can, however, formulate a simpler, necessary criterion thereof: among all LOCC operations, the local unitary transformations $\varrho \rightarrow \mathcal{U}_1 \otimes \mathcal{U}_2 \varrho \mathcal{U}_1^\dagger \otimes \mathcal{U}_2^\dagger$ are special since they have an inverse that is again LOCC. If one applies some arbitrary local unitary in a first step, and its inverse in a second step, then a monotone $\mathcal{M}$ cannot increase after either step

$$\mathcal{M}(\varrho) \geq \mathcal{M}(\mathcal{U}_1 \otimes \mathcal{U}_2 \, \varrho \, \mathcal{U}_1^\dagger \otimes \mathcal{U}_2^\dagger) \geq \mathcal{M}(\varrho) . \qquad (2.43)$$

However, because initial and final states are equal, so is their entanglement, and one necessarily concludes that any entanglement monotone is invariant under local unitaries

$$\mathcal{M}(\varrho) = \mathcal{M}(\mathcal{U}_1 \otimes \mathcal{U}_2 \, \varrho \, \mathcal{U}_1^\dagger \otimes \mathcal{U}_2^\dagger) . \qquad (2.44)$$

This invariance is significantly easier to check than monotonicity under LOCC. However, as mentioned above, it provides only a necessary, but not a sufficient condition.

Schmidt Coefficients and Majorization

Invariance under local unitary transformations is not only a simple test to rule out potential candidates for entanglement monotones as non-monotonous under LOCC, but indeed it has much deeper implications. It implies that any entanglement monotone can be expressed as a function only of invariants under local unitaries. Consequently, if one can identify these invariants, one proceeds a big step forward, toward the systematic construction of entanglement monotones. Although the exhaustive search for such invariants turns out to be a very intricate task for a general state, it has a surprisingly simple answer in the case of pure states of bipartite systems. There, the Schmidt coefficients introduced earlier in Sect. 2.3.2 provide a complete set of invariants, and all entanglement properties can be expressed in terms of only those quantities.

Majorization

One very useful application of the characterization of entanglement in terms of Schmidt coefficients is a simple test that allows to check whether one state $|\Phi\rangle$ can be prepared by LOCC starting from another state $|\Psi\rangle$. This is possible [16, 17] if and only if their Schmidt coefficients, ordered decreasingly (i.e., $\lambda_1 \geq \lambda_2 \geq \ldots$), satisfy the set of inequalities

$$\lambda_1^{(\Phi)} \geq \lambda_1^{(\Psi)}$$

$$\sum_{i=1}^{2} \lambda_i^{(\Phi)} \geq \sum_{i=1}^{2} \lambda_i^{(\Psi)}$$

$$\sum_{i=1}^{3} \lambda_i^{(\Phi)} \geq \sum_{i=1}^{3} \lambda_i^{(\Psi)} \qquad (2.45)$$

$$\vdots \quad \vdots$$

This set of conditions is often expressed in short-hand notation $\boldsymbol{\lambda}^{(\Phi)} \succ \boldsymbol{\lambda}^{(\Psi)}$ in terms of the *Schmidt vectors* $\boldsymbol{\lambda}^{(\Phi)} = [\lambda_1^{(\Phi)}, \lambda_2^{(\Phi)}, \ldots]$ and similarly for $\boldsymbol{\lambda}^{(\Psi)}$, and reads '$\boldsymbol{\lambda}^{(\Phi)}$ majorizes $\boldsymbol{\lambda}^{(\Psi)}$', or, also '$\boldsymbol{\lambda}^{(\Psi)}$ is majorized by $\boldsymbol{\lambda}^{(\Phi)}$'.

An Example

In order to get a bit better idea of how such an LOCC transformation works, let us look at the exemplary case to start out with the state $|\Psi\rangle = (|00\rangle + |11\rangle)/\sqrt{2}$, and aim at the preparation of the state $|\Phi\rangle = \sqrt{\lambda_1}|00\rangle + \sqrt{\lambda_2}|11\rangle$ using only LOCC operations. This is possible since $\boldsymbol{\lambda}^{(\Phi)}$ actually majorizes $\boldsymbol{\lambda}^{(\Psi)}$. However, this majorization criterion does not give a prescription on how such a transformation can be achieved. Therefore, we will content ourselves with verifying that the LOCC operation that is comprised of the operators

$$a_1 = \sqrt{\lambda_1}|0\rangle\langle 0| + \sqrt{\lambda_2}|1\rangle\langle 1| \,, \qquad b_{11} = |0\rangle\langle 0| + |1\rangle\langle 1| \,,$$
$$a_2 = \sqrt{\lambda_1}|0\rangle\langle 1| + \sqrt{\lambda_2}|1\rangle\langle 0| \,, \qquad b_{21} = |0\rangle\langle 1| + |1\rangle\langle 0| \,, \tag{2.46}$$

indeed transforms $|\Psi\rangle$ to $|\Phi\rangle$. First, however, one should verify that the resolutions to identity $\sum_i a_i^\dagger a_i = \mathbb{1}$ and $b_{11}^\dagger b_{11} = b_{21}^\dagger b_{21} = \mathbb{1}$ are given. Then, consider the action of these operators onto the state $|\Psi\rangle$. First the a_i:

$$a_1|\Psi\rangle = \sqrt{\frac{\lambda_1}{2}}|00\rangle + \sqrt{\frac{\lambda_2}{2}}|11\rangle = \frac{1}{\sqrt{2}}|\Phi\rangle, \tag{2.47}$$

$$a_2|\Psi\rangle = \sqrt{\frac{\lambda_1}{2}}|01\rangle + \sqrt{\frac{\lambda_2}{2}}|10\rangle \,. \tag{2.48}$$

The first term is already proportional to $|\Phi\rangle$, so that in the next step the identity operation b_{11} is applied. But, the second term does not have the correct form yet. Here, one needs to transform $|0\rangle$ of the second subsystem into $|1\rangle$ and vice versa, what is exactly what b_{21} does. Thus, one obtains $b_{11}a_1|\Psi\rangle = b_{21}a_2|\Psi\rangle = 1/\sqrt{2}|\Phi\rangle$. So all together, the final state reads

$$\sum_{ij} a_i \otimes b_{ij} |\Psi\rangle\langle\Psi| a_i^\dagger \otimes b_{ij}^\dagger = |\Phi\rangle\langle\Phi| \,. \tag{2.49}$$

And, this is exactly what we were aiming at.

Inequivalent Entanglement Properties

So far, we found a criterion that excludes some quantities from the list of potential quantifiers of entanglement, but does not yet define one *unique* entanglement measure. Whether such a *unique* measure exists is still a subject of debate, and beyond the scope of the present introduction. Let us however briefly illustrate why the characterization of entanglement by a simple scalar quantity might reveal problematic.

The entanglement of a pure state of two qubits is characterized by a single independent Schmidt coefficient due to the normalization of the reduced density matrix. Therefore, the set of majorization conditions (2.45) reduces to its first line. For two arbitrary pure states $|\Psi_1\rangle$ and $|\Psi_2\rangle$ either both Schmidt vectors coincide, i.e., $\boldsymbol{\lambda}_1^{(\Psi_1)} = \boldsymbol{\lambda}_1^{(\Psi_2)}$, or one majorizes the other. That is, there is an unambiguous order of pure states with respect to their degree of entanglement, and any entanglement monotone will respect this order. The situation is different in higher dimensional systems as one can see in the exemplary case of the following two states

$$|\Psi_1\rangle = \frac{1}{\sqrt{2}}|00\rangle + \frac{1}{\sqrt{2}}|11\rangle \,,$$

$$|\Psi_2\rangle = \sqrt{\frac{3}{5}}|00\rangle + \sqrt{\frac{1}{5}}|11\rangle + \sqrt{\frac{1}{5}}|22\rangle \,. \tag{2.50}$$

The Schmidt vectors are $\boldsymbol{\lambda}^{(\Psi_1)} = [1/2, 1/2, 0]$ and $\boldsymbol{\lambda}^{(\Psi_2)} = [3/5, 1/5, 1/5]$, respectively, and neither does $\boldsymbol{\lambda}^{(\Psi_1)}$ majorize $\boldsymbol{\lambda}^{(\Psi_2)}$, nor vice versa

$$\lambda_1^{(\Psi_1)} = \tfrac{1}{2} < \tfrac{3}{5} = \lambda_1^{(\Psi_2)}$$
$$\sum_{i=1}^{2} \lambda_i^{(\Psi_1)} = 1 > \tfrac{4}{5} = \sum_{i=1}^{2} \lambda_i^{(\Psi)}$$
$$\sum_{i=1}^{3} \lambda_i^{(\Psi_1)} = 1 = 1 = \sum_{i=1}^{3} \lambda_i^{(\Psi)} \ .$$

Thus neither can $|\Psi_1\rangle$ be prepared by LOCC from $|\Psi_2\rangle$, nor is there an LOCC operation that takes $|\Psi_2\rangle$ to $|\Psi_1\rangle$. This implies that the two states have non-equivalent entanglement properties, and it is not obvious that either one can be considered more entangled than the other. In particular, the use of different entanglement monotones may lead to contradictory conclusions on the relative entanglement content of both states.

Entanglement Measures

So far, we required only monotonicity under LOCC for a potential entanglement quantifier. There are additional axioms that qualify a monotone as an *entanglement measure*. While there is no general agreement on the complete list of axioms, we list some important ones:

- Mixing two states ϱ and σ probabilistically can increase only classical correlations. Therefore, one expects that a probabilistic mixture $p\varrho + (1 - p)\sigma$, $(0 \leq p \leq 1)$, should be no more entangled than the two individual states on average. This implies *convexity* of an entanglement measure, i.e., $\mathcal{M}(p\varrho + (1 - p)\sigma) \leq p\mathcal{M}(\varrho) + (1 - p)\mathcal{M}(\sigma)$.
- Assume one is given n copies of a state ϱ on $\mathcal{H}_1 \otimes \mathcal{H}_2$. This is equivalent to a single n-fold state $\varrho^{\otimes n} = \varrho \otimes \ldots \otimes \varrho$, and one wants to quantify the entanglement between the subsystems associated with the larger Hilbert spaces $\mathcal{H}_1^{\otimes n}$ and $\mathcal{H}_2^{\otimes n}$. An entanglement monotone that fulfills $\mathcal{M}(\varrho^{\otimes n}) = n\mathcal{M}(\varrho)$ is called *additive*.
- Similarly, one can consider two different states ϱ and σ on $\mathcal{H}_1 \otimes \mathcal{H}_2$ and evaluate the entanglement of the joint state $\varrho \otimes \sigma$ on $\mathcal{H}_1^{\otimes 2} \otimes \mathcal{H}_2^{\otimes 2}$. A monotone $\mathcal{M}$ that satisfies the inequality $\mathcal{M}(\varrho \otimes \sigma) \leq \mathcal{M}(\varrho) + \mathcal{M}(\sigma)$ is called *subadditive*.

2.4.2 Some Specific Monotones and Measures

In the above, we discussed very general properties of entanglement quantifiers. Now we will discuss some more specific entanglement monotones and measures that are frequently used in the literature.

Pure States

We saw earlier that any entanglement monotone or measure can be expressed in terms of invariants under local unitary transformations, and that, in the case of bipartite pure states, the Schmidt coefficients provide a complete set thereof. Therefore, we can restrict our discussion to functions $\mathcal{F}(\boldsymbol{\lambda})$ of the Schmidt coefficients only. But not every such function is also an entanglement monotone, i.e., non-increasing under LOCC. The following criterion allows to verify this property: A function $\mathcal{F}(\boldsymbol{\lambda})$ is monotonously decreasing under LOCC if $\mathcal{F}$ is invariant under any permutation of the Schmidt coefficients λ_i, and if $\mathcal{F}$ is Schur concave, i.e., [18]

$$(\lambda_1 - \lambda_2)\left(\frac{\partial \mathcal{F}}{\partial \lambda_1} - \frac{\partial \mathcal{F}}{\partial \lambda_2}\right) \leq 0 \qquad (2.51)$$

It suffices to express the condition for Schur concavity in terms of only the first two Schmidt coefficients because of the required permutation invariance. We now evaluate this criterion for a few specific monotones and measures.

Entanglement Entropy

The *entanglement entropy*, which is the von Neumann entropy of the reduced density matrix,

$$E(\Psi) = S(\varrho_r) = -\mathrm{tr}\varrho_r \ln \varrho_r = -\sum_i \lambda_i \ln \lambda_i \ , \qquad (2.52)$$

is indeed invariant under permutation of the λ_i, satisfies

$$(\lambda_1 - \lambda_2)\left(\frac{\partial S(\rho_r)}{\partial \lambda_1} - \frac{\partial S(\rho_r)}{\partial \lambda_2}\right) = (\lambda_1 - \lambda_2)\ln\frac{\lambda_2}{\lambda_1} \leq 0 \ , \qquad (2.53)$$

and thus is a valid entanglement monotone.

Concurrence

Another frequently used monotone is *concurrence c*. For bipartite systems, c is often defined in terms of the local Pauli matrices

$$\sigma_y = \begin{bmatrix} 0 & -i \\ i & 0 \end{bmatrix} \qquad (2.54)$$

represented in a given orthonormal basis $\{|0\rangle, |1\rangle\}$ of the factor spaces $\mathcal{H}_1$, and $\mathcal{H}_2$ of $\mathcal{H}$ [19],

$$c(\Psi) = |\langle \Psi^* | \sigma_y \otimes \sigma_y | \Psi \rangle| \ . \qquad (2.55)$$

$\langle \Psi^* |$ denotes the complex conjugate of $\langle \Psi |$, with the conjugation performed in the same basis. That is, if $\langle \Psi |$ reads $\langle \Psi | = \sum_{ij} \beta_{ij} \langle ij |$, then $\langle \Psi^* |$ reads $\langle \Psi^* | = \sum_{ij} \beta_{ij}^* \langle ij |$. Equivalently, $\langle \Psi^* |$ is the transpose of $|\Psi\rangle$, whereas $\langle \Psi |$ is the adjoint of $|\Psi\rangle$.

A possible generalization of the above definition for higher dimensional systems (see e.g., [20]) reads [21]

$$c(\Psi) = \sqrt{2(1 - \mathrm{tr}\varrho_r^2)} \,, \tag{2.56}$$

and is equivalent to (2.55), for two-level systems. In terms of the Schmidt coefficients, concurrence reads

$$c(\Psi) = \sqrt{2 \sum_{i \neq j} \lambda_i \lambda_j} \,. \tag{2.57}$$

This is invariant under permutations of the λ_i, and since

$$(\lambda_1 - \lambda_2) \left(\frac{\partial c}{\partial \lambda_1} - \frac{\partial c}{\partial \lambda_2} \right) = \frac{(\lambda_1 - \lambda_2)}{2c} \left(\sum_{i \neq 1} \lambda_i - \sum_{i \neq 2} \lambda_i \right) = -\frac{(\lambda_1 - \lambda_2)^2}{2c} \leq 0 \,, \tag{2.58}$$

concurrence is a valid monotone.

Mixed States

For pure states, we were able to give constructive definitions for some entanglement measures. In the case of mixed states, however, it turns out to be much more involved to find a quantity that is monotonously decreasing under LOCC. The basic difference between mixed and pure states in this specific context is that pure states bear no classical correlations. These need to be distinguished from genuine quantum correlations by a mixed state entanglement monotone.

Negativity

So far, only very few constructively defined quantities were proved to be non-increasing under LOCC. The most prominent example is *negativity* [22]. Earlier, in Sect. 2.3.2, we saw that the partial transpose ϱ^{pt} of a mixed state ϱ can be very helpful to decide on the separability of ϱ: if one of the eigenvalues λ_i of ϱ^{pt} is negative, then ϱ is entangled. This inspired the definition of negativity as

$$\mathcal{N}(\varrho) = \frac{(\sum_i |\lambda_i|) - 1}{2} \,, \tag{2.59}$$

what was proved to be monotonously decreasing under LOCC [22]. If ϱ^{pt} is positive semi-definite, $\mathcal{N}$ vanishes, but takes positive values if ϱ^{pt} has one, or more negative eigenvalues. In comparison to virtually all other mixed state entanglement monotoness, $\mathcal{N}$ can be evaluated easily, since it is an algebraic function of the spectrum of ϱ^{pt}. This advantage, however, comes at the price that negativity assigns non-vanishing entanglement only to those states that are detected via their negative partial transpose. Therefore, much as for the ppt-criterion itself, negativity is fully reliable only for 2×2 or 2×3 system.

Convex Roofs

The failure to detect all entangled states finds its remedy with the so-called convex roof measures. However, the solution to this issue comes at the expense of an additional optimization problem that prevents the explicit algebraic evaluation in most cases. Since any mixed state can be decomposed into a probabilistic mixture of pure states

$$\varrho = \sum_i p_i |\Psi_i\rangle\langle\Psi_i| \,, \tag{2.60}$$

with positive prefactors p_i, one can characterize the entanglement properties of ϱ in terms of those of its pure state components. A very suggestive generalization of a pure state monotone for mixed states is the *average* value $\sum_i p_i \mathcal{M}(\Psi_i)$ of the monotone $\mathcal{M}$. However, a mixed state does not have a unique pure state decomposition, and different decompositions typically yield different average values. A valid mixed state generalization that is monotonously decreasing under LOCC is the infimum over all pure state decompositions, i.e., the minimal average value

$$\mathcal{M}(\varrho) = \inf_{\{p_i, |\Psi_i\rangle\}} \sum_i p_i \mathcal{M}(\Psi_i) \,, \tag{2.61}$$

what is called the *convex roof*. To solve the optimization problem implicit in the convex roof definition (2.61), one needs a systematic way to explore all pure state decompositions of ϱ. Given the eigenstates $|\Phi_j\rangle$ of ϱ, together with the associated eigenvalues μ_i, any linear combination of the eigenstates

$$\sqrt{p_i}|\Psi_i\rangle = \sum_j V_{ij}\sqrt{\mu_j}|\Phi_j\rangle \,, \tag{2.62}$$

defines another valid decomposition [23], provided $\sum_k V_{ik}^\dagger V_{kj} = \delta_{jk}$, i.e., for a left-unitary coefficient matrix V (with adjoint $V^\dagger$):

$$\begin{aligned}
\sum_i p_i |\Psi_i\rangle\langle\Psi_i| &= \sum_{ijk} V_{ij}\sqrt{\mu_j}|\Phi_j\rangle\langle\Phi_k|\sqrt{\mu_k}V_{ik}^* \\
&= \sum_{ijk} V_{ki}^\dagger V_{ij}\sqrt{\mu_j\mu_k}|\Phi_j\rangle\langle\Phi_k| \\
&= \sum_{jk} \delta_{jk}\sqrt{\mu_j\mu_k}|\Phi_j\rangle\langle\Phi_k| \\
&= \sum_j \mu_j|\Phi_j\rangle\langle\Phi_j| = \varrho \,;
\end{aligned} \tag{2.63}$$

and *any* pure state decomposition of ϱ can be obtained in this fashion [23].

Concurrence of Mixed States

With this characterization of pure state decompositions at hand, we can now focus on the evaluation of concurrence for mixed states. So far, concurrence is virtually the only quantity for which the convex roof can be evaluated algebraically. Later in Sect. 2.4.2, we will see that also the convex roof of the entanglement entropy has an algebraic solution. This solution, however follows from the known solution for concurrence.

A crucial property of concurrence in contrast to other monotones is the homogeneity,

$$c(\eta|\Psi\rangle\langle\Psi|) = \eta\,c(|\Psi\rangle)\ ,\ \text{for}\ \eta \geq 0, \tag{2.64}$$

which allows to rewrite the convex roof expression above as

$$c(\varrho) = \inf_{\{|\psi_i\rangle\}} \sum_i c(\psi_i)\ , \tag{2.65}$$

where everything is expressed in terms of *subnormalized* states $|\psi_i\rangle = \sqrt{p_i}|\Psi_i\rangle$, and the probabilities p_i do not enter explicitly any more.

This allows to reformulate (2.61) in the following closed form

$$
\begin{aligned}
c(\varrho) &= \inf_{\{|\psi_i\rangle\}} \sum_i c(\psi_i) \\
&= \inf_{\{|\psi_i\rangle\}} \sum_i |\langle\psi_i^*|\sigma_y \otimes \sigma_y|\psi_i\rangle| \\
&= \inf_V \sum_i \left|\sum_{jk} V_{ij}\langle\phi_j^*|\sigma_y \otimes \sigma_y|\phi_k\rangle V_{ki}^T\right| \\
&= \inf_V \sum_i \left|[V\tau V^T]_{ii}\right|\ ,
\end{aligned}
\tag{2.66}
$$

where we used (2.55) and (2.62). In the last line, we introduced a short-hand notation, where τ is a complex symmetric matrix, $\tau = \tau^T$, with elements

$$\tau_{ij} = \langle\phi_i^*|\sigma_y \otimes \sigma_y|\phi_j\rangle\ . \tag{2.67}$$

Equation (2.66) resembles the diagonalization of a hermitean matrix H through a unitary transformation $\mathcal{U}H\mathcal{U}^\dagger$, where $\mathcal{U}$ is unitary. The difference resides, however, in the fact that τ is symmetric and not hermitean, and that the transpose of a unitary, respectively left unitarys, enters instead of its adjoint. But also a symmetric matrix can be diagonalized in a similar fashion. Already earlier, in (2.13) we have been invoking the singular value decomposition of a matrix. It stated that any matrix A could be diagonalized with two unitary transformations u_1 and u_2 as $u_1 A u_2$. This, of course, also holds for the particular case of a symmetric matrix that we are facing here. However, in this specific case, u_2 is equal to u_1^T. Therefore, we can rephrase the infimum to be evaluated as

$$c(\varrho) = \inf_{V} \sum_{i} \left|[VU^{\dagger}U\tau U^T U^* V^T]_{ii}\right| = \inf_{\tilde{V}} \sum_{i} \left|[\tilde{V}\tau_d \tilde{V}^T]_{ii}\right|, \tag{2.68}$$

where $\tilde{V} = VU^{\dagger}$, and $\tau_d = U\tau U^T = \mathrm{diag}[\mathcal{S}_1, \mathcal{S}_2, \mathcal{S}_3, \mathcal{S}_4]$ is the diagonal form of τ. The order of the diagonal elements is not determined and can be chosen arbitrarily. But in the following, we will use the convention that $\mathcal{S}_1$ is the largest of all diagonal entries.

With the diagonal form τ_d of τ, we have simplified the problem a lot: instead of 20 real parameter that characterize a general complex symmetric matrix, we are left with only four real parameters. But it is still not straightforward to derive an optimal matrix $\tilde{V}$ that achieves the infimum. Instead of a systematic derivation, we are going to take an Ansatz that eventually will turn out to do the job. Let us take $\tilde{V}$ equal to $\mathcal{V}$ with

$$\mathcal{V} = \frac{1}{2} \begin{bmatrix} 1 & e^{i\varphi_2} & e^{i\varphi_3} & e^{i\varphi_4} \\ 1 & e^{i\varphi_2} & -e^{i\varphi_3} & -e^{i\varphi_4} \\ 1 & -e^{i\varphi_2} & e^{i\varphi_3} & -e^{i\varphi_4} \\ 1 & -e^{i\varphi_2} & -e^{i\varphi_3} & e^{i\varphi_4} \end{bmatrix}, \tag{2.69}$$

where we still have the free phases φ_2, φ_3, and φ_4 that we can adjust. With this choice, we obtain

$$\sum_{i} \left|\mathcal{V}\tau_d \mathcal{V}^T\right| = \left|\mathcal{S}_1 + \sum_{i>1} e^{2i\varphi_i} \mathcal{S}_i\right|. \tag{2.70}$$

Now, we can minimize this expression by proper choices of the free phases, what is most conveniently done by distinguishing two cases. In the former case, where $\mathcal{S}_1 \geq \sum_{i>1} \mathcal{S}_i$, it is optimal to take $\varphi_2 = \varphi_3 = \varphi_4 = \pi/2$, what leads to $\sum_{i} |\mathcal{V}\tau_d \mathcal{V}^T| = \mathcal{S}_1 - \sum_{i>1} \mathcal{S}_i$. In the latter case, $\mathcal{S}_1 < \sum_{i>1} \mathcal{S}_i$, one can always find a choice of phases such that $\sum_{i} |\mathcal{V}\tau_d \mathcal{V}^T| = 0$, as depicted in Fig. 2.2. That is, we found a pure state decomposition in which the average

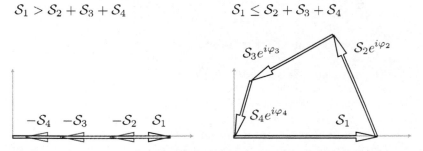

Fig. 2.2. Schematic drawing of the singular values $\mathcal{S}_i$ added up with adjustable phases $e^{2i\varphi_i}$ in the complex plane. If $\mathcal{S}_1 > \sum_{i>1} \mathcal{S}_i$, as depicted on the left, the optimal choice to minimize $|\mathcal{S}_1 + \sum_{i>1} \mathcal{S}_i e^{2i\varphi_i}|$ of the phases is $\varphi_i = \pi/2$. If, on the other hand, $\mathcal{S}_1 < \sum_{i>1} \mathcal{S}_i$ as depicted on the right, then one can always find phases φ_i such that $|\mathcal{S}_1 + \sum_{i>1} \mathcal{S}_i e^{2i\varphi_i}|$ vanishes

concurrence reads $\max(\mathcal{S}_1 - \sum_{i>1}\mathcal{S}_i, 0)$. However, we still do not know, if this is optimal, or if there are decompositions that yield a smaller value.

For answering this question, we can restrict ourselves to the case $\mathcal{S}_1 \geq \sum_{i>1}\mathcal{S}_i$. In the other case, we found a vanishing value for concurrence, which obviously is the infimum, since concurrence cannot be negative. Now, let us start out not with $\tilde{\tau} = \mathcal{V}\tau_d\mathcal{V}^T$ with the choice $\varphi_2 = \varphi_3 = \varphi_4 = \pi/2$ that we found optimal above. We now show that there is no left-unitary W that could yield a smaller value than what we have found so far.

$$
\begin{aligned}
\sum_i \left|[W\tau_d W^T]_{ii}\right| &= \sum_i \left|\sum_j W_{ij}^2 \mathcal{S}_j\right| \\
&= \sum_i \left|W_{i1}^2 \mathcal{S}_1 + \sum_{j>1} W_{ij}^2 \mathcal{S}_j\right| \\
&\geq \sum_i \left(\left|W_{i1}^2\right|\mathcal{S}_1 - \left|\sum_{j>1} W_{ij}^2 \mathcal{S}_j\right|\right) \\
&= \mathcal{S}_1 - \sum_i \left|\sum_{j>1} W_{ij}^2 \mathcal{S}_j\right| \\
&\geq \mathcal{S}_1 - \sum_i \sum_{j>1} \left|W_{ij}\right|^2 \mathcal{S}_j \\
&= \mathcal{S}_1 - \sum_{j>1} \mathcal{S}_j ,
\end{aligned}
\tag{2.71}
$$

where going from the second to the third line we used $|a + b| \geq |a| - |b|$ with $a = W_{i1}^2\mathcal{S}_1$, and $b = \sum_{j>1} W_{ij}^2\mathcal{S}_j$, and in the fourth line, we used the left-unitarity condition of W, i.e., $\sum_i |W_{ij}|^2 = 1$. We obtained the fifth line using $-|\sum a_j| \geq -\sum |a_j|$, with $a_j = W_{ij}^2\mathcal{S}_j$, and the last line followed again from the left-unitarirty of W. Thus, we found the algebraic solution

$$
c(\varrho) = \max(\mathcal{S}_1 - \sum_{i>1} \mathcal{S}_i, 0)
\tag{2.72}
$$

for the concurrence of an arbitrary mixed state of a bipartite two-level system.

Entanglement of Formation of Mixed States

With this solution for concurrence, we can now proceed and consider entanglement of formation, which is the convex roof extension of the entanglement entropy. Here we will make use of the fact that for *pure* states in bipartite two-level systems, there is only one independent Schmidt coefficient, since they sum up to unity. Therefore, one can determine both Schmidt coefficients in terms of the concurrence:

$$
\lambda_\pm = \frac{1 \pm \sqrt{1 - c^2}}{2} .
\tag{2.73}
$$

And, since the entanglement entropy is a function of λ, it can also be expressed in terms of concurrence via

$$E(\Psi) = -\frac{1+\sqrt{1-c^2}}{2}\ln\frac{1+\sqrt{1-c^2}}{2} - \frac{1-\sqrt{1-c^2}}{2}\ln\frac{1-\sqrt{1-c^2}}{2} \quad (2.74)$$
$$\equiv \mathcal{E}(c) ,$$

where we introduced the function $\mathcal{E}(c)$. One easily convinces oneself that $\mathcal{E}(c)$ is monotonously increasing $\partial\mathcal{E}(c)/\partial c \geq 0$, and convex $\partial^2\mathcal{E}(c)/\partial c^2 \geq 0$, for $c \geq 0$. Convexity can equivalently be expressed as $\sum_i p_i\mathcal{E}(q_i) \geq \mathcal{E}(\sum_i p_iq_i)$. With the help of these properties, we arrive at the following reasoning:

$$E(\varrho) = \inf\sum_i p_iE(\Psi_i)$$
$$= \inf\sum_i p_i\mathcal{E}(c(\Psi_i))$$
$$\geq \inf\mathcal{E}(\sum_i p_ic(\Psi_i)) \quad (2.75)$$
$$= \mathcal{E}(\inf\sum_i p_ic(\Psi_i))$$
$$= \mathcal{E}(c(\varrho)) .$$

Here, in going from the second to the third line, we used the convexity of $\mathcal{E}$, and from the third to the fourth its monotonicity. Thus, we found that entanglement of formation is bounded from below by $\mathcal{E}(c(\varrho))$. But, we are close to seeing that this is indeed not only a bound but rather the exact result. The crucial feature here is the fact that there is not a single optimal decomposition of a mixed state ϱ into pure states that yields the actual value of concurrence, but there is actually a continuum of optimal decompositions. And, in particular, there is one, $\varrho = \sum_i \tilde{p}_i|\tilde{\Psi}_i\rangle\langle\tilde{\Psi}_i|$ in which all pure states do have the same value of concurrence, i.e., $c(\tilde{\Psi}_i) = c(\varrho)$ [19]. With the help of this particular decomposition, we can now show that $\mathcal{E}(c(\varrho))$ is not only a lower bound on entanglement of formation, but actually its exact value: due to its definition as convex roof, $E(\varrho)$ is bounded from above by its average value evaluated in any decomposition – in particular $\{\tilde{p}_i, |\tilde{\Psi}_i\rangle\}$, i.e., $E(\varrho) \leq \sum_i \tilde{p}_i\mathcal{E}(c(\tilde{\Psi}_i))$. Now, we can replace $c(\tilde{\Psi}_i)$ by $c(\varrho)$, so that we end up with $E(\varrho) \leq \sum_i \tilde{p}_i\mathcal{E}(c(\varrho))$. And, finally, since the probabilities add up to 1, we arrive at the conclusion that entanglement of formation is bounded from above by $\mathcal{E}(c(\varrho))$. Since we found above in (2.75) that it is also bounded from below by the same quantity, we necessarily need to conclude that these two quantities coincide:

$$E(\varrho) = \mathcal{E}(c(\varrho)) . \quad (2.76)$$

Therefore, once one has evaluated concurrence for a mixed state – what can be done algebraically (2.72) – one can easily also obtain entanglement of formation.

References

1. A. Einstein, B. Podolsky, and N. Rosen: *Can quantum-mechanical description of physical reality be considered complete?*. Phys. Rev. **47**, 777 (1935)
2. R. F. Werner: *Quantum states with Einstein-Podolsky-Rosen correlations admitting a hidden-variable model*. Phys. Rev. A **40**, 4277 (1989)
3. E. Schmidt: *Zur Theorie der linearen und nichtlinearen Integralgleichungen*. Math. Ann. **63**, 433 (1907)
4. P. A. Horn and C. R. Johnson: *Matrix Analysis* (Cambridge University Press, New York 1985)
5. M. Horodecki, P. Horodecki, and R. Horodecki: *Separability of mixed states: necessary and sufficient conditions*. Phys. Lett. A **223**, 1 (1996)
6. B. M. Terhal: *Bell inequalities and the separability criterion*. Phys. Lett. A **271**, 319 (2001)
7. M. Reed and B. Simon: *Analysis of Operators*. (Elsevier, Amsterdam 1978)
8. A. Peres: *Separability criterion for density matrices*. Phys. Rev. Lett. **77**, 1413 (1996)
9. E. Størmer: *Positive linear maps of operator algebras*. Acta. Math. **110**, 233 (1963)
10. S. L. Woronowicz: *Positive maps of low dimensional matrix algebras*. Rep. Math. Phys. **10**, 165 (1976)
11. K. Kraus: *States, Effects and Operations: Fundamental Notions of Quantum Theory* (Springer, New York 1983)
12. W. F. Stinespring: *Positive functions on c*-algebras*. Proc. Am. Math. Soc. **6**, 211 (1955)
13. Man-Duen Choi: *Completely positive linear maps on complex matrices*. Linear Algebra Appl. **10**, 285 (1975)
14. V. Vedral, M. B. Plenio, M. A. Rippin, and P. L. Knight: *Quantifying entanglement*. Phys. Rev. Lett. **78**, 2275 (1997)
15. G. Vidal. *Entanglement monotones*. J. Mod. Opt. **47**, 355 (2000)
16. M. A. Nielsen: *Conditions for a class of entanglement transformations*. Phys. Rev. Lett. **83**, 436 (1999)
17. M. A. Nielsen and G. Vidal: *Majorization and the interconversion of bipartite states*. Quant. Inf. Comp. **1**, 76 (2001)
18. T. Ando: *Majorization, doubly stochastic matrices, and comparison of eigenvalues*. Lin. Alg. Appl. **118**, 163 (1989)
19. W. K. Wootters: *Entanglement of formation of an arbitrary state of two qubits*. Phys. Rev. Lett. **80**, 2245 (1998)
20. A. Uhlmann: *Fidelity and concurrence of conjugated states*. Phys. Rev. A **62**, 032307 (2000)
21. P. Rungta, V. Bužek, C. M. Caves, M. Hillery, and G. J. Milburn: *Universal state inversion and concurrence in arbitrary dimensions*. Phys. Rev. A **64**, 042315 (2001)
22. G. Vidal and R. F. Werner: *Computable measure of entanglement*. Phys. Rev. A **65**, 032314 (2002)
23. E. Schrödinger. *Probability relations between separated systems*. Proc. Cambridge Philos. Soc. **32**, 446 (1936)

3 Topology and Quantum Computing

L.H. Kauffman[1] and S.J. Lomonaco Jr.[2]

[1] Department of Mathematics, Statistics and Computer Science (m/c 249),
   851 South Morgan Street, University of Illinois, Chicago, ILL 60607-7045, USA
[2] Department of Computer Science and Electrical Engineering,
   University of Maryland Baltimore County,
   1000 Hilltop Circle, Baltimore, MD 21250, USA

3.1 Introduction

This chapter describes relationships between topology and quantum computing. It is fruitful to move back and forth between topological methods and the techniques of quantum information theory.[1]

We sketch the background topology, discuss analogies (such as topological entanglement and quantum entanglement), show direct correspondences between certain topological operators (solutions to the Yang–Baxter equation), and universal quantum gates. We then describe the background for topological quantum computing in terms of Temperley–Lieb recoupling theory. This is a recoupling theory that generalizes standard angular momentum recoupling theory, generalizes the Penrose theory of spin networks and is inherently topological. Temperley–Lieb recoupling theory is based on the bracket polynomial model [2, 3] for the Jones polynomial. It is built in terms of diagrammatic combinatorial topology. The same structure can be explained in terms of the $SU(2)_q$ quantum group and has relationships with functional integration and Witten's approach to topological quantum field theory. Nevertheless, the approach given here will be unrelentingly elementary. Elementary, does not necessarily mean simple. In this case, an architecture is built from simple beginnings, and this architecture and its recoupling language can be applied to many things including, e.g., colored Jones polynomials, Witten–Reshetikhin–Turaev invariants of three manifolds, topological quantum field theory, and quantum computing.

In quantum computing, the application of topology is most interesting because the simplest non-trivial example of the Temperley–Lieb recoupling theory gives the so-called Fibonacci model. The recoupling theory yields representations of the Artin braid group into unitary groups $U(n)$, where n is a Fibonacci number. These representations are *dense* in the unitary group, and it can be used to model quantum computation universally in terms of representations of the braid group. Hence the term topological quantum

---

[1] This paper is an expanded version of joint work of the authors [1], and it includes expository and background material.

Kauffman, L.H., Lomonaco, S.J.: *Topology and Quantum Computing*. Lect. Notes Phys. **768**, 87–156 (2009)
DOI 10.1007/978-3-540-88169-8_3   © Springer-Verlag Berlin Heidelberg 2009

computation. In this work, we outline the basics of the TL-Recoupling Theory, and show explicitly how the Fibonacci model arises from it.

While this chapter attempts to be self-contained, and hence has some expository material, most of the results are either new or new points of view on the known results. The material on $SU(2)$ representations of the Artin braid group is new, and the relationship of this material to the recoupling theory is new. The treatment of elementary cobordism categories is well known, but new in the context of quantum information theory. The reformulation of Temperley–Lieb recoupling theory for the purpose of producing unitary braid group representations is new for quantum information theory and is directly related to much of the recent work of Freedman and his collaborators. The treatment of the Fibonacci model in terms of two-strand recoupling theory is new and at the same time, the most elementary non-trivial example of the recoupling theory. The models in Sect. 3.11 for quantum computation of colored Jones polynomials and for quantum computation of the Witten–Reshetikhin–Turaev invariant are new in this form of the recoupling theory. They take a particularly simple aspect in this context.

Here is a very condensed presentation of how unitary representations of the braid group are constructed via topological quantum field theoretic methods. One has a mathematical particle with label P that can interact with itself to produce either itself labeled P or itself with the null label $*$. When $*$ interacts with P, the result is always P. When $*$ interacts with $*$, the result is always $*$. One considers process spaces where a row of particles labeled P can successively interact, subject to the restriction that the end result is P. For example, the space $V[(ab)c]$ denotes the space of interactions of three particles labeled P. The particles are placed in the positions a, b, c. Thus we begin with $(PP)P$. In a typical sequence of interactions, the first two Ps interact to produce a $*$, and the $*$ interacts with P to produce P.

$$(PP)P \longrightarrow (*)P \longrightarrow P.$$

In another possibility, the first two Ps interact to produce a P, and the P interacts with P to produce P.

$$(PP)P \longrightarrow (P)P \longrightarrow P.$$

It follows from this analysis that the space of linear combinations of processes $V[(ab)c]$ is two dimensional. The two processes we have just described can be taken to be the qubit basis for this space. One obtains a representation of the three-strand Artin braid group on $V[(ab)c]$ by assigning appropriate phase changes to each of the generating processes. One can think of these phases as corresponding to the interchange of the particles labeled a and b in the association $(ab)c$. The other operator for this representation corresponds to the interchange of b and c. This interchange is accomplished by a *unitary change of basis mapping*

$$F : V[(ab)c] \longrightarrow V[a(bc)].$$

If
$$A : V[(ab)c] \longrightarrow V[(ba)c]$$

is the first braiding operator (corresponding to an interchange of the first two particles in the association), then the second operator

$$B : V[(ab)c] \longrightarrow V[(ac)b]$$

is accomplished via the formula $B = F^{-1}RF$, where the R acts in the second vector space $V[a(bc)]$ to apply the phases for the interchange of b and c. These issues are illustrated in Fig. 3.1, where the parenthesization of the particles is indicated by circles and also by trees. The trees can be taken to indicate patterns of particle interaction, where two particles interact at the branch of a binary tree to produce the particle product at the root. See also Fig. 3.27 for an illustration of the braiding $B = F^{-1}RF$.

In this scheme, vector spaces corresponding to the associated strings of particle interactions are interrelated by *recoupling transformations* that generalize the mapping F indicated above. A full representation of the Artin braid group on each space is defined in terms of the local interchange phase gates and the recoupling transformations. These gates and transformations have to satisfy a number of identities in order to produce a well-defined representation of the braid group. These identities were discovered originally

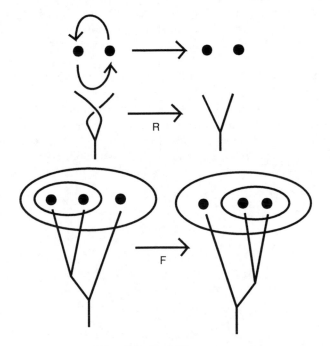

Fig. 3.1. Braiding anyons

in relation to topological quantum field theory. In our approach, the structure of phase gates and recoupling transformations arise naturally from the structure of the bracket model for the Jones polynomial. Thus we obtain a knot-theoretic basis for topological quantum computing.

Aspects of the quantum Hall effect are related to topological quantum field theory [4–7], where, in two dimensional space, the braiding of quasi-particles or collective excitations leads to non-trival representations of the Artin braid group. Such particles are called *Anyons*. It is hoped that the mathematics we explain here will form the bridge between theoretical models of anyons and their applications to quantum computing.

3.2 Knots and Braids

The purpose of this section is to give a quick introduction to the diagrammatic theory of knots, links, and braids. A *knot* is an embedding of a circle in three-dimensional space, taken up to ambient isotopy. The problem of deciding whether two knots are isotopic is an example of a *placement problem*, a problem of studying the topological forms that can be made by placing one space inside another. In the case of knot theory, we consider the placements of a circle inside three-dimensional space. That is, two knots are regarded as equivalent if one embedding can be obtained from the other through a continuous family of embeddings of circles in three space. A *link* is an embedding of a disjoint collection of circles, taken up to ambient isotopy. Figure 3.2 illustrates a diagram for a knot. The diagram is regarded both as a schematic picture of the knot and as a plane graph with extra structure at the nodes (indicating how the curve of the knot passes over or under itself by standard pictorial conventions).

There are many applications of the theory of knots. Topology is a background for the physical structure of real knots made from rope of cable. As a result, the field of practical knot tying is a field of applied topology that existed well before the mathematical discipline of topology arose. Then again long molecules such as rubber molecules and DNA molecules can be knotted

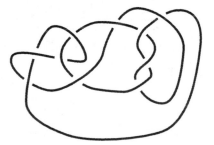

Fig. 3.2. A knot diagram

and linked. There have been a number of intense applications of knot theory
to the study of DNA [8] and to polymer physics [9]. Knot theory is closely
related to theoretical physics as well with applications in quantum gravity
[10–12] and many applications of ideas in physics to the topological structure
of knots themselves [3].

Quantum topology is the study and invention of topological invariants via
the use of analogies and techniques from mathematical physics. Many invariants such as the Jones polynomial are constructed via partition functions and
generalized quantum amplitudes. As a result, one expects to see relationships
between knot theory and physics. In this lecture, we will study how knot theory can be used to produce unitary representations of the braid group. Such
representations can play a fundamental role in quantum computing.

Ambient isotopy is mathematically the same as the equivalence relation
generated on diagrams by the *Reidemeister moves*. These moves are illustrated in Fig. 3.3. Each move is performed on a local part of the diagram
that is topologically identical to the part of the diagram illustrated in this
figure (these figures are representative examples of the types of Reidemeister
moves), without changing the rest of the diagram. The Reidemeister moves
are useful in doing combinatorial topology with knots and links, notably in
working out the behavior of knot invariants. A *knot invariant* is a function defined from knots and links to some other mathematical object (such as groups
or polynomials or numbers) such that equivalent diagrams are mapped to
equivalent objects (isomorphic groups, identical polynomials, identical numbers). The Reidemeister moves are of great use for analyzing the structure of
knot invariants, and they are closely related to the *Artin braid group*, which
we discuss below.

Another significant structure related to knots and links is the Artin braid
group. A *braid* is an embedding of a collection of strands that have their ends
top and bottom row points in two rows of points that are set one above the
other with respect to a choice of vertical. The strands are not individually

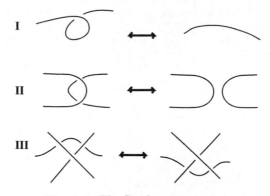

Fig. 3.3. The Reidemeister moves

knotted, and they are disjoint from one another. See Figs. 3.4, 3.5, and 3.6, for illustrations of braids and moves on braids. Braids can be multiplied by attaching the bottom row of one braid to the top row of the other braid.

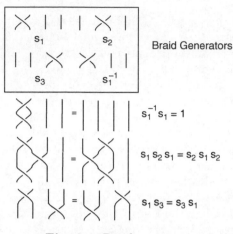

Fig. 3.4. Braid generators

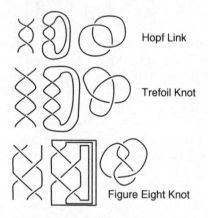

Fig. 3.5. Closing braids to form knots and links

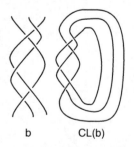

Fig. 3.6. Borromean rings as a braid closure

Taken up to ambient isotopy, fixing the endpoints, the braids form a group under this notion of multiplication. In Fig. 3.4, we illustrate the form of the basic generators of the braid group and the form of the relations among these generators. Figure 3.5 illustrates how to close a braid by attaching the top strands to the bottom strands by a collection of parallel arcs. A key theorem of Alexander states that every knot or link can be represented as a closed braid. Thus the theory of braids is critical to the theory of knots and links. Figure 3.6 illustrates the famous Borromean rings (a link of three unknotted loops such that any two of the loops are unlinked) as the closure of a braid.

Let B_n denote the Artin braid group on n strands. We recall here that B_n is generated by elementary braids $\{s_1, \cdots, s_{n-1}\}$ with relations

1. $s_i s_j = s_j s_i$ for $|i - j| > 1$,
2. $s_i s_{i+1} s_i = s_{i+1} s_i s_{i+1}$ for $i = 1, \cdots n - 2$.

See Fig. 3.4 for an illustration of the elementary braids and their relations. Note that the braid group has a diagrammatic topological interpretation, where a braid is an intertwining of strands that lead from one set of n points to another set of n points. The braid generators s_i are represented by diagrams, where the ith and $(i + 1)$th strands wind around one another by a single half-twist (the sense of this turn is shown in Fig. 3.4), and all other strands drop straight to the bottom. Braids are diagrammed vertically as in Fig. 3.4, and the products are taken in order from top to bottom. The product of two braid diagrams is accomplished by adjoining the top strands of one braid to the bottom strands of the other braid.

In Fig. 3.4, we have restricted the illustration to the four-stranded braid group B_4. In that figure, the three braid generators of B_4 are shown, and then the inverse of the first generator is drawn. Following this, one sees the identities $s_1 s_1^{-1} = 1$ (where the identity element in B_4 consists in four vertical strands), $s_1 s_2 s_1 = s_2 s_1 s_2$, and finally $s_1 s_3 = s_3 s_1$.

Braids are a key structure in mathematics. It is not just that they are a collection of groups with a vivid topological interpretation. From the algebraic point of view, the braid groups B_n are important extensions of the symmetric groups S_n. Recall that the symmetric group S_n of all permutations of n distinct objects has presentation as shown below:

1. $s_i^2 = 1$ for $i = 1, \cdots n - 1$,
2. $s_i s_j = s_j s_i$ for $|i - j| > 1$,
3. $s_i s_{i+1} s_i = s_{i+1} s_i s_{i+1}$ for $i = 1, \cdots n - 2$.

Thus S_n is obtained from B_n by setting the square of each braiding generator equal to one. We have an exact sequence of groups

$$1 \longrightarrow B_n \longrightarrow S_n \longrightarrow 1,$$

exhibiting the Artin braid group as an extension of the symmetric group.

In the next sections, we shall show how representations of the Artin braid group are rich enough to provide a dense set of transformations in the unitary groups. Thus the braid groups are *in principle* fundamental to quantum computation and quantum information theory.

3.3 Quantum Mechanics and Quantum Computation

We shall quickly indicate the basic principles of quantum mechanics. The quantum information context encapsulates a concise model of quantum theory:

The initial state of a quantum process is a vector $|v\rangle$ in a complex vector space H. Measurement returns basis elements β of H with probability

$$|\langle \beta | v\rangle|^2 / \langle v | v\rangle,$$

where $\langle v | w\rangle = v^\dagger w$ with $v^\dagger$ the conjugate transpose of v. A physical process occurs in steps $|v\rangle \longrightarrow U|v\rangle = |Uv\rangle$ where U is a unitary linear transformation.

Note that since $\langle Uv | Uw\rangle = \langle v | U^\dagger U | w\rangle = \langle v | w\rangle$ when U is unitary, it follows that probability is preserved in the course of a quantum process.

One of the details for any specific quantum problem is the nature of the unitary evolution. This is specified by knowing appropriate information about the classical physics that supports the phenomena. This information is used to choose an appropriate Hamiltonian through which the unitary operator is constructed via a correspondence principle that replaces classical variables with appropriate quantum operators. (In the path integral approach, one needs a Langrangian to construct the action on which the path integral is based.) One needs to know certain aspects of classical physics to solve any specific quantum problem.

A key concept in the quantum information viewpoint is the notion of the superposition of states. If a quantum system has two distinct states $|v\rangle$ and $|w\rangle$, then it has infinitely many states of the form $a|v\rangle + b|w\rangle$, where a and b are complex numbers taken up to a common multiple. States are "really" in the projective space associated with H. There is only one superposition of a single state $|v\rangle$ with itself.

Dirac [13] introduced the "bra-(c)-ket" notation $\langle A | B\rangle = A^\dagger B$ for the inner product of complex vectors $A, B \in H$. He also separated the parts of the bracket into the *bra* $\langle A |$ and the *ket* $|B\rangle$. Thus

$$\langle A | B\rangle = \langle A | \, | B\rangle.$$

In this interpretation, the ket $|B\rangle$ is identified with the vector $B \in H$, while the bra $\langle A |$ is regarded as the element dual to A in the dual space H^*.

The dual element to A corresponds to the conjugate transpose $A^\dagger$ of the vector A, and the inner product is expressed in conventional language by the matrix product $A^\dagger B$ (which is a scalar since B is a column vector). Having separated the bra and the ket, Dirac can write the "ket-bra" $|A\rangle\langle B| = AB^\dagger$. In conventional notation, the ket-bra is a matrix, not a scalar, and we have the following formula for the square of $P = |A\rangle\langle B|$:

$$P^2 = |A\rangle\langle B \,||A\rangle\langle B| = A(B^\dagger A)B^\dagger = (B^\dagger A)AB^\dagger = \langle B\,|A\rangle P\,.$$

The standard example is a ket-bra $P = |A\rangle\langle A|$, where $\langle A\,|A\rangle = 1$ so that $P^2 = P$. Then P is a projection matrix, projecting to the subspace of H that is spanned by the vector $|A\rangle$. In fact, for any vector $|B\rangle$, we have

$$P|B\rangle = |A\rangle\langle A\,||B\rangle = |A\rangle\langle A\,|B\rangle = \langle A\,|B\rangle|A\rangle\,.$$

If $\{|C_1\rangle, |C_2\rangle, \cdots |C_n\rangle\}$ is an orthonormal basis for H, and

$$P_i = |C_i\rangle\langle C_i|\,,$$

then for any vector $|A\rangle$, we have

$$|A\rangle = \langle C_1\,|A\rangle|C_1\rangle + \cdots + \langle C_n\,|A\rangle|C_n\rangle\,.$$

Hence

$$\langle B\,|A\rangle = \langle B\,|C_1\rangle\langle C_1\,|A\rangle + \cdots + \langle B\,|C_n\rangle\langle C_n\,|A\rangle\,.$$

One wants the probability of starting in state $|A\rangle$ and ending in state $|B\rangle$. The probability for this event is equal to $|\langle B\,|A\rangle|^2$. This can be refined if we have more knowledge. If the intermediate states $|C_i\rangle$ are a complete set of orthonormal alternatives, then we can assume that $\langle C_i\,|C_i\rangle = 1$ for each i and that $\Sigma_i|C_i\rangle\langle C_i| = 1$. This identity now corresponds to the fact that 1 is the sum of the probabilities of an arbitrary state being projected into one of these intermediate states.

If there are intermediate states between the intermediate states, this formulation can be continued until one is summing over all possible paths from A to B. This becomes the path integral expression for the amplitude $\langle B|A\rangle$.

3.3.1 What Is a Quantum Computer?

A *quantum computer* is, abstractly, a composition U of unitary transformations, together with an initial state and a choice of measurement basis. One runs the computer by repeatedly initializing it, and then measuring the result of applying the unitary transformation U to the initial state. The results of these measurements are then analyzed for the desired information that the computer was set to determine. The key to using the computer is the design of the initial state and the design of the composition of unitary transformations. The reader should consult [14] for more specific examples of quantum algorithms.

Let H be a given finite-dimensional vector space over the complex numbers C. Let $\{W_0, W_1, \ldots, W_n\}$ be an orthonormal basis for H so that with $|i\rangle := |W_i\rangle$ denoting W_i and $\langle i|$ denoting the conjugate transpose of $|i\rangle$, we have

$$\langle i|j\rangle = \delta_{ij},$$

where δ_{ij} denotes the Kronecker delta (equal to one when its indices are equal to one another, and equal to zero otherwise). Given a vector v in H let $|v|^2 := \langle v|v\rangle$. Note that $\langle i|v$ is the ith coordinate of v. A *measurement of v* returns one of the coordinates $|i\rangle$ of v with probability $|\langle i|v|^2$. This model of measurement is a simple instance of the situation with a quantum mechanical system that is in a mixed state until it is observed. The result of observation is to put the system into one of the basis states.

When the dimension of the space H is two ($n = 2$), a vector in the space is called a *qubit*. A qubit represents one quantum of binary information. On measurement, one obtains either the ket $|0\rangle$ or the ket $|1\rangle$. This constitutes the binary distinction that is inherent in a qubit. Note however that the information obtained is probabilistic. If the qubit is

$$|\psi\rangle = \alpha|0\rangle + \beta\,|1\rangle,$$

then the ket $|0\rangle$ is observed with probability $|\alpha|^2$, and the ket $|1\rangle$ is observed with probability $|\beta|^2$. In speaking of an idealized quantum computer, we do not specify the nature of measurement process beyond these probability postulates.

In the case of general dimension n of the space H, we will call the vectors in H *qunits*. It is quite common to use spaces H that are tensor products of two-dimensional spaces (so that all computations are expressed in terms of qubits), but this is not necessary in principle. One can start with a given space, and later work out factorizations into qubit transformations.

A *quantum computation* consists in the application of a unitary transformation U to an initial qunit $\psi = a_0|0\rangle + \ldots + a_n|n\rangle$ with $|\psi|^2 = 1$, plus a measurement of $U\psi$. A measurement of $U\psi$ returns the ket $|i\rangle$ with probability $|\langle i|U\psi|^2$. In particular, if we start the computer in the state $|i\rangle$, then the probability that it will return the state $|j\rangle$ is $|\langle j|U|i\rangle|^2$.

It is the necessity for writing a given computation in terms of unitary transformations, and the probabilistic nature of the result that characterizes quantum computation. Such computation could be carried out by an idealized quantum mechanical system. It is hoped that such systems can be physically realized.

3.4 Braiding Operators and Universal Quantum Gates

A class of invariants of knots and links called quantum invariants can be constructed using representations of the Artin braid group, and more specifically using solutions to the Yang–Baxter equation [15], first discovered in relation to $1+1$-dimensional quantum field theory, and 2-dimensional statistical mechanics. Braiding operators feature in constructing representations of the Artin braid group, and in the construction of invariants of knots and links.

A key concept in the construction of quantum link invariants is the association of a Yang–Baxter operator R to each elementary crossing in a link diagram. The operator R is a linear mapping

$$: V \otimes V \longrightarrow V \otimes V$$

defined on the 2-fold tensor product of a vector space V, generalizing the permutation of the factors (i.e., generalizing a swap gate when V represents one qubit). Such transformations are not necessarily unitary in topological applications. It is useful to understand when they can be replaced by unitary transformations for the purpose of quantum computing. Such unitary R-matrices can be used to make unitary representations of the Artin braid group.

A solution to the Yang–Baxter equation, as described in the last paragraph is a matrix R, regarded as a mapping of a 2-fold tensor product of a vector space $V \otimes V$ to itself that satisfies the equation

$$(R \otimes I)(I \otimes R)(R \otimes I) = (I \otimes R)(R \otimes I)(I \otimes R).$$

From the point of view of topology, the matrix R is regarded as representing an elementary bit of braiding represented by one string crossing over another. In Fig. 3.7, we have illustrated the braiding identity that corresponds to the Yang–Baxter equation. Each braiding picture with its three input lines (below) and output lines (above) corresponds to a mapping of the 3-fold tensor product of the vector space V to itself, as required by the algebraic equation quoted above. The pattern of placement of the crossings in the diagram corresponds to the factors $R \otimes I$ and $I \otimes R$. This crucial topological move has an algebraic expression in terms of such a matrix R. Our approach in this section to relate topology, quantum computing, and quantum entanglement is through the use of the Yang–Baxter equation. In order to accomplish this

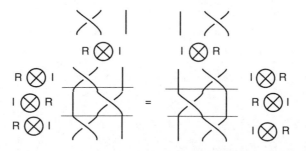

Fig. 3.7. The Yang–Baxter equation $- (R \otimes I)(I \otimes R)(R \otimes I) = (I \otimes R)(R \otimes I)(I \otimes R)$

aim, *we need to study solutions of the Yang–Baxter equation that are unitary.*
Then the R matrix can be seen *either* as a braiding matrix *or* as a quantum
gate in a quantum computer.

The problem of finding solutions to the Yang–Baxter equation that are
unitary turns out to be surprisingly difficult. Dye [16] has classified all such
matrices of size 4×4. A rough summary of her classification is that all
4×4 unitary solutions to the Yang–Baxter equation are similar to one of the
following types of matrix:

$$R = \begin{pmatrix} 1/\sqrt{2} & 0 & 0 & 1/\sqrt{2} \\ 0 & 1/\sqrt{2} & -1/\sqrt{2} & 0 \\ 0 & 1/\sqrt{2} & 1/\sqrt{2} & 0 \\ -1/\sqrt{2} & 0 & 0 & 1/\sqrt{2} \end{pmatrix},$$

$$R' = \begin{pmatrix} a & 0 & 0 & 0 \\ 0 & 0 & b & 0 \\ 0 & c & 0 & 0 \\ 0 & 0 & 0 & d \end{pmatrix},$$

$$R'' = \begin{pmatrix} 0 & 0 & 0 & a \\ 0 & b & 0 & 0 \\ 0 & 0 & c & 0 \\ d & 0 & 0 & 0 \end{pmatrix},$$

where a, b, c, and d are unit complex numbers.

For the purpose of quantum computing, one should regard each matrix
as acting on the standard basis $\{|00\rangle, |01\rangle, |10\rangle, |11\rangle\}$ of $H = V \otimes V$, where
V is a two-dimensional complex vector space. Then, for example we have

$$R|00\rangle = (1/\sqrt{2})|00\rangle - (1/\sqrt{2})|11\rangle,$$
$$R|01\rangle = (1/\sqrt{2})|01\rangle + (1/\sqrt{2})|10\rangle,$$
$$R|10\rangle = -(1/\sqrt{2})|01\rangle + (1/\sqrt{2})|10\rangle,$$
$$R|11\rangle = (1/\sqrt{2})|00\rangle + (1/\sqrt{2})|11\rangle.$$

The reader should note that R is the familiar change-of-basis matrix from
the standard basis to the Bell basis of entangled states. In the case of R', we
have

$$R'|00\rangle = a|00\rangle, \qquad\qquad R'|01\rangle = c|10\rangle,$$
$$R'|10\rangle = b|01\rangle, \qquad\qquad R'|11\rangle = d|11\rangle.$$

Note that R' can be regarded as a diagonal phase gate P, composed with a
swap gate S,

$$P = \begin{pmatrix} a & 0 & 0 & 0 \\ 0 & b & 0 & 0 \\ 0 & 0 & c & 0 \\ 0 & 0 & 0 & d \end{pmatrix}, \qquad\qquad S = \begin{pmatrix} 1 & 0 & 0 & 0 \\ 0 & 0 & 1 & 0 \\ 0 & 1 & 0 & 0 \\ 0 & 0 & 0 & 1 \end{pmatrix}.$$

Compositions of solutions of the (Braiding) Yang–Baxter equation with the swap gate S are called *solutions to the algebraic Yang–Baxter equation*. Thus the diagonal matrix P is a solution to the algebraic Yang–Baxter equation.

Remark 1. Another avenue related to unitary solutions to the Yang–Baxter equation as quantum gates comes from using extra physical parameters in this equation (the rapidity parameter) that are related to statistical physics. In [17] we discovered that solutions to the Yang–Baxter equation with the rapidity parameter allow many new unitary solutions. The significance of these gates for quantum computing is still under investigation.

3.4.1 Universal Gates

A *two-qubit gate* G is a unitary linear mapping $G : V \otimes V \longrightarrow V$, where V is a two complex dimensional vector space. We say that the gate G is *universal for quantum computation* (or just *universal*) if G together with local unitary transformations (unitary transformations from V to V) generates all unitary transformations of the complex vector space of dimension 2^n to itself. It is well known [14] that *CNOT* is a universal gate.[2]

A gate G, as above, is said to be *entangling* if there is a vector

$$|\alpha\beta\rangle = |\alpha\rangle \otimes |\beta\rangle \in V \otimes V$$

such that $G|\alpha\beta\rangle$ is not decomposable as a tensor product of two qubits. Under these circumstances, one says that $G|\alpha\beta\rangle$ is *entangled*.

In [18], the Brylinskis give a general criterion of G to be universal. They prove that *a two-qubit gate G is universal if and only if it is entangling*.

Remark 2. A two-qubit pure state

$$|\phi\rangle = a|00\rangle + b|01\rangle + c|10\rangle + d|11\rangle$$

is entangled exactly when $(ad - bc) \neq 0$. It is easy to use this fact to check when a specific matrix is, or is not, entangling.

Remark 3. There are many gates other than *CNOT* that can be used as universal gates in the presence of local unitary transformations. Some of these are themselves topological (unitary solutions to the Yang–Baxter equation, see [19]) and themselves generate representations of the Artin braid Group. Replacing *CNOT* by a solution to the Yang–Baxter equation does not place the local unitary transformations as part of the corresponding representation of the braid group. Thus such substitutions connote only a partial solution to creating topological quantum computation. In this lecture, we are concerned

---

[2] On the standard basis, *CNOT* is the identity when the first qubit is 0, and it flips the second qubit, leaving the first alone, when the first qubit is 1.

with braid group representations that include all aspects of the unitary group. Accordingly, in the next section we shall first examine, how the braid group on three strands can be represented as local unitary transformations.

Theorem 1. *Let D denote the phase gate shown below. D is a solution to the algebraic Yang–Baxter equation (see the earlier discussion in this section). Then D is a universal gate.*

$$D = \begin{pmatrix} 1 & 0 & 0 & 0 \\ 0 & 1 & 0 & 0 \\ 0 & 0 & 1 & 0 \\ 0 & 0 & 0 & -1 \end{pmatrix}.$$

Proof. It follows at once from the Brylinski Theorem that D is universal. For a more specific proof, note that $CNOT = QDQ^{-1}$, where $Q = H \otimes I$, H is the 2×2 Hadamard matrix. The conclusion then follows at once from this identity and the discussion above. We illustrate the matrices involved in this proof below:

$$H = \frac{1}{\sqrt{2}} \begin{pmatrix} 1 & 1 \\ 1 & -1 \end{pmatrix}, \quad Q = \frac{1}{\sqrt{2}} \begin{pmatrix} 1 & 1 & 0 & 0 \\ 1 & -1 & 0 & 0 \\ 0 & 0 & 1 & 1 \\ 0 & 0 & 1 & -1 \end{pmatrix}, \quad D = \begin{pmatrix} 1 & 0 & 0 & 0 \\ 0 & 1 & 0 & 0 \\ 0 & 0 & 1 & 0 \\ 0 & 0 & 0 & -1 \end{pmatrix},$$

$$QDQ^{-1} = QDQ = \begin{pmatrix} 1 & 0 & 0 & 0 \\ 0 & 1 & 0 & 0 \\ 0 & 0 & 0 & 1 \\ 0 & 0 & 1 & 0 \end{pmatrix} = CNOT.$$

This completes the proof of the Theorem.[3]   □

Theorem 2. *The matrix solutions R' and R'' to the Yang–Baxter equation, described above, are universal gates exactly when $ad - bc \neq 0$ for their internal parameters a, b, c, d. In particular, let R_0 denote the solution R' (above) to the Yang–Baxter equation with $a = b = c = 1$, $d = -1$,*

$$R_0 = \begin{pmatrix} 1 & 0 & 0 & 0 \\ 0 & 0 & 1 & 0 \\ 0 & 1 & 0 & 0 \\ 0 & 0 & 0 & -1 \end{pmatrix}.$$

Then R_0 is a universal gate.

Proof. The first part follows at once from the Brylinski Theorem. In fact, letting H be the Hadamard matrix as before, and

---

[3] We thank Martin Roetteler (private conversation, fall 2003) for pointing out the specific factorization of *CNOT* used in this proof.

$$\sigma = \begin{pmatrix} 1/\sqrt{2} & i/\sqrt{2} \\ i/\sqrt{2} & 1/\sqrt{2} \end{pmatrix} , \quad \lambda = \begin{pmatrix} 1/\sqrt{2} & 1/\sqrt{2} \\ i/\sqrt{2} & -i/\sqrt{2} \end{pmatrix} , \quad \mu = \begin{pmatrix} (1-i)/2 & (1+i)/2 \\ (1-i)/2 & (-1-i)/2 \end{pmatrix} .$$

Then
$$CNOT = (\lambda \otimes \mu)(R_0(I \otimes \sigma)R_0)(H \otimes H) .$$

This gives an explicit expression for $CNOT$ in terms of R_0 and local unitary transformations (for which we thank Ben Reichardt). □

Remark 4. Let $SWAP$ denote the Yang–Baxter Solution R' with $a = b = c = d = 1$,

$$SWAP = \begin{pmatrix} 1 & 0 & 0 & 0 \\ 0 & 0 & 1 & 0 \\ 0 & 1 & 0 & 0 \\ 0 & 0 & 0 & 1 \end{pmatrix} .$$

$SWAP$ is the standard swap gate. Note that $SWAP$ is not a universal gate. This also follows from the Brylinski Theorem, since $SWAP$ is not entangling. Note also that R_0 is the composition of the phase gate D with this swap gate.

Theorem 3. *Let*

$$R = \begin{pmatrix} 1/\sqrt{2} & 0 & 0 & 1/\sqrt{2} \\ 0 & 1/\sqrt{2} & -1/\sqrt{2} & 0 \\ 0 & 1/\sqrt{2} & 1/\sqrt{2} & 0 \\ -1/\sqrt{2} & 0 & 0 & 1/\sqrt{2} \end{pmatrix}$$

be the unitary solution to the Yang–Baxter equation discussed above. Then R is a universal gate. The proof below gives a specific expression for CNOT in terms of R.

Proof. This result follows at once from the Brylinksi Theorem, since R is highly entangling. For a direct computational proof, it suffices to show that $CNOT$ can be generated from R and local unitary transformations. Let

$$\alpha = \begin{pmatrix} 1/\sqrt{2} & 1/\sqrt{2} \\ 1/\sqrt{2} & -1/\sqrt{2} \end{pmatrix} , \qquad \beta = \begin{pmatrix} -1/\sqrt{2} & 1/\sqrt{2} \\ i/\sqrt{2} & i/\sqrt{2} \end{pmatrix} ,$$

$$\gamma = \begin{pmatrix} 1/\sqrt{2} & i/\sqrt{2} \\ 1/\sqrt{2} & -i/\sqrt{2} \end{pmatrix} , \qquad \delta = \begin{pmatrix} -1 & 0 \\ 0 & -i \end{pmatrix} .$$

Let $M = \alpha \otimes \beta$ and $N = \gamma \otimes \delta$. Then it is straightforward to verify that

$$CNOT = MRN .$$

This completes the proof.[4] □

---

[4] See [19] for more information about these calculations.

3.5 A Remark About EPR, Entanglement and Bell's Inequality

It is remarkable that the simple algebraic situation of an element in a tensor product that is not itself a tensor product of elements of the factors corresponds to subtle nonlocality in physics. It helps to place this algebraic structure in the context of a gedanken experiment to see where the physics comes in. Consider

$$S = (|0\rangle|1\rangle + |1\rangle|0\rangle)/\sqrt{2}.$$

In an EPR thought experiment, we think of two "parts" of this state that are separated in space. We want a notation for these parts and suggest the following:

$$L = (\{|0\rangle\}|1\rangle + \{|1\rangle\}|0\rangle)/\sqrt{2},$$
$$R = (|0\rangle\{|1\rangle\} + |1\rangle\{|0\rangle\})/\sqrt{2}.$$

In the left state L, an observer can only observe the left-hand factor. In the right state R, an observer can only observe the right-hand factor. These "states" L and R together comprise the EPR state S, but they are accessible individually just as are the two photons in the usual thought experiment. One can transport L and R individually and we shall write

$$S = L * R$$

to denote that they are the "parts" (but not tensor factors) of S.

The curious thing about this formalism is that it includes a little bit of macroscopic physics implicitly, and so it makes it a bit more apparent what EPR was concerned about. After all, lots of things that we can do to L or R do not affect S. For example, transporting L from one place to another, as in the original experiment where the photons separate. On the other hand, if Alice has L and Bob has R and Alice performs a local unitary transformation on "her" tensor factor, this applies to both L and R since the transformation is actually being applied to the state S. This is also a "spooky action at a distance" whose consequence does not appear until a measurement is made.

To go a bit deeper, it is worthwhile seeing what entanglement, in the sense of tensor indecomposability, has to do with the structure of the EPR thought experiment. To this end, we look at the structure of the Bell inequalities using the Clauser, Horne, Shimony, and Holt (CHSH) formalism as explained in the book by Nielsen and Chuang [14]. For this we use the following observables with eigenvalues ± 1:

$$Q = \begin{pmatrix} 1 & 0 \\ 0 & -1 \end{pmatrix}_1, \qquad R = \begin{pmatrix} 0 & 1 \\ 1 & 0 \end{pmatrix}_1,$$
$$S = \frac{1}{\sqrt{2}} \begin{pmatrix} -1 & -1 \\ -1 & 1 \end{pmatrix}_2, \qquad T = \frac{1}{\sqrt{2}} \begin{pmatrix} 1 & -1 \\ -1 & -1 \end{pmatrix}_2.$$

The subscripts 1 and 2 on these matrices indicate that they are to operate on the first- and second-tensor factors, respectively, of a quantum state of the form

$$\phi = a|00\rangle + b|01\rangle + c|10\rangle + d|11\rangle.$$

To simplify the results of this calculation, we shall here assume that the coefficients a, b, c, and d are real numbers. We calculate the quantity

$$\Delta = \langle\phi|QS|\phi\rangle + \langle\phi|RS|\phi\rangle + \langle\phi|RT|\phi\rangle - \langle\phi|QT|\phi\rangle,$$

finding that

$$\Delta = (2 - 4(a + d)^2 + 4(ad - bc))/\sqrt{2}.$$

Classical probability calculation with random variables of value ± 1 gives the value of $QS + RS + RT - QT = \pm 2$ (with each of Q, R, S, and T equal to ± 1). Hence the classical expectation satisfies the Bell inequality

$$E(QS) + E(RS) + E(RT) - E(QT) \leq 2.$$

That quantum expectation is not classical and is embodied in the fact that Δ can be greater than 2. The classic case is that of the Bell state

$$\phi = (|01\rangle - |10\rangle)/\sqrt{2}.$$

Here

$$\Delta = 6/\sqrt{2} > 2.$$

In general, we see that the following inequality is needed in order to violate the Bell inequality

$$(2 - 4(a + d)^2 + 4(ad - bc))/\sqrt{2} > 2.$$

This is equivalent to

$$(\sqrt{2} - 1)/2 < (ad - bc) - (a + d)^2.$$

Since we know that ϕ is entangled exactly when $ad - bc$ is non-zero, this shows that an unentangled state cannot violate the Bell inequality. This formula *also* shows that it is possible for a state to be entangled and yet not violate the Bell inequality. For example, if

$$\phi = (|00\rangle - |01\rangle + |10\rangle + |11\rangle)/2,$$

then $\Delta(\phi)$ satisfies Bell's inequality, but ϕ is an entangled state. We see from this calculation that entanglement in the sense of tensor indecomposability and entanglement in the sense of Bell inequality violation for a given choice of Bell operators are not equivalent concepts. On the other hand, Benjamin Schumacher has pointed out [20] that any entangled two-qubit state will violate Bell inequalities for an appropriate choice of operators. This deepens the context for our question of the relationship between topological entanglement and quantum entanglement. The Bell inequality violation is an indication of quantum mechanical entanglement. One's intuition suggests that it is *this* sort of entanglement that should have a topological context.

3.6 The Aravind Hypothesis

Link diagrams can be used as graphical devices and holders of information. In this vein, Aravind [21] proposed that the entanglement of a link should correspond to the entanglement of a state. *Measurement of a link would be modeled by deleting one component of the link.* A key example is the Borromean rings (see Fig. 3.8).

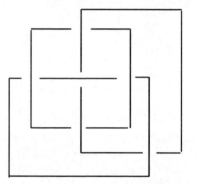

Fig. 3.8. Borromean rings

Deleting any component of the Borromean rings yields a remaining pair of unlinked rings. The Borromean rings are entangled, but any two of them are unentangled. In this sense, the Borromean rings are analogous to the *GHZ* state $|GHZ\rangle = (1/\sqrt{2})(|000\rangle + |111\rangle)$. Measurement in any factor of the *GHZ* yields an unentangled state. Aravind points out that this property is basis dependent. *We point out that there are states whose entanglement after a measurement is a matter of probability (via quantum amplitudes).* Consider for example the state

$$|\psi\rangle = (1/2)(|000\rangle + |001\rangle + |101\rangle + |110\rangle).$$

Measurement in any coordinate yields an entangled or an unentangled state with equal probability. For example

$$|\psi\rangle = (1/2)(|0\rangle(|00\rangle + |01\rangle) + |1\rangle(|01\rangle + |10\rangle)),$$

so that projecting to $|0\rangle$ in the first coordinate yields an unentangled state, while projecting to $|1\rangle$ yields an entangled state, each with equal probability.

New ways to use link diagrams must be invented to map the properties of such states (see [22]).

3.7 $SU(2)$ Representations of the Artin Braid Group

The purpose of this section is to determine all the representations of the three strand Artin braid group B_3 to the special unitary group $SU(2)$ and concomitantly to the unitary group $U(2)$. One regards the groups $SU(2)$ and $U(2)$ as acting on a single qubit, and so $U(2)$ is usually regarded as the group of local unitary transformations in a quantum information setting. If one is looking for a coherent way to represent all unitary transformations by way of braids, then $U(2)$ is the place to start. Here we will show that there are many representations of the three-strand braid group that generate a dense subset of $U(2)$. Thus it is a fact that local unitary transformations can be "generated by braids" in many ways.

We begin with the structure of $SU(2)$. A matrix in $SU(2)$ has the form

$$M = \begin{pmatrix} z & w \\ -\bar{w} & \bar{z} \end{pmatrix},$$

where z and w are complex numbers, and $\bar{z}$ denotes the complex conjugate of z. To be in $SU(2)$, it is required that $\det(M) = 1$ and that $M^\dagger = M^{-1}$ where det denotes determinant, and $M^\dagger$ is the conjugate transpose of M. Thus if $z = a + bi$ and $w = c + di$ where a, b, c, d are real numbers, and $i^2 = -1$, then

$$M = \begin{pmatrix} a + bi & c + di \\ -c + di & a - bi \end{pmatrix}$$

with $a^2 + b^2 + c^2 + d^2 = 1$. It is convenient to write

$$M = a \begin{pmatrix} 1 & 0 \\ 0 & 1 \end{pmatrix} + b \begin{pmatrix} i & 0 \\ 0 & -i \end{pmatrix} + c \begin{pmatrix} 0 & 1 \\ -1 & 0 \end{pmatrix} + d \begin{pmatrix} 0 & i \\ i & 0 \end{pmatrix},$$

and to abbreviate this decomposition as

$$M = a + bi + cj + dk,$$

where

$$1 \equiv \begin{pmatrix} 1 & 0 \\ 0 & 1 \end{pmatrix}, \quad i \equiv \begin{pmatrix} i & 0 \\ 0 & -i \end{pmatrix}, \quad j \equiv \begin{pmatrix} 0 & 1 \\ -1 & 0 \end{pmatrix}, \quad k \equiv \begin{pmatrix} 0 & i \\ i & 0 \end{pmatrix},$$

so that

$$i^2 = j^2 = k^2 = ijk = -1$$

and

$$ij = k, \qquad jk = i, \qquad ki = j$$
$$ji = -k, \qquad kj = -i, \qquad ik = -j.$$

The algebra of 1, i, j, k is called the *quaternions* after William Rowan Hamilton who discovered this algebra prior to the discovery of matrix algebra. Thus the units are identified with $SU(2)$ in this way. We shall use this identification, and some facts about the quaternions to find the $SU(2)$ representations of braiding. First we recall some facts about the quaternions.

1. Note that if $q = a + bi + cj + dk$ (as above), then $q^\dagger = a - bi - cj - dk$ so that $qq^\dagger = a^2 + b^2 + c^2 + d^2 = 1$.
2. A general quaternion has the form $q = a + bi + cj + dk$, where the value of $qq^\dagger = a^2 + b^2 + c^2 + d^2$ is not fixed to unity. The *length* of q is by definition $\sqrt{qq^\dagger}$.
3. A quaternion of the form $ri + sj + tk$ for real numbers r, s, t is said to be a *pure* quaternion. We identify the set of pure quaternions with the vector space of triples (r, s, t) of real numbers R^3.
4. Thus a general quaternion has the form $q = a + bu$, where u is a pure quaternion of unit length, and a and b are arbitrary real numbers. A unit quaternion (element of $SU(2)$) has the addition property that $a^2 + b^2 = 1$.
5. If u is a pure unit length quaternion, then $u^2 = -1$. Note that the set of pure unit quaternions forms the two-dimensional sphere $S^2 = \{(r, s, t) | r^2 + s^2 + t^2 = 1\}$ in R^3.
6. If u, v are pure quaternions, then

$$uv = -u \cdot v + u \times v\,,$$

where $u \cdot v$ is the dot product of the vectors u and v, and $u \times v$ is the vector cross product of u and v. In fact, one can take the definition of quaternion multiplication as

$$(a + bu)(c + dv) = ac + bc(u) + ad(v) + bd(-u \cdot v + u \times v)\,,$$

and all the above properties are consequences of this definition. Note that quaternion multiplication is associative.

7. Let $g = a + bu$ be a unit length quaternion so that $u^2 = -1$ and $a = cos(\theta/2)$, $b = sin(\theta/2)$ for a chosen angle θ. Define $\phi_g : R^3 \longrightarrow R^3$ by the equation $\phi_g(P) = gPg^\dagger$, for P any point in R^3, regarded as a pure quaternion. Then ϕ_g is an orientation preserving rotation of R^3 (hence an element of the rotation group $SO(3)$). Specifically, ϕ_g is a rotation about the axis u by the angle θ. The mapping

$$\phi : SU(2) \longrightarrow SO(3)$$

is a two-to-one surjective map from the special unitary group to the rotation group. In quaternionic form, this result was proved by Hamilton and by Rodrigues in the middle of the nineteenth century. The specific formula for $\phi_g(P)$ is shown below:

$$\phi_g(P) = gPg^{-1} = (a^2 - b^2)P + 2ab(P \times u) + 2(P \cdot u)b^2 u\,.$$

We want a representation of the three-strand braid group in $SU(2)$. This means that we want a homomorphism $\rho : B_3 \longrightarrow SU(2)$, and hence we want elements $g = \rho(s_1)$ and $h = \rho(s_2)$ in $SU(2)$ representing the braid group generators s_1 and s_2. Since $s_1 s_2 s_1 = s_2 s_1 s_2$ is the generating relation for B_3, the only requirement on g and h is that $ghg = hgh$. We rewrite this relation as $h^{-1}gh = ghg^{-1}$ and analyze its meaning in the unit quaternions.

Suppose that $g = a + bu$ and $h = c + dv$, where u and v are unit pure quaternions so that $a^2 + b^2 = 1$ and $c^2 + d^2 = 1$, then $ghg^{-1} = c + d\phi_g(v)$ and $h^{-1}gh = a + b\phi_{h^{-1}}(u)$. Thus it follows from the braiding relation that $a = c$, $b = \pm d$, and that $\phi_g(v) = \pm\phi_{h^{-1}}(u)$. However, in the case where there is a minus sign, we have $g = a + bu$ and $h = a - bv = a + b(-v)$. Thus we can now prove the following theorem.

Theorem 4. *If $g = a + bu$ and $h = c + dv$ are pure unit quaternions, then, without loss of generality, the braid relation $ghg = hgh$ is true if and only if $h = a + bv$, and $\phi_g(v) = \phi_{h^{-1}}(u)$. Furthermore, given that $g = a + bu$ and $h = a + bv$, the condition $\phi_g(v) = \phi_{h^{-1}}(u)$ is satisfied if and only if $u \cdot v = \frac{a^2 - b^2}{2b^2}$ when $u \neq v$. If $u = v$, then $g = h$, and the braid relation is trivially satisfied.*

Proof. We have proved the first sentence of the theorem in the discussion prior to its statement. Therefore assume that $g = a + bu$, $h = a + bv$, and $\phi_g(v) = \phi_{h^{-1}}(u)$. We have already stated the formula for $\phi_g(v)$ in the discussion about quaternions:

$$\phi_g(v) = gvg^{-1} = (a^2 - b^2)v + 2ab(v \times u) + 2(v \cdot u)b^2 u\,.$$

By the same token, we have

$$\begin{aligned}
\phi_{h^{-1}}(u) &= h^{-1}uh \\
&= (a^2 - b^2)u + 2ab(u \times -v) + 2(u \cdot (-v))b^2(-v) \\
&= (a^2 - b^2)u + 2ab(v \times u) + 2(v \cdot u)b^2(v)\,.
\end{aligned}$$

Hence we require that

$$(a^2 - b^2)v + 2(v \cdot u)b^2 u = (a^2 - b^2)u + 2(v \cdot u)b^2(v)\,.$$

This equation is equivalent to

$$2(u \cdot v)b^2(u - v) = (a^2 - b^2)(u - v)\,.$$

If $u \neq v$, then this implies that

$$u \cdot v = \frac{a^2 - b^2}{2b^2}\,.$$

This completes the proof of the Theorem.   □

Example 1. Let

$$g = e^{i\theta} = a + bi\,,$$

where $a = \cos(\theta)$ and $b = \sin(\theta)$. Let

$$h = a + b[(c^2 - s^2)i + 2csk],$$

where $c^2 + s^2 = 1$ and $c^2 - s^2 = \frac{a^2 - b^2}{2b^2}$. Then we can reexpress g and h in matrix form as the matrices G and $\tilde{H}$. Instead of writing the explicit form of H, we write $H = FGF^\dagger$, where F is an element of $SU(2)$ as shown below:

$$G = \begin{pmatrix} e^{i\theta} & 0 \\ 0 & e^{-i\theta} \end{pmatrix}, \qquad\qquad F = i \begin{pmatrix} c & s \\ s & -c \end{pmatrix}.$$

This representation of braiding, where one generator G is a simple matrix of phases, while the other generator $H = FGF^\dagger$ is derived from G by conjugation by a unitary matrix, has the possibility for generalization to representations of braid groups (on greater than three strands) to $SU(n)$ or $U(n)$ for n greater than 2. In fact we shall see just such representations constructed later in this paper, using a version of topological quantum field theory. The simplest example is given by

$$g = e^{7\pi i/10},$$
$$f = i\tau + k\sqrt{\tau},$$
$$h = frf^{-1},$$

where $\tau^2 + \tau = 1$. Then g and h satisfy $ghg = hgh$ and generate a representation of the three-strand braid group that is dense in $SU(2)$. We shall call this the *Fibonacci* representation of B_3 to $SU(2)$.

Remark 5 (Density). Consider representations of B_3 into $SU(2)$ produced by the method of this section. That is consider the subgroup $SU[G, H]$ of $SU(2)$ generated by a pair of elements $\{g, h\}$ such that $ghg = hgh$. We wish to understand when such a representation will be dense in $SU(2)$. We need the following lemma.

Lemma 1. $e^{ai}e^{bj}e^{ci} = \cos(b)e^{i(a+c)} + \sin(b)e^{i(a-c)}j$. *Hence any element of $SU(2)$ can be written in the form $e^{ai}e^{bj}e^{ci}$ for appropriate choices of angles a, b, c. In fact, if u and v are linearly independent unit vectors in R^3, then any element of $SU(2)$ can be written in the form*

$$e^{au}e^{bv}e^{cu}$$

for appropriate choices of the real numbers a, b, c.

Proof. It is easy to check that

$$e^{ai}e^{bj}e^{ci} = \cos(b)e^{i(a+c)} + \sin(b)e^{i(a-c)}j.$$

This completes the verification of the identity in the statement of the Lemma. Let v be any unit direction in R^3 and λ an arbitrary angle. We have

$$e^{v\lambda} = \cos(\lambda) + \sin(\lambda)v\,,$$

and

$$v = r + si + (p+qi)j\,,$$

where $r^2 + s^2 + p^2 + q^2 = 1$. So

$$\begin{aligned}
e^{v\lambda} &= \cos(\lambda) + \sin(\lambda)(r + si) + \sin(\lambda)(p + qi)j \\
&= [(\cos(\lambda) + \sin(\lambda)r) + \sin(\lambda)si] + [\sin(\lambda)p + \sin(\lambda)qi]\,j\,.
\end{aligned}$$

By the identity just proved, we can choose angles a, b, c so that

$$e^{v\lambda} = e^{ia}e^{jb}e^{ic}\,.$$

Hence

$$\cos(b)e^{i(a+c)} = (\cos(\lambda) + \sin(\lambda)r) + \sin(\lambda)si$$

and

$$\sin(b)e^{i(a-c)} = \sin(\lambda)p + \sin(\lambda)qi\,.$$

Suppose we keep v fixed and vary λ, then the last equations show that this will result in a full variation of b.

Now consider

$$e^{ia'}e^{v\lambda}e^{ic'} = e^{ia'}e^{ia}e^{jb}e^{ic}e^{ib'} = e^{i(a'+a)}e^{jb}e^{i(c+c')}\,.$$

By the basic identity, this shows that any element of $SU(2)$ can be written in the form

$$e^{ia'}e^{v\lambda}e^{ic'}\,.$$

Then, by applying a rotation, we finally conclude that if u and v are linearly independent unit vectors in R^3, then any element of $SU(2)$ can be written in the form

$$e^{au}e^{bv}e^{cu}$$

for appropriate choices of the real numbers a, b, c. $\square$

This Lemma can be used to verify density of a representation, by finding two elements A and B in the representation such that the powers of A are dense in the rotations about its axis, and the powers of B are dense in the rotations about its axis, and such that the axes of A and B are linearly independent in R^3. Then by the Lemma, the set of elements $A^{a+c}B^bA^{a-c}$ are dense in $SU(2)$. It follows, for example, that the Fibonacci representation described above is dense in $SU(2)$, and indeed the generic representation of B_3 into $SU(2)$ will be dense in $SU(2)$. Our next task is to describe representations of the higher braid groups that will extend some of these unitary representations of the three-strand braid group. For this, we need more topology.

3.8 The Bracket Polynomial and the Jones Polynomial

We now discuss the Jones polynomial. We shall construct the Jones polynomial by using the bracket state summation model [2]. The bracket polynomial, invariant under Reidmeister moves II and III, can be normalized to give an invariant of all three Reidemeister moves (see Fig. 3.3). This normalized invariant, with a change of variable, is the Jones polynomial [23, 24]. The Jones polynomial was originally discovered by a different method than the one given here.

The *bracket polynomial*, $< K > = < K > (A)$, assigns to each unoriented link diagram K a Laurent polynomial in the variable A, such that

1. If K and K' are regularly isotopic diagrams, then $< K > = < K' >$.
2. If $K \sqcup O$ denotes the disjoint union of K with an extra unknotted and unlinked component O (also called "loop" or "simple closed curve" or "Jordan curve"), then

$$< K \sqcup O > = \delta < K >,$$

where

$$\delta = -A^2 - A^{-2}.$$

3. $< K >$ satisfies the following formulas

$$< \chi > = A < \asymp > + A^{-1} <)(>,$$
$$< \overline{\chi} > = A^{-1} < \asymp > + A <)(>,$$

where the small diagrams represent parts of larger diagrams that are identical except at the site indicated in the bracket. We take the convention that the letter chi, χ, denotes a crossing where *the curved line is crossing over the straight segment*. The barred letter denotes the switch of this crossing, where *the curved line is undercrossing the straight segment*. See Fig. 3.9 for a graphic illustration of this relation, and an indication of the convention for choosing the labels A and A^{-1} at a given crossing.

It is easy to see that Properties 2 and 3 define the calculation of the bracket on arbitrary link diagrams. The choices of coefficients (A and A^{-1}) and the value of δ make the bracket invariant under the Reidemeister moves II and III. Thus Property 1 is a consequence of the other two properties.

In computing the bracket, one finds the following behavior under Reidemeister move I:

$$< \gamma > = -A^3 < \smile >$$

and

$$< \overline{\gamma} > = -A^{-3} < \smile >,$$

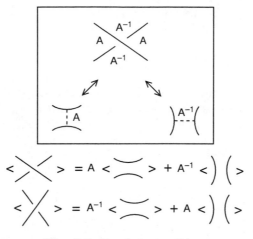

Fig. 3.9. Bracket smoothings

where γ denotes a curl of positive type as indicated in Fig. 3.10, and $\bar{\gamma}$ indicates a curl of negative type, as also seen in this figure. The type of a curl is the sign of the crossing when we orient it locally. Our convention of signs is also given in Fig. 3.10. Note that the type of a curl does not depend on the orientation we choose. The small arcs on the right-hand side of these formulas indicate the removal of the curl from the corresponding diagram.

The bracket is invariant under regular isotopy and can be normalized to an invariant of ambient isotopy by the definition

$$f_K(A) = (-A^3)^{-w(K)} < K > (A),$$

where we chose an orientation for K, and where $w(K)$ is the sum of the crossing signs of the oriented link K. $w(K)$ is called the *writhe* of K. The convention for crossing signs is shown in Fig. 3.10.

One useful consequence of these formulas is the following *switching formula*

$$A < \chi > -A^{-1} < \bar{\chi} >= (A^2 - A^{-2}) < \asymp > .$$

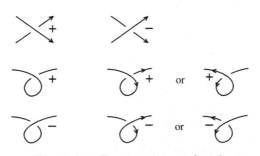

Fig. 3.10. Crossing signs and curls

Note that in these conventions, the A-smoothing of χ is $\asymp$, while the A-smoothing of $\overline{\chi}$ is $)($. Properly interpreted, the switching formula above says that you can switch a crossing and smooth it either way and obtain a three-diagram relation. This is useful since some computations will simplify quite quickly with the proper choices of switching and smoothing. Remember that it is necessary to keep track of the diagrams up to regular isotopy (the equivalence relation generated by the second and third Reidemeister moves). Here is an example: Figure 3.11 shows a trefoil diagram K, an unknot diagram U, and another unknot diagram U'. Applying the switching formula, we have

$$A^{-1} < K > -A < U >= (A^{-2} - A^2) < U' >,$$

$< U >= -A^3$, and $< U' >= (-A^{-3})^2 = A^{-6}$. Thus

$$A^{-1} < K > -A(-A^3) = (A^{-2} - A^2)A^{-6}.$$

Hence

$$A^{-1} < K > = -A^4 + A^{-8} - A^{-4}.$$

Thus

$$< K > = -A^5 - A^{-3} + A^{-7}.$$

This is the bracket polynomial of the trefoil diagram K.

Since the trefoil diagram K has writhe $w(K) = 3$, we have the normalized polynomial

$$f_K(A) = (-A^3)^{-3} < K >= -A^{-9}(-A^5 - A^{-3} + A^{-7}) = A^{-4} + A^{-12} - A^{-16}.$$

The bracket model for the Jones polynomial is quite useful both theoretically and in terms of practical computations. One of the neatest applications is to simply compute, as we have done, $f_K(A)$ for the trefoil knot K and determine that $f_K(A)$ is not equal to $f_K(A^{-1}) = f_{-K}(A)$. This shows that the trefoil is not ambient isotopic to its mirror image, a fact that is much harder to prove by classical methods.

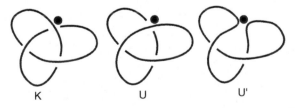

K   U   U'

Fig. 3.11. Trefoil and two relatives

3.8.1 The State Summation

In order to obtain a closed formula for the bracket, we now describe it as a state summation. Let K be any unoriented link diagram. Define a *state*, S, of K to be a choice of smoothing for each crossing of K. There are two choices for smoothing a given crossing, and thus there are 2^N states of a diagram with N crossings. In a state, we label each smoothing with A or A^{-1} according to the left–right convention discussed in Property 3 (see Fig. 3.9). The label is called a *vertex weight* of the state. There are two evaluations related to a state. The first one is the product of the vertex weights, denoted

$$< K|S > .$$

The second evaluation is the number of loops in the state S, denoted

$$||S|| .$$

Define the *state summation*, $< K >$, by the formula

$$< K > = \sum_S < K|S > \delta^{||S||-1} .$$

It follows from this definition that $< K >$ satisfies the equations

$$< \chi > = A < \asymp > + A^{-1} <)(> ,$$
$$< K \sqcup O > = \delta < K > ,$$
$$< O > = 1 .$$

The first equation expresses the fact that the entire set of states of a given diagram is the union, with respect to a given crossing, of those states with an A-type smoothing and those with an A^{-1}-type smoothing at that crossing. The second and the third equation are clear from the formula defining the state summation. Hence this state summation produces the bracket polynomial as we have described it at the beginning of the section.

Remark 6. By a change of variables, one obtains the original Jones polynomial, $V_K(t)$, for oriented knots and links from the normalized bracket:

$$V_K(t) = f_K(t^{-\frac{1}{4}}) .$$

Remark 7. The bracket polynomial provides a connection between knot theory and physics, in that the state summation expression for it exhibits it as a generalized partition function defined on the knot diagram. Partition functions are ubiquitous in statistical mechanics, where they express the summation over all states of the physical system of probability weighting functions for the individual states. Such physical partition functions contain large amounts of information about the corresponding physical system. Some of this information is directly present in the properties of the function, such as the location of critical points and phase transition. Some of the information can be obtained by differentiating the partition function, or performing other mathematical operations on it.

There is much more in this connection with statistical mechanics in that the local weights in a partition function are often expressed in terms of solutions to a matrix equation, the Yang–Baxter equation, that turns out to fit perfectly invariance under the third Reidemeister move. As a result, there are many ways to define partition functions of knot diagrams that give rise to invariants of knots and links. The subject is intertwined with the algebraic structure of Hopf algebras and quantum groups, useful for producing systematic solutions to the Yang–Baxter equation. In fact Hopf algebras are deeply connected with the problem of constructing invariants of three-dimensional manifolds in relation to invariants of knots. We have chosen, in this lecture, to not discuss the details of these approaches, but rather to proceed to Vassiliev invariants and the relationships with Witten's functional integral. The reader is referred to [2, 3, 23–34] for more information about relationships of knot theory with statistical mechanics, Hopf algebras, and quantum groups. For topology, the key point is that Lie algebras can be used to construct invariants of knots and links.

3.8.2 Quantum Computation of the Jones Polynomial

Can the invariants of knots and links such as the Jones polynomial be configured as quantum computers? This is an important question because the algorithms to compute the Jones polynomial are known to be *NP*-hard, and so corresponding quantum algorithms may shed light on the relationship of this level of computational complexity with quantum computing. Such models can be formulated in terms of the Yang–Baxter equation [2, 3, 25, 35]. The next paragraph explains how this comes about.

In Fig. 3.12, we indicate how topological braiding plus maxima (caps) and minima (cups) can be used to configure the diagram of a knot or link. This also can be translated into algebra by the association of a Yang–Baxter matrix R (not necessarily the R of the previous sections) to each crossing and other matrices to the maxima and minima. There are models of very effective invariants of knots and links such as the Jones polynomial that can be put into this form [35]. In this way of looking at things, the knot diagram can be viewed as a picture, with time as the vertical dimension, of particles arising from the vacuum, interacting (in a two-dimensional space) and finally annihilating one another. The invariant takes the form of an amplitude for this process that is computed through the association of the Yang–Baxter solution R as the scattering matrix at the crossings and the minima and maxima as creation and annihilation operators. Thus we can write the amplitude in the form

$$Z_K = \langle CUP|M|CAP \rangle,$$

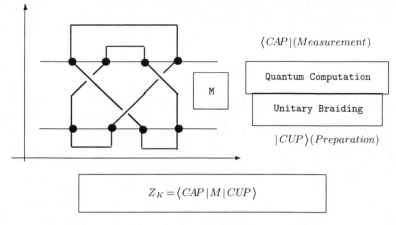

$$Z_K = \langle CAP \,|\, M \,|\, CUP \rangle$$

Fig. 3.12. A knot quantum computer

where $\langle CUP|$ denotes the composition of cups, M is the composition of elementary braiding matrices, and $|CAP\rangle$ is the composition of caps. We regard $\langle CUP|$ as the preparation of this state, and $|CAP\rangle$ as the measurement of this state. In order to view Z_K as a quantum computation, M must be a unitary operator. This is the case when the R-matrices (the solutions to the Yang–Baxter equation used in the model) are unitary. Each R-matrix is viewed as a quantum gate (or possibly a composition of quantum gates), and the vacuum–vacuum diagram for the knot is interpreted as a quantum computer. This quantum computer will probabilistically (via quantum amplitudes) compute the values of the states in the state sum for Z_K.

We should remark, however, that it is not necessary that the invariant be modeled via solutions to the Yang–Baxter equation. One can use unitary representations of the braid group that are constructed in other ways. In fact, the presently successful quantum algorithms for computing knot invariants indeed use such representations of the braid group, and we shall see this below. Nevertheless, it is useful to point out this analogy between the structure of the knot invariants and quantum computation.

Quantum algorithms for computing the Jones polynomial have been discussed elsewhere. See [19, 35–39]. Here, as an example, we give a local unitary representation that can be used to compute the Jones polynomial for closures of 3-braids. We analyze this representation by making explicit how the bracket polynomial is computed from it and showing how the quantum computation devolves to finding the trace of a unitary transformation.

The idea behind the construction of this representation depends upon the algebra generated by two single-qubit density matrices (ket-bras). Let $|v\rangle$ and $|w\rangle$ be two qubits in V, a complex vector space of dimension two over the complex numbers. Let $P = |v\rangle\langle v|$ and $Q = |w\rangle\langle w|$ be the corresponding ket-bras. Note that

$$P^2 = |v|^2 P \,,$$
$$Q^2 = |w|^2 Q \,,$$
$$PQP = |\langle v|w\rangle|^2 P \,,$$
$$QPQ = |\langle v|w\rangle|^2 Q \,.$$

P and Q generate a representation of the Temperley–Lieb algebra (see Sect. 3.6 of the present paper). One can adjust parameters to make a representation of the three-strand braid group in the form

$$s_1 \longmapsto rP + sI \,,$$
$$s_2 \longmapsto tQ + uI \,,$$

where I is the identity mapping on V, and r, s, t, u are suitably chosen scalars. In the following, we use this method to adjust such a representation so that it is unitary. Note also that this is a local unitary representation of B_3 to $U(2)$. We leave it as an exercise for the reader to verify that it fits into our general classification of such representations as given in Sect. 3.4.

The representation depends on two symmetric but non-unitary matrices U_1 and U_2 with

$$U_1 = \begin{pmatrix} d & 0 \\ 0 & 0 \end{pmatrix} = d|w\rangle\langle w|$$

and

$$U_2 = \begin{pmatrix} d^{-1} & \sqrt{1 - d^{-2}} \\ \sqrt{1 - d^{-2}} & d - d^{-1} \end{pmatrix} = d|v\rangle\langle v| \,,$$

where $w = (1, 0)$ and $v = (d^{-1}, \sqrt{1 - d^{-2}})$, assuming the entries of v are real. Note that $U_1^2 = dU_1$ and $U_2^2 = dU_1$. Moreover, $U_1 U_2 U_1 = U_1$ and $U_2 U_1 U_2 = U_1$. This is an example of a specific representation of the Temperley–Lieb algebra [2, 35]. The desired representation of the Artin braid group is given on the two braid generators for the three-strand braid group by the equations:

$$\Phi(s_1) = AI + A^{-1}U_1 \,,$$
$$\Phi(s_2) = AI + A^{-1}U_2 \,.$$

Here I denotes the 2×2 identity matrix. For any A with $d = -A^2 - A^{-2}$ these formulas define a representation of the braid group. With $A = e^{i\theta}$, we have $d = -2cos(2\theta)$. We find a specific range of angles θ in the following disjoint union of angular intervals

$$\theta \in [0, \pi/6] \sqcup [\pi/3, 2\pi/3] \sqcup [5\pi/6, 7\pi/6] \sqcup [4\pi/3, 5\pi/3] \sqcup [11\pi/6, 2\pi]$$

that give unitary representations of the three-strand braid group. Thus a specialization of a more general represention of the braid group gives rise to a continuous family of unitary representations of the braid group.

Note that the traces of these matrices are given by the formulas $\text{tr}(U_1) = \text{tr}(U_2) = d$ while $\text{tr}(U_1 U_2) = \text{tr}(U_2 U_1) = 1$. If b is any braid, let $I(b)$ denote the sum of the exponents in the braid word that expresses b. For b a three-strand braid, it follows that

$$\Phi(b) = A^{I(b)} I + \Pi(b),$$

where I is the 2×2-identity matrix, and $\Pi(b)$ is a sum of products in the Temperley–Lieb algebra involving U_1 and U_2. Since the Temperley–Lieb algebra in this dimension is generated by I, U_1, U_2, $U_1 U_2$, and $U_2 U_1$, it follows that the value of the bracket polynomial of the closure of the braid b, denoted $< \overline{b} >$, can be calculated directly from the trace of this representation, except for the part involving the identity matrix. The result is the equation

$$< \overline{b} >= A^{I(b)} d^2 + \text{tr}(\Pi(b)),$$

where $\overline{b}$ denotes the standard braid closure of b, and the sharp brackets denote the bracket polynomial. From this, we see at once that

$$< \overline{b} >= \text{tr}(\Phi(b)) + A^{I(b)}(d^2 - 2).$$

It follows from this calculation that the question of computing the bracket polynomial for the closure of the three-strand braid b is mathematically equivalent to the problem of computing the trace of the unitary matrix $\Phi(b)$.

3.8.3 The Hadamard Test

In order to (quantum) compute the trace of a unitary matrix U, one can use the *Hadamard test* to obtain the diagonal matrix elements $\langle \psi | U | \psi \rangle$ of U. The trace is then the sum of these matrix elements as $|\psi\rangle$ runs over an orthonormal basis for the vector space. We first obtain

$$\frac{1}{2} + \frac{1}{2}\text{Re}\langle \psi | U | \psi \rangle$$

as an expectation by applying the Hadamard gate H

$$H|0\rangle = \frac{1}{\sqrt{2}}(|0\rangle + |1\rangle)$$

$$H|1\rangle = \frac{1}{\sqrt{2}}(|0\rangle - |1\rangle)$$

to the first qubit of

$$C_U \circ (H \otimes 1)|0\rangle|\psi\rangle = \frac{1}{\sqrt{2}}(|0\rangle \otimes |\psi\rangle + |1\rangle \otimes U|\psi\rangle).$$

Here C_U denotes controlled U, acting as U when the control bit is $|1\rangle$ and the identity mapping when the control bit is $|0\rangle$. We measure the expectation for the first qubit $|0\rangle$ of the resulting state

$$\frac{1}{\sqrt{2}}(H|0\rangle \otimes |\psi\rangle + H|1\rangle \otimes U|\psi\rangle)$$

$$= \frac{1}{2}((|0\rangle + |1\rangle) \otimes |\psi\rangle + (|0\rangle - |1\rangle) \otimes U|\psi\rangle)$$

$$= \frac{1}{2}(|0\rangle \otimes (|\psi\rangle + U|\psi\rangle) + |1\rangle \otimes (|\psi\rangle - U|\psi\rangle)) .$$

This expectation is

$$\frac{1}{2}((\langle\psi| + \langle\psi|U^{\dagger})(|\psi\rangle + U|\psi\rangle)) = \frac{1}{2} + \frac{1}{2}\mathrm{Re}\langle\psi|U|\psi\rangle .$$

The imaginary part is obtained by applying the same procedure to

$$\frac{1}{\sqrt{2}}(|0\rangle \otimes |\psi\rangle - i|1\rangle \otimes U|\psi\rangle) .$$

This is the method used in [36], and the reader may wish to contemplate its efficiency in the context of this simple model. Note that the Hadamard test enables this quantum computation to estimate the trace of any unitary matrix U by repeated trials that estimate individual matrix entries $\langle\psi|U|\psi\rangle$.

3.9 Quantum Topology, Cobordism Categories, Temperley–Lieb Algebra and Topological Quantum Field Theory

The purpose of this section is to discuss the general idea behind topological quantum field theory and to illustrate its application to basic quantum mechanics and quantum mechanical formalism. It is useful in this regard to have available the concept of *category*, and we shall begin the section by discussing this far-reaching mathematical concept.

Definition 1. *A category Cat consists in two related collections:*

1. *Obj(Cat), the objects of Cat, and*
2. *Morph(Cat), the morphisms of Cat.*

satisfying the following axioms:

1. *Each morphism f is associated to two objects of Cat, the domain of f and the codomain of f. Letting A denote the domain of f and B denote the codomain of f, it is customary to denote the morphism f by the arrow notation $f : A \longrightarrow B$.*
2. *Given $f : A \longrightarrow B$ and $g : B \longrightarrow C$, where A, B, and C are objects of Cat, then there exists an associated morphism $g \circ f : A \longrightarrow C$ called the composition of f and g.*

3. *To each object A of Cat, there is a unique identity morphism $1_A : A \longrightarrow A$ such that $1_A \circ f = f$ for any morphism f with codomain A, and $g \circ 1_A = g$ for any morphism g with domain A.*
4. *Given three morphisms $f : A \longrightarrow B$, $g : B \longrightarrow C$, and $h : C \longrightarrow D$, then composition is associative. That is*

$$(h \circ g) \circ f = h \circ (g \circ f).$$

If Cat_1 and Cat_2 are two categories, then a *functor* $F : Cat_1 \longrightarrow Cat_2$ consists in functions $F_O : Obj(Cat_1) \longrightarrow Obj(Cat_2)$ and $F_M : Morph(Cat_1) \longrightarrow Morph(Cat_2)$ such that identity morphisms and composition of morphisms are preserved under these mappings. That is (writing just F for F_O and F_M),

1. $F(1_A) = 1_{F(A)}$,
2. $F(f : A \longrightarrow B) = F(f) : F(A) \longrightarrow F(B)$,
3. $F(g \circ f) = F(g) \circ F(f)$.

A functor $F : Cat_1 \longrightarrow Cat_2$ is a structure-preserving mapping from one category to another. It is often convenient to think of the image of the functor F as an *interpretation* of the first category in terms of the second. We shall use this terminology below and sometimes refer to an interpretation without specifying all the details of the functor that describes it.

The notion of category is a broad mathematical concept, encompassing many fields of mathematics. Thus one has the category of sets where the objects are sets (collections) and the morphisms are mappings between sets. One has the category of topological spaces where the objects are spaces and the morphisms are continuous mappings of topological spaces. One has the category of groups where the objects are groups and the morphisms are homomorphisms of groups. Functors are structure-preserving mappings from one category to another. For example, the fundamental group is a functor from the category of topological spaces with base point, to the category of groups. In all the examples mentioned so far, the morphisms in the category are restrictions of mappings in the category of sets, but this is not necessarily the case. For example, any group G can be regarded as a category, $Cat(G)$, with one object $*$. The morphisms from $*$ to itself are the elements of the group and composition is group multiplication. In this example, the object has no internal structure, and all the complexity of the category is in the morphisms.

The Artin braid group B_n can be regarded as a category whose single object is an ordered row of points $[n] = \{1, 2, 3, \ldots, n\}$. The morphisms are the braids themselves, and the composition is the multiplication of the braids. The ordered row of points is interpreted as the starting and ending row of points at the bottom and the top of the braid. In the case of the braid category, the morphisms have both external and internal structure. Each morphism produces a permutation of the ordered row of points (corresponding to the beginning and ending points of the individual braid strands), and weaving of the braid is extra structure beyond the object that is its domain and codomain. Finally, for this example, we can take all the braid groups B_n (n a positive integer) under the wing of a single category, $Cat(B)$, whose objects are all ordered rows of points $[n]$, and whose morphisms are of the form $b : [n] \longrightarrow [n]$ where b is a braid in B_n. The reader may wish to have morphisms between objects with different n. We will have this shortly in the Temperley–Lieb category and in the category of tangles.

The n-*Cobordism Category*, $Cob[n]$, has as its objects smooth manifolds of dimension n, and as its morphisms, smooth manifolds M^{n+1} of dimension $n + 1$ with a partition of the boundary, ∂M^{n+1}, into two collections of n-manifolds that we denote by $L(M^{n+1})$ and $R(M^{n+1})$. We regard M^{n+1} as a morphism from $L(M^{n+1})$ to $R(M^{n+1})$

$$M^{n+1} : L(M^{n+1}) \longrightarrow R(M^{n+1}).$$

As we shall see, these cobordism categories are highly significant for quantum mechanics, and the simplest one, $Cob[0]$ is directly related to the Dirac notation of bras and kets and to the Temperley–Lieb algebra. We shall concentrate in this section on these cobordism categories, and their relationships with quantum mechanics.

One can choose to consider either oriented or non-oriented manifolds, and within unoriented manifolds, there are those that are orientable and those that are not orientable. In this section, we will implicitly discuss only orientable manifolds, but we shall not specify an orientation. In the next section, with the standard definition of topological quantum field theory, the manifolds will be oriented. The definitions of the cobordism categories for oriented manifolds go over mutatis mutandis.

Lets begin with $Cob[0]$. Zero-dimensional manifolds are just collections of points. The simplest zero-dimensional manifold is a single point p. We take p to be an object of this category and also $*$, where $*$ denotes the empty manifold (i.e., the empty set in the category of manifolds). The object $*$ occurs in $Cob[n]$ for every n, since it is possible that either the left set or the right set of a morphism is empty. A line segment S with boundary points p and q is a morphism from p to q.

$$S : p \longrightarrow q.$$

In Fig. 3.13, we have illustrated the morphism from p to p. The simplest convention for this category is to take this morphism to be the identity. Thus if we look at the subcategory of $Cob[0]$ whose only object is p, then the only morphism is the identity morphism. Two points occur as the boundary of an interval. The reader will note that $Cob[0]$ and the usual arrow notation for morphisms are very closely related. This is a place where notation and mathematical structure share common elements. In general, the objects of $Cob[0]$ consist in the empty object $*$ and non-empty rows of points, symbolized by

$$p \otimes p \otimes \cdots \otimes p \otimes p.$$

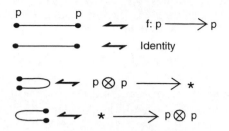

Fig. 3.13. Elementary cobordisms

Figure 3.13 also contains a morphism

$$p \otimes p \longrightarrow *$$

and the morphism

$$* \longrightarrow p \otimes p.$$

The first represents a cobordism of two points to the empty set (via the bounding curved interval). The second represents a cobordism from the empty set to two points.

In Fig. 3.14, we have indicated more morphisms in $Cob[0]$, and we have named the morphisms just discussed as

$$|\Omega\rangle : p \otimes p \longrightarrow *,$$
$$\langle\Theta| : * \longrightarrow p \otimes p.$$

The point to notice is that the usual conventions for handling Dirac bra–kets are essentially the same as the composition rules in this topological category. Thus, in Fig. 3.14, we have that

$$\langle\Theta| \circ |\Omega\rangle = \langle\Theta|\Omega\rangle : * \longrightarrow *,$$

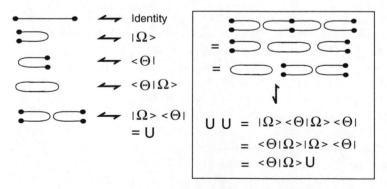

Fig. 3.14. Bras, kets, and projectors

which represents a cobordism from the empty manifold to itself. This cobordism is topologically a circle, and in the Dirac formalism is interpreted as a scalar. In order to interpret the notion of scalar, we would have to map the cobordism category to the category of vector spaces and linear mappings. We shall discuss this after describing the similarities with quantum mechanical formalism. Nevertheless, the reader should note that if V is a vector space over the complex numbers C, then a linear mapping from C to C is determined by the image of 1, and hence is characterized by the scalar that is the image of 1. In this sense, a mapping $C \longrightarrow C$ can be regarded as a possible image in vector spaces of the abstract structure $\langle \Theta | \Omega \rangle : * \longrightarrow *$. It is therefore assumed that in $Cob[0]$, the composition with the morphism $\langle \Theta | \Omega \rangle$ commutes with any other morphism. In that way, $\langle \Theta | \Omega \rangle$ behaves like a scalar in the cobordism category. In general, an $n + 1$ manifold without boundary behaves as a scalar in $Cob[n]$, and if a manifold M^{n+1} can be written as a union of two submanifolds L^{n+1} and R^{n+1} so that that an n-manifold W^n is their common boundary:

$$M^{n+1} = L^{n+1} \cup R^{n+1}$$

with

$$L^{n+1} \cap R^{n+1} = W^n$$

then, we can write

$$\langle M^{n+1} \rangle = \langle L^{n+1} \cup R^{n+1} \rangle = \langle L^{n+1} | R^{n+1} \rangle ,$$

and $\langle M^{n+1} \rangle$ will be a scalar (morphism that commutes with all other morphisms) in the category $Cob[n]$.

Getting back to the contents of Fig. 3.14, note how the zero-dimensional cobordism category has structural parallels to the Dirac ket–bra formalism

$$U = | \Omega \rangle \langle \Theta |$$
$$UU = | \Omega \rangle \langle \Theta | \Omega \rangle \langle \Theta | = \langle \Theta | \Omega \rangle | \Omega \rangle \langle \Theta | = \langle \Theta | \Omega \rangle U .$$

In the cobordism category, the bra–ket and ket–bra formalism is seen as patterns of connection of the one manifold that realize the cobordisms.

Now, Fig. 3.15 illustrates a morphism S in $Cob[0]$ that requires two crossed line segments for its planar representation. Thus S can be regarded as a nontrivial permutation, and $S^2 = I$, where I denotes the identity morphisms for a two-point row. From this example, it is clear that $Cob[0]$ contains the structure of all the symmetric groups and more. In fact, if we take the subcategory of $Cob[0]$ consisting of all morphisms from $[n]$ to $[n]$ for a fixed positive integer n, then this gives the well-known *Brauer algebra* (see [40]) extending the symmetric group by allowing any connections among the points in the two rows. In this sense, one could call $Cob[0]$ the *Brauer category*. We shall return to this point of view later.

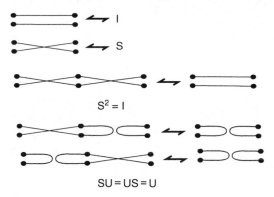

Fig. 3.15. Permutations

In this section, we shall be concentrating on the part of $Cob[0]$ that does not involve permutations. This part can be characterized by those morphisms that can be represented by planar diagrams without crossings between any of the line segments (the one manifolds). We shall call this crossingless sub-category of $Cob[0]$ the *Temperley–Lieb Category* and denote it by $CatTL$. In $CatTL$, we have the subcategory $TL[n]$ whose only objects are the row of n points and the empty object $*$, and whose morphisms can all be represented by configurations that embed in the plane as in the morphisms P and Q in Fig. 3.16. Note that with the empty object $*$, the morphism whose diagram is a single loop appears in $TL[n]$ and is taken to commute with all other morphisms.

The *Temperley–Lieb Algebra*, $AlgTL[n]$, is generated by the morphisms in $TL[n]$ that go from $[n]$ to itself. Up to multiplication by the loop, the product (composition) of two such morphisms is another flat morphism from $[n]$ to itself. For algebraic purposes, the loop $* \longrightarrow *$ is taken to be a scalar algebraic variable δ that commutes with all elements in the algebra. Thus the equation

$$UU = \langle \Theta | \Omega \rangle U$$

becomes

$$UU = \delta U$$

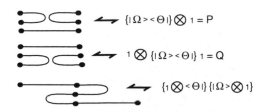

Fig. 3.16. Projectors in tensor lines and elementary topology

in the algebra. In the algebra, we are allowed to add morphisms formally, and this addition is taken to be commutative. Initially the algebra is taken with coefficients in the integers, but a different commutative ring of coefficients can be chosen and the value of the loop may be taken in this ring. For example, for quantum mechanical applications, it is natural to work over the complex numbers. The multiplicative structure of $AlgTL[n]$ can be described by generators and relations as follows: Let I_n denote the identity morphism from $[n]$ to $[n]$. Let U_i denote the morphism from $[n]$ to $[n]$ that connects k with k for $k < i$ and $k > i + 1$ from one row to the other and connects i to $i + 1$ in each row. Then the algebra $AlgTL[n]$ is generated by $\{I_n, U_1, U_2, \cdots, U_{n-1}\}$ with relations

$$U_i^2 = \delta U_i \,,$$
$$U_i U_{i+1} U_i = U_i \,,$$
$$U_i U_j = U_j U_i : |i - j| > 1 \,.$$

These relations are illustrated for three strands in Fig. 3.16. We leave the commuting relation for the reader to draw in the case where n is four or greater. For a proof that these are indeed all the relations, see [41].

Figures 3.16 and 3.17 indicate how the zero-dimensional cobordism category contains structure that goes well beyond the usual Dirac formalism. By tensoring the ket–bra on one side or another by identity morphisms, we obtain the beginnings of the Temperley–Lieb algebra and the Temperley–Lieb category. Thus Fig. 3.17 illustrates the morphisms P and Q obtained by such tensoring, and the relation $PQP = P$ which is the same as $U_1 U_2 U_1 = U_1$.

Note the composition at the bottom of the Fig. 3.17. Here we see a composition of the identity tensored with a ket, followed by a bra tensored with the identity. The diagrammatic for this association involves "straightening" the curved structure of the morphism to a straight line. In Fig. 3.18, we have elaborated this situation even further, pointing out that in this category each

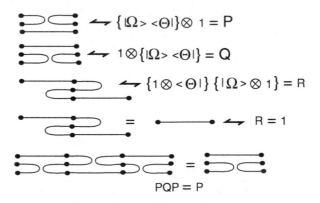

Fig. 3.17. The basic Temperley–Lieb relation

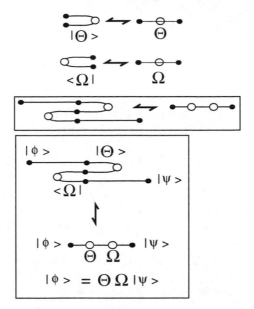

Fig. 3.18. The key to teleportation

of the morphisms $\langle \Theta |$ and $| \Omega \rangle$ can be seen, by straightening, as mappings from the generating object to itself. We have denoted these corresponding morphisms by Θ and Ω, respectively. In this way, there is a correspondence between morphisms $p \otimes p \longrightarrow *$ and morphism $p \longrightarrow p$.

In Fig. 3.18, we have illustrated the generalization of the straightening procedure of Fig. 3.17. In Fig. 3.17, the straightening occurs because the connection structure in the morphism of $Cob[0]$ does not depend on the wandering of curves in diagrams for the morphisms in that category. Nevertheless, one can envisage a more complex interpretation of the morphisms where each one manifold (line segment) has a label, and a multiplicity of morphisms can correspond to a single-line segment. This is exactly what we expect in interpretations. For example, we can interpret the line segment $[1] \longrightarrow [1]$ as a mapping from a vector space V to itself. Then $[1] \longrightarrow [1]$ is the diagrammatic abstraction for $V \longrightarrow V$, and there are many instances of linear mappings from V to V.

At the vector space level, there is a duality between mappings $V \otimes V \longrightarrow C$ and linear maps $V \longrightarrow V$. Specifically, let

$$\{|0\rangle, \cdots, |m\rangle\}$$

be a basis for V. Then $\Theta : V \longrightarrow V$ is determined by

$$\Theta |i\rangle = \Theta_{ij} |j\rangle$$

(where we have used the Einstein summation convention on the repeated index j), which corresponds to the bra

$$\langle \Theta | : V \otimes V \longrightarrow C$$

defined by

$$\langle \Theta | ij \rangle = \Theta_{ij} \, .$$

Given $\langle \Theta | : V \otimes V \longrightarrow C$, we associate $\Theta : V \longrightarrow V$ in this way.

Comparing with the diagrammatic for the category $Cob[0]$, we say that $\Theta : V \longrightarrow V$ is obtained by *straightening* the mapping

$$\langle \Theta | : V \otimes V \longrightarrow C.$$

Note that in this interpretation, the bras and kets are defined relative to the tensor product of V with itself and $[2]$ is interpreted as $V \otimes V$. If we interpret $[2]$ as a single vector space W, then the usual formalisms of bras and kets still pass over from the cobordism category.

Figure 3.18 illustrates the straightening of $|\Theta\rangle$ and $\langle \Omega |$, and the straightening of a composition of these applied to $|\psi\rangle$, resulting in $|\phi\rangle$. In the left-hand part of the bottom of Fig. 3.18, we illustrate the preparation of the tensor product $|\Theta\rangle \otimes |\psi\rangle$ followed by a successful measurement by $\langle \Omega |$ in the second two tensor factors. The resulting single qubit state, as seen by straightening, is $|\phi\rangle = \Theta \circ \Omega |\psi\rangle$.

From this, we see that it is possible to reversibly, indeed unitarily, transform a state $|\psi\rangle$ via a combination of preparation and measurement just so long as the straightenings of the preparation and measurement (Θ and Ω) are each invertible (unitary). This is the key to teleportation [42–44]. In the standard teleportation procedure, one chooses the preparation Θ to be (up to normalization) the two-dimensional identity matrix so that $|\theta\rangle = |00\rangle + |11\rangle$. If the successful measurement Ω is also the identity, then the transmitted state $|\phi\rangle$ will be equal to $|\psi\rangle$. In general, we will have $|\phi\rangle = \Omega |\psi\rangle$. One can then choose a basis of measurements $|\Omega\rangle$, each corresponding to a unitary transformation Ω so that the recipient of the transmission can rotate the result by the inverse of Ω to reconstitute $|\psi\rangle$ if given the requisite information. This is the basic design of the teleportation procedure.

There is much more to say about the category $Cob[0]$ and its relationship with quantum mechanics. We will stop here, and invite the reader to explore further. Later in the text, we shall use these ideas in formulating our representations of the braid group. For now, we point out how things look as we move upward to $Cob[n]$ for $n > 0$. In Fig. 3.19, we show typical cobordisms (morphisms) in $Cob[1]$ from two circles to one circle and from one circle to two circles. These are often called "pairs of pants". Their composition is a surface of genus one seen as a morphism from two circles to two circles. The bottom of the figure indicates a ket–bra in this dimension in the form of a mapping from one circle to one circle as a composition of a cobordism of a circle to the empty set and a cobordism from the empty set to a circle (circles bounding disks). As we go to higher dimensions, the structure of cobordisms becomes more interesting and more complicated. It is remarkable that there is so much structure in the lowest dimensions of these categories.

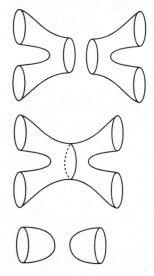

Fig. 3.19. Corbordisms of 1-manifolds are surfaces

3.10 Braiding and Topological Quantum Field Theory

The purpose of this section is to discuss in a very general way how braiding is related to topological quantum field theory. In the section to follow, we will use the Temperley–Lieb recoupling theory to produce specific unitary representations of the Artin braid group.

The ideas in the subject of topological quantum field theory (TQFT) are well expressed in the book [45] by Michael Atiyah and the paper [46] by Edward Witten. Here is Atiyah's definition:

Definition 2. *A TQFT in dimension d is a functor $Z(\Sigma)$ from the cobordism category Cob[d] to the category Vect of vector spaces and linear mappings which assigns*

1. *a finite-dimensional vector space $Z(\Sigma)$ to each compact, oriented d-dimensional manifold Σ,*
2. *a vector $Z(Y) \in Z(\Sigma)$ for each compact, oriented $(d+1)$-dimensional manifold Y with boundary Σ.*
3. *a linear mapping $Z(Y) : Z(\Sigma_1) \longrightarrow Z(\Sigma_2)$ when Y is a $(d+1)$-manifold that is a cobordism between Σ_1 and Σ_2 (whence the boundary of Y is the union of Σ_1 and $-\Sigma_2$).*

The functor satisfies the following axioms.

1. *$Z(\Sigma^\dagger) = Z(\Sigma)^\dagger$ where $\Sigma^\dagger$ denotes the manifold Σ with the opposite orientation, and $Z(\Sigma)^\dagger$ is the dual vector space.*
2. *$Z(\Sigma_1 \cup \Sigma_2) = Z(\Sigma_1) \otimes Z(\Sigma_2)$, where $\cup$ denotes disjoint union.*

3. If Y_1 is a cobordism from Σ_1 to Σ_2, Y_2 is a cobordism from Σ_2 to Σ_3 and Y is the composite cobordism $Y = Y_1 \cup_{\Sigma_2} Y_2$, then

$$Z(Y) = Z(Y_2) \circ Z(Y_1) : Z(\Sigma_1) \longrightarrow Z(\Sigma_2)$$

is the composite of the corresponding linear mappings.

4. $Z(\phi) = C$ (C denotes the complex numbers) for the empty manifold ϕ.

5. With $\Sigma \times I$ (where I denotes the unit interval) denoting the identity cobordism from Σ to Σ, $Z(\Sigma \times I)$ is the identity mapping on $Z(\Sigma)$.

Note that in this view a TQFT is basically a functor from the cobordism categories defined in the last section to vector spaces over the complex numbers. We have already seen that in the lowest dimensional case of cobordisms of zero-dimensional manifolds, this gives rise to a rich structure related to quantum mechanics and quantum information theory. The remarkable fact is that the case of three dimensions is also related to quantum theory and to the lower dimensional versions of the TQFT. This gives a significant way to think about three manifold invariants in terms of lower dimensional patterns of interaction. Here follows a brief description.

Regard the three manifold as a union of two handlebodies with boundary an orientable surface S_g of genus g. The surface is divided up into trinions as illustrated in Fig. 3.20. A *trinion* is a surface with boundary that is topologically equivalent to a sphere with three punctures. The trinion constitutes in itself a cobordism in $Cob[1]$ from two circles to a single circle, or from a single circle to two circles, or from three circles to the empty set. The *pattern* of a trinion is a trivalent graphical vertex, as illustrated in Fig. 3.21. In that

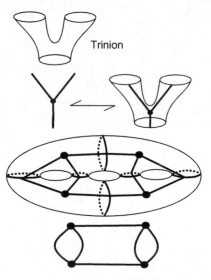

Fig. 3.20. Decomposition of a surface into trinions

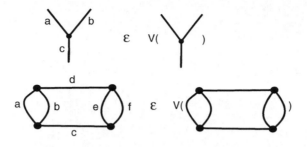

Fig. 3.21. Trivalent vectors

figure, we show the trivalent vertex graphical pattern drawn on the surface of the trinion, forming a graphical pattern for this combordism. It should be clear from this figure that any cobordism in $Cob[1]$ can be diagrammed by a trivalent graph, so that the category of trivalent graphs (as morphisms from ordered sets of points to ordered sets of points) has an image in the category of cobordisms of compact one-dimensional manifolds. Given a surface S (possibly with boundary) and a decomposition of that surface into trinions, we associate to it a trivalent graph $G(S,t)$, where t denotes the particular trinion decomposition.

In this correspondence, distinct graphs can correspond to topologically identical cobordisms of circles, as illustrated in Fig. 3.22. It turns out that the graphical structure is important and that it is extraordinarily useful to articulate transformations between the graphs that correspond to the homeomorphisms of the corresponding surfaces. The beginning of this structure is indicated in the bottom part of Fig. 3.22.

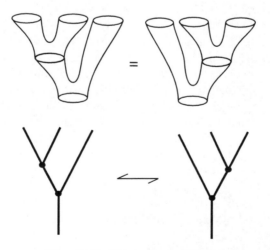

Fig. 3.22. Trinion associativity

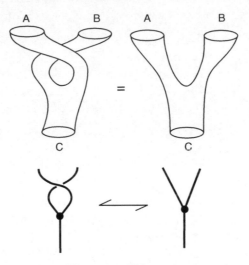

Fig. 3.23. Tube twist

In Fig. 3.23, we illustrate another feature of the relationship between surfaces and graphs. At the top of the figure, we indicate a homeomorphism between a twisted trinion and a standard trinion. The homeomorphism leaves the ends of the trinion (denoted A, B, and C) fixed while undoing the internal twist. This can be accomplished as an ambient isotopy of the embeddings in three-dimensional space that are indicated by this figure. Below this isotopy, we indicate the corresponding graphs. In the graph category, there will have to be a transformation between a braided and an unbraided trivalent vertex that corresponds to this homeomorphism.

From the point of view that we shall take in this paper, the key to the mathematical structure of three-dimensional TQFT lies in the trivalent graphs, including the braiding of graphical arcs. We can think of these braided graphs as representing idealized Feynman diagrams, with the trivalent vertex as the basic particle interaction vertex, and the braiding of lines representing an interaction resulting from an exchange of particles. In this view, one thinks of the particles as moving in a two-dimensional medium, and the diagrams of braiding and trivalent vertex interactions as indications of the temporal events in the system, with time indicated in the direction of the morphisms in the category. Adding such graphs to the category of knots and links is an extension of the *tangle category* where one has already extended braids to allow any embedding of strands and circles that start in n-ordered points and end in m-ordered points. The tangle category includes the braid category and the Temperley–Lieb category. Both are included in the category of braided trivalent graphs.

Thinking of the basic trivalent vertex as the form of a particle interaction, there will be a set of particle states that can label each arc incident to the vertex. In Fig. 3.21, we illustrate the labeling of the trivalent graphs by such particle states. In the next two sections, we will see specific rules for labeling such states. Here it suffices to note that there will be some restrictions on these labels, so that a trivalent vertex has a set of possible labelings. Similarly, any trivalent graph will have a set of admissible labelings. These are the possible particle processes that this graph can support. We take the set of admissible labelings of a given graph G as a basis for a vector space $V(G)$ over the complex numbers. This vector space is the space of *processes* associated with the graph G. Given a surface S and a decomposition t of the surface into trinions, we have the associated graph $G(S,t)$ and hence a vector space of processes $V(G(S,t))$. It is desirable to have this vector space independent of the particular decomposition into trinions. If this can be accomplished, then the set of vector spaces and linear mappings associated to the surfaces can constitute a functor from the category of cobordisms of one manifold to vector spaces, and hence gives rise to a one-dimensional topological quantum field theory. To this end, we need some properties of the particle interactions that will be described below.

A *spin network* is, by definition, a labeled trivalent graph in a category of graphs that satisfy the properties outlined in the previous paragraph. We shall detail the requirements below.

The simplest case of this idea is C. N. Yang's original interpretation of the Yang–Baxter equation [47]. Yang articulated a quantum field theory in one dimension of space and one dimension of time in which the R-matrix giving the scattering amplitudes for an interaction of two particles whose (let us say) spins corresponded to the matrix indices so that R_{ab}^{cd} is the amplitude for particles of spin a and spin b to interact and produce particles of spin c and d. Since these interactions are between particles in a line, one takes the convention that the particle with spin a is to the left of the particle with spin b, and the particle with spin c is to the left of the particle with spin d. If one follows the concatenation of such interactions, then there is an underlying permutation that is obtained by following strands from the bottom to the top of the diagram (thinking of time as moving up the page). Yang designed the Yang–Baxter equation for R so that *the amplitudes for a composite process depend only on the underlying permutation corresponding to the process and not on the individual sequences of interactions.*

In taking over the Yang–Baxter equation for topological purposes, we can use the same interpretation, but think of the diagrams with their under- and over-crossings as modeling events in a spacetime with two dimensions of space and one dimension of time. The extra spatial dimension is taken in displacing the woven strands perpendicular to the page and allows to use braiding operators R and R^{-1} as scattering matrices. Taking this picture to heart, one can add other particle properties to the idealized theory. In particular, one can add fusion and creation vertices where in fusion two particles interact to become a single particle and in creation one particle changes (decays)

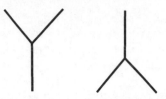

Fig. 3.24. Creation and fusion

into two particles. These are the trivalent vertices discussed above. Matrix elements corresponding to trivalent vertices can represent these interactions (see Fig. 3.24).

Once one introduces trivalent vertices for fusion and creation, there is the question how these interactions will behave in respect to the braiding operators. There will be a matrix expression for the compositions of braiding and fusion or creation as indicated in Fig. 3.25. Here we will restrict ourselves to showing the diagrammatics with the intent of giving the reader a flavor of these structures. It is natural to assume that braiding intertwines with creation as shown in Fig. 3.26 (similarly with fusion). This intertwining identity is clearly the sort of thing that a topologist will love, since it indicates that the diagrams can be interpreted as embeddings of graphs in three-dimensional space, and it fits with our interpretation of the vertices in terms of trinions. The intertwining identity is an assumption like the Yang–Baxter equation itself (see Fig. 3.7), which simplifies the mathematical structure of the model.

It is to be expected that there will be an operator that expresses the recoupling of vertex interactions as shown in Fig. 3.27 and labeled by Q. This corresponds to the associativity at the level of trinion combinations shown in Fig. 3.22. The actual formalism of such an operator will parallel the mathematics of recoupling for angular momentum (see e.g., [26]). If one

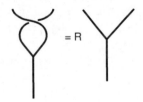

Fig. 3.25. Braiding

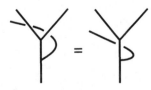

Fig. 3.26. Interwining

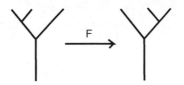

Fig. 3.27. Recoupling

just considers the abstract structure of recoupling, then one sees that for trees with four branches (each with a single root), there is a cycle of length five as shown in Fig. 3.28. One can start with any pattern of three-vertex interactions and go through a sequence of five recouplings that bring one back to the same tree from which one started. *It is a natural simplifying axiom to assume that this composition is the identity mapping.* This axiom is called the *pentagon identity.*

Finally there is a hexagonal cycle of interactions between braiding, recoupling, and intertwining identity as shown in Fig. 3.29. One says that the interactions satisfy the *hexagon identity* if this composition is the identity.

A *graphical three-dimensional topological quantum field theory* is an algebra of interactions that satisfies the Yang–Baxter equation, the intertwining identity, the pentagon identity, and the hexagon identity. There is no room in this summary to detail the way that these properties fit into the topology of knots and three-dimensional manifolds, but a sketch is in order. For the case of topological quantum field theory related to the group $SU(2)$, there is a construction based entirely on the combinatorial topology of the bracket polynomial (see Sects. 3.7, 3.9, and 3.10). See [3, 26] for more information on this approach.

Now return to Fig. 3.20 where we illustrate trinions, shown in relation to a trivalent vertex, and a surface of genus three that is decomposed into four trinions. It turns out that the vector space $V(S_g) = V(G(S_g, t))$ to a surface with a trinion decomposition as t described above, and defined in

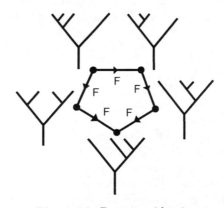

Fig. 3.28. Pentagon identity

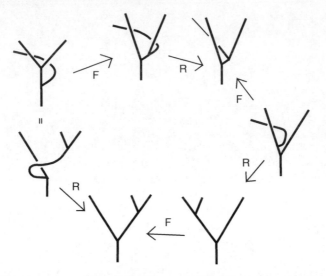

Fig. 3.29. Hexagon identity

terms of the graphical topological quantum field theory, does not depend on the choice of trinion decomposition. This independence is guaranteed by the braiding, hexagon, and pentagon identities. One can then associate a well-defined vector $|M\rangle$ in $V(S_g)$ whenever M is a three manifold whose boundary is S_g. Furthermore, if a closed three manifold M^3 is decomposed along a surface S_g into the union of M_- and M_+ where these parts are otherwise disjoint three manifolds with boundary S_g, then the inner product $I(M) = \langle M_-|M_+\rangle$ is, up to normalization, an invariant of the three manifold M_3. With the definition of graphical topological quantum field theory given above, knots and links can be incorporated as well, so that one obtains a source of invariants $I(M^3, K)$ of knots and links in orientable three manifolds. Here we see the uses of the relationships that occur in the higher dimensional cobordism categories, as described in the previous section.

The invariant $I(M^3, K)$ can be formally compared with the Witten [46] integral

$$Z(M^3, K) = \int DA \, e^{(ik/4\pi)S(M,A)} W_K(A) \, .$$

It can be shown that up to limits of the heuristics, $Z(M, K)$ and $I(M^3, K)$ are essentially equivalent for appropriate choice of gauge group and corresponding spin networks.

By these graphical reformulations, a three-dimensional TQFT is, at base, a highly simplified theory of point particle interactions in $2 + 1$-dimensional spacetime. It can be used to articulate invariants of knots and links and invariants of three manifolds. The reader interested in the $SU(2)$ case of this structure and its implications for invariants of knots and three manifolds can

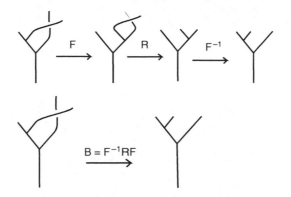

Fig. 3.30. A more complex braiding operator

consult [3, 26, 48–50]. One expects that physical situations involving $2 + 1$ spacetime will be approximated by such an idealized theory. There are also applications to $3 + 1$ quantum gravity [12, 51, 52]. Aspects of the quantum Hall effect may be related to topological quantum field theory [4]. One can study a physics in two-dimensional space, where the braiding of particles or collective excitations leads to non-trival representations of the Artin braid group. Such particles are called *Anyons*. Such TQFT models would describe applicable physics. One can think about applications of anyons to quantum computing along the lines of the topological models described here.

A key point in the application of TQFT to quantum information theory is contained in the structure illustrated in Fig. 3.30. A more complex braiding operator is shown, based on the composition of recoupling with the elementary braiding at a vertex. (This structure is implicit in the Hexagon identity of Fig. 3.29.) The new braiding operator is a source of unitary representations of braid group in situations (which exist mathematically) where the recoupling transformations are themselves unitary. This kind of pattern is utilized in the work of Freedman and collaborators [53–57] and in the case of classical angular momentum formalism has been dubbed a "spin-network quantum simulator" by Rasetti and collaborators [58, 59]. In the next section, we show how certain natural deformations [26] of Penrose spin networks [60] can be used to produce these unitary representations of the Artin braid group and the corresponding models for anyonic topological quantum computation.

3.11 Spin Networks and Temperley–Lieb Recoupling Theory

In this section, we discuss a combinatorial construction for spin networks that generalizes the original construction of Roger Penrose. The result of this generalization is a structure that satisfies all the properties of a graphical TQFT as described in the previous section, and specializes to classical angular momentum recoupling theory in the limit of its basic variable. The construction is based on the properties of the bracket polynomial (as already described in Sect. 3.4). A complete description of this theory can be found in the book "Temperley–Lieb Recoupling Theory and Invariants of Three-Manifolds" by Kauffman and Lins [26].

The "q-deformed" spin networks that we construct here are based on the bracket polynomial relation (see Figs. 3.31 and 3.32).

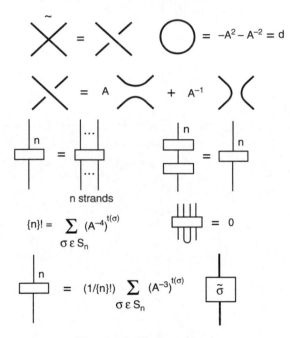

Fig. 3.31. Basic projectors

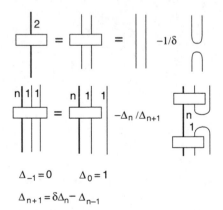

Fig. 3.32. Two strand projector

In Fig. 3.31, we indicate how the basic projector (symmetrizer, Jones-Wenzl projector) is constructed on the basis of the bracket polynomial expansion. In this technology, a symmetrizer is a sum of tangles on n strands (for a chosen integer n). The tangles are made by summing over braid lifts of permutations in the symmetric group on n letters, as indicated in Fig. 3.31. Each elementary braid is then expanded by the bracket polynomial relation as indicated in Fig. 3.31 so that the resulting sum consists of flat tangles without any crossings (these can be viewed as elements in the Temperley–Lieb algebra). The projectors have the property that the concatenation of a projector with itself is just that projector, and if you tie two lines on the top or the bottom of a projector together, then the evaluation is zero. This general definition of projectors is very useful for this theory. The two-strand projector is shown in Fig. 3.32. Here the formula for that projector is particularly simple. It is the sum of two parallel arcs and two turn-around arcs (with coefficient $-1/d$, with $d = -A^2 - A^{-2}$ is the loop value for the bracket polynomial). Figure 3.32 also shows the recursion formula for the general projector. This recursion formula is due to Jones and Wenzl and the projector in this form, developed as a sum in the Temperley–Lieb algebra (see Sect. 3.9), is usually known as the *Jones–Wenzl projector*.

The projectors are combinatorial analogs of irreducible representations of a group (the original spin nets were based on $SU(2)$ and these deformed nets are based on the corresponding quantum group to SU(2)). As such the reader can think of them as "particles". The interactions of these particles are governed by how they can be tied together into three vertices (see Fig. 3.33). In Fig. 3.33, we show how to tie three projectors, of a, b, c strands, respectively, together to form a three vertex. In order to accomplish this interaction, we must share lines between them as shown in that figure so that there are non-negative integers i, j, k so that $a = i + j, b = j + k$, and $c = i + k$. This is equivalent to the condition that $a + b + c$ is even and that the sum of any two of a, b, c is greater than or equal to the third. For example $a + b \geq c$. One can think of the vertex as a possible particle interaction where $[a]$ and $[b]$ interact to produce $[c]$. That is, any two of the legs of the vertex can be regarded as interacting to produce the third leg.

There is a basic orthogonality of three vertices as shown in Fig. 3.34. Here if we tie two three-vertices together so that they form a "bubble" in

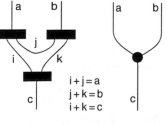

$$i + j = a$$
$$j + k = b$$
$$i + k = c$$

Fig. 3.33. Vertex

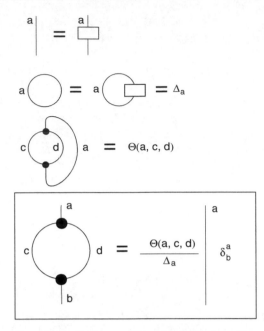

Fig. 3.34. Orthogonality of trivalent vertices

the middle, then the resulting network with labels a and b on its free ends is a multiple of an a-line (meaning a line with an a-projector on it) or zero (if a is not equal to b). The multiple is compatible with the results of closing the diagram in the equation of Fig. 3.34 so the two free ends are identified with one another. On closure, as shown in the figure, the left-hand side of the equation becomes a theta graph and the right-hand side becomes a multiple of a "delta" where Δ_a denotes the bracket polynomial evaluation of the a-strand loop with a projector on it. The $\Theta(a, b, c)$ denotes the bracket evaluation of a theta graph made from three trivalent vertices and labeled with a, b, c on its edges.

There is a recoupling formula in this theory in the form shown in Fig. 3.35. Here there are "6-j symbols", recoupling coefficients that can be expressed, as shown in Fig. 3.35, in terms of tetrahedral graph evaluations and theta graph evaluations. The tetrahedral graph is shown in Fig. 3.36. One derives

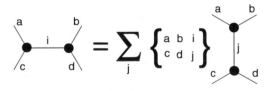

Fig. 3.35. Recoupling formula

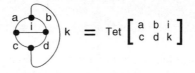

$$\left(\; \right) k \;=\; \text{Tet} \begin{bmatrix} a & b & i \\ c & d & k \end{bmatrix}$$

Fig. 3.36. Tetrahedron network

the formulas for these coefficients directly from the orthogonality relations for the trivalent vertices by closing the left-hand side of the recoupling formula and using orthogonality to evaluate the right-hand side. This is illustrated in Fig. 3.37. Finally, there is the braiding relation, as illustrated in Fig. 3.38.

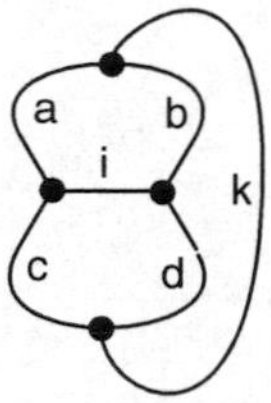

$$\left(\; \right) k \;=\; \sum_j \left\{ \begin{matrix} a & b & i \\ c & d & j \end{matrix} \right\} \left(\; \right) k$$

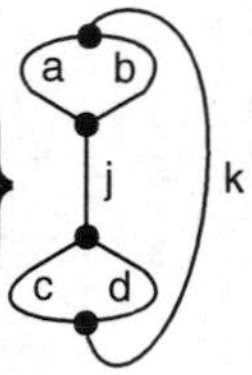

$$=\; \sum_j \left\{ \begin{matrix} a & b & i \\ c & d & j \end{matrix} \right\} \frac{\Theta(a,b,j)}{\Delta_j} \frac{\Theta(c,d,j)}{\Delta_j} \Delta_j \delta_j^k$$

$$=\; \left\{ \begin{matrix} a & b & i \\ c & d & k \end{matrix} \right\} \frac{\Theta(a,b,k)\;\Theta(c,d,k)}{\Delta_k}$$

$$\boxed{\left\{ \begin{matrix} a & b & i \\ c & d & k \end{matrix} \right\} \;=\; \frac{\text{Tet} \begin{bmatrix} a & b & i \\ c & d & k \end{bmatrix} \Delta_k}{\Theta(a,b,k)\;\Theta(c,d,k)}}$$

Fig. 3.37. Tetrahedron formula for recoupling coefficients

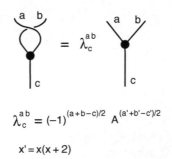

$$\left(\; \right) \;=\; \lambda_c^{ab} \left(\; \right)$$

$$\lambda_c^{ab} \;=\; (-1)^{(a+b-c)/2} \, A^{(a'+b'-c')/2}$$

$$x' = x(x+2)$$

Fig. 3.38. Local braiding formula

With the braiding relation in place, this q-deformed spin network theory satisfies the pentagon, hexagon, and braiding naturality identities needed for a topological quantum field theory. All these identities follow naturally from the basic underlying topological construction of the bracket polynomial. One can apply the theory to many different situations.

3.11.1 Evaluations

In this section, we discuss the structure of the evaluations for Δ_n and the theta and tetrahedral networks. We refer to [26] for the details behind these formulas. Recall that Δ_n is the bracket evaluation of the closure of the n-strand projector, as illustrated in Fig. 3.34. For the bracket variable A, one finds that

$$\Delta_n = (-1)^n \frac{A^{2n+2} - A^{-2n-2}}{A^2 - A^{-2}}.$$

One sometimes writes the *quantum integer*

$$[n] = (-1)^{n-1} \Delta_{n-1} = \frac{A^{2n} - A^{-2n}}{A^2 - A^{-2}}.$$

If

$$A = e^{i\pi/2r},$$

where r is a positive integer, then

$$\Delta_n = (-1)^n \frac{\sin((n+1)\pi/r)}{\sin(\pi/r)}.$$

Here the corresponding quantum integer is

$$[n] = \frac{\sin(n\pi/r)}{\sin(\pi/r)}.$$

Note that $[n+1]$ is a positive real number for $n = 0, 1, 2, \ldots r-2$ and that $[r-1] = 0$.

The evaluation of the theta net is expressed in terms of quantum integers by the formula

$$\Theta(a, b, c) = (-1)^{m+n+p} \frac{[m+n+p+1]![n]![m]![p]!}{[m+n]![n+p]![p+m]!},$$

where

$$a = m + p, \ b = m + n, \ c = n + p.$$

Note that

$$(a + b + c)/2 = m + n + p.$$

When $A = e^{i\pi/2r}$, the recoupling theory becomes finite with the restriction that only three vertices (labeled with a, b, c) are *admissible* when $a + b + c \leq 2r - 4$. All the summations in the formulas for recoupling are restricted to admissible triples of this form.

3.11.2 Symmetry and Unitarity

The formula for the recoupling coefficients given in Fig. 3.37 has less symmetry than is actually inherent in the structure of the situation. By multiplying all the vertices by an appropriate factor, we can reconfigure the formulas in this theory so that the revised recoupling transformation is orthogonal, in the sense that its transpose is equal to its inverse. This is a very useful fact. It means that when the resulting matrices are real, then the recoupling transformations are unitary. We shall see particular applications of this viewpoint later in the lecture.

Figure 3.39 illustrates this modification of the three vertex. Let $Vert[a, b, c]$ denote the original three vertex of the Temperley–Lieb recoupling theory. Let $ModVert[a, b, c]$ denote the modified vertex. Then we have the formula

$$ModVert[a, b, c] = \frac{\sqrt{\sqrt{\Delta_a \Delta_b \Delta_c}}}{\sqrt{\Theta(a, b, c)}} \, Vert[a, b, c] \, .$$

Lemma 2. *For the bracket evaluation at the root of unity* $A = e^{i\pi/2r}$, *the factor*

$$f(a, b, c) = \frac{\sqrt{\sqrt{\Delta_a \Delta_b \Delta_c}}}{\sqrt{\Theta(a, b, c)}}$$

is real, and can be taken to be a positive real number for (a, b, c) *admissible (i.e.,* $a + b + c \leq 2r - 4$*).*

Proof. By the results from the previous subsection,

$$\Theta(a, b, c) = (-1)^{(a+b+c)/2} \hat{\Theta}(a, b, c),$$

where $\hat{\Theta}(a, b, c)$ is positive real, and

$$\Delta_a \Delta_b \Delta_c = (-1)^{(a+b+c)} [a + 1][b + 1][c + 1],$$

where the quantum integers in this formula can be taken to be positive real. It follows from this that

$$f(a, b, c) = \sqrt{\frac{\sqrt{[a + 1][b + 1][c + 1]}}{\hat{\Theta}(a, b, c)}} \, ,$$

showing that this factor can be taken to be positive real.   □

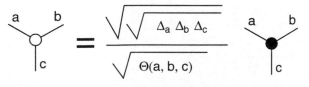

Fig. 3.39. Modified three vertex

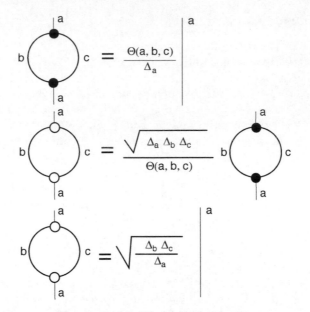

Fig. 3.40. Modified bubble identity

In Fig. 3.40, we show how this modification of the vertex affects the non-zero term of the orthogonality of trivalent vertices (cf. Fig. 3.34). We refer to this as the "modified bubble identity". The coefficient in the modified bubble identity is

$$\sqrt{\frac{\Delta_b \Delta_c}{\Delta_a}} = (-1)^{(b+c-a)/2} \sqrt{\frac{[b+1][c+1]}{[a+1]}},$$

where (a, b, c) form an admissible triple. In particular $b + c - a$ is even and hence this factor can be taken to be real.

We rewrite the recoupling formula in this new basis and emphasize that the recoupling coefficients can be seen (for fixed external labels a, b, c, d) as a matrix transforming the horizontal "double-Y" basis to a vertically disposed double-Y basis. In Figs. 3.41, 3.42, and 3.43, we have shown the form of this transformation, using the matrix notation

$$M[a, b, c, d]_{ij}$$

for the modified recoupling coefficients. In Fig. 3.41, we derive an explicit formula for these matrix elements. The proof of this formula follows directly from trivalent–vertex orthogonality (see Figs. 3.34 and 3.37) and is given in Fig. 3.41. The result shown in Figs. 3.41 and 3.42 is the following formula for the recoupling matrix elements

$$M[a, b, c, d]_{ij} = ModTet \begin{pmatrix} a & b & i \\ c & d & j \end{pmatrix} / \sqrt{\Delta_a \Delta_b \Delta_c \Delta_d},$$

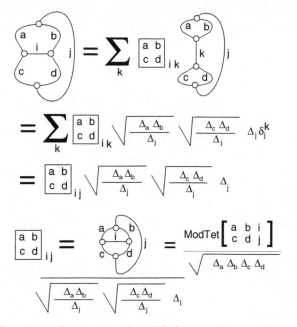

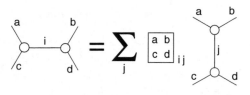

Fig. 3.41. Derivation of modified recoupling coefficients

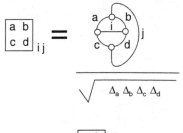

Fig. 3.42. Modified recoupling formula

Fig. 3.43. Modified recoupling matrix

where $\sqrt{\Delta_a \Delta_b \Delta_c \Delta_d}$ is short hand for the product

$$\sqrt{\frac{\Delta_a \Delta_b}{\Delta_j}} \sqrt{\frac{\Delta_c \Delta_d}{\Delta_j}} \Delta_j$$

$$= (-1)^{(a+b-j)/2}(-1)^{(c+d-j)/2}(-1)^j \sqrt{\frac{[a+1][b+1]}{[j+1]}} \sqrt{\frac{[c+1][d+1]}{[j+1]}}[j+1]$$

$$= (-1)^{(a+b+c+d)/2}\sqrt{[a+1][b+1][c+1][d+1]}.$$

In this form, since (a, b, j) and (c, d, j) are admissible triples, we see that this coefficient can be taken to be real, and its value is independent of the choice of i and j. The matrix $M[a, b, c, d]$ is real valued.

It follows from Fig. 3.35 (turn the diagrams by 90°) that

$$M[a, b, c, d]^{-1} = M[b, d, a, c].$$

In Fig. 3.44, we illustrate the formula

$$M[a, b, c, d]^T = M[b, d, a, c].$$

It follows from this formula that

$$M[a, b, c, d]^T = M[a, b, c, d]^{-1}.$$

Hence $M[a, b, c, d]$ is an orthogonal, real-valued matrix.

Theorem 5. *In the Temperley–Lieb theory we obtain unitary (in fact real orthogonal) recoupling transformations when the bracket variable A has the form $A = e^{i\pi/2r}$ for r a positive integer. Thus we obtain families of unitary representations of the Artin braid group from the recoupling theory at these roots of unity.*

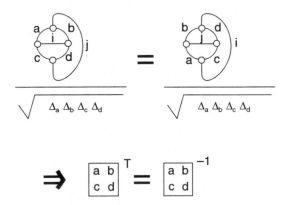

Fig. 3.44. Modified matrix transpose

Proof. The proof is given in the discussion above.   □

In Sect. 3.12, we shall show explicitly how these methods work in the case of the Fibonacci model, where $A = e^{3i\pi/5}$.

3.12 Fibonacci Particles

In this section and the next, we detail how the Fibonacci model for anyonic quantum computing [61, 62] can be constructed by using a version of the two-stranded bracket polynomial and a generalization of Penrose spin networks. This is a fragment of the Temperly–Lieb recoupling theory [26]. We already gave in the preceding sections a general discussion of the theory of spin networks and their relationship with quantum computing.

The Fibonacci model is a TQFT that is based on a single "particle" with two states that we shall call the *marked state* and the *unmarked state*. The particle in the marked state can interact with itself either to produce a single particle in the marked state, or to produce a single particle in the unmarked state. The particle in the unmarked state has no influence in interactions (an unmarked state interacting with any state S yields that state S). One way to indicate these two interactions symbolically is to use a box, for the marked state and a blank space for the unmarked state. Then one has two modes of interaction of a box with itself:

1. Adjacency: □   □
   and
2. Nesting: ☐ .

With this convention, we take the adjacency interaction to yield a single box, and the nesting interaction to produce nothing:

We take the notational opportunity to denote nothing by an asterisk (*). The syntactical rules for operating the asterisk are the following: the asterisk is a stand-in for no mark at all, and it can be erased or placed wherever it is convenient to do so. Thus

We shall make a recoupling theory based on this particle, but it is worth noting some of its purely combinatorial properties first. The arithmetic of combining boxes (standing for acts of distinction) according to these rules has been studied and formalized in [63] and correlated with Boolean algebra and classical logic. Here *within* and *next to* are ways to refer to the two sides delineated by the given distinction. From this point of view, there are two modes of relationship (adjacency and nesting) that arise at once in the presence of a distinction.

From here on, we shall denote the Fibonacci particle by the letter P. Thus the two possible interactions of P with itself are as follows:

1. $P, P \longrightarrow *$
2. $P, P \longrightarrow P$

In Fig. 3.45, we indicate in small tree diagrams the two possible interactions of the particle P with itself. In the first interaction, the particle vanishes, producing the asterix. In the second interaction, the particle a single copy of P is produced. These are the two basic actions of a single distinction relative to itself, and they constitute our formalism for this very elementary particle.

In Fig. 3.46, we have indicated the different results of particle processes, where we begin with a left-associated tree structure with three branches, all marked and then four branches all marked. In each case, we demand that the particles interact successively to produce an unmarked particle in the end, at the root of the tree. More generally one can consider a left-associated tree with n upward branches and one root. Let $T(a_1, a_2, \cdots, a_n : b)$ denote

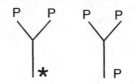

Fig. 3.45. Fibonacci particle interaction

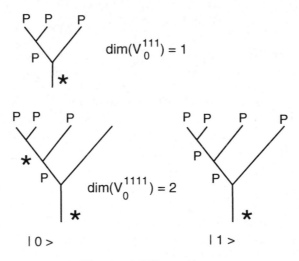

Fig. 3.46. Fibonacci tree

such a tree with particle labels $a_1, \cdots, a_n$ on the top and root label b at the bottom of the tree. We consider all possible processes (sequences of particle interactions) that start with the labels at the top of the tree, and end with the labels at the bottom of the tree. Each such sequence is regarded as a basis vector in a complex vector space

$$V_b^{a_1, a_2, \cdots, a_n}$$

associated with the tree. In the case where all the labels are marked at the top and the bottom label is unmarked, we shall denote this tree by

$$V_0^{111 \cdots 11} = V_0^{(n)},$$

where n denotes the number of upward branches in the tree. We see from Fig. 3.46 that the dimension of $V_0^{(3)}$ is 1, and that

$$dim(V_0^{(4)}) = 2.$$

This means that $V_0^{(4)}$ is a natural candidate in this context for the two-qubit space.

Given the tree $T(1, 1, 1, \cdots, 1 : 0)$ (n marked states at the top, an unmarked state at the bottom), a process basis vector in $V_0^{(n)}$ is in direct correspondence with a string of boxes and asterisks (1s and 0s) of length $n - 2$ with no repeated asterisks and ending in a marked state. See Fig. 3.46 for an illustration of the simplest cases. It follows from this that

$$dim(V_0^{(n)}) = f_{n-2},$$

where f_k denotes the kth Fibonacci number:

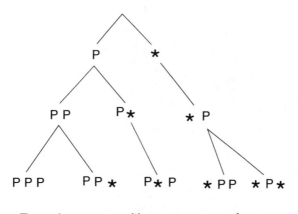

Tree of squences with no occurence of ★ ★

Fig. 3.47. Fibonacci sequence

$$f_0 = 1\,, f_1 = 1\,, f_2 = 2\,, f_3 = 3\,, f_4 = 5\,, f_5 = 8, \cdots\,,$$

where

$$f_{n+2} = f_{n+1} + f_n\,.$$

The dimension formula for these spaces follows from the fact that there are f_n sequences of length $n-1$ of marked and unmarked states with no repetition of an unmarked state. This fact is illustrated in Fig. 3.47.

3.13 The Fibonacci Recoupling Model

We now show how to make a model for recoupling the Fibonacci particle by using the Temperley–Lieb recoupling theory and the bracket polynomial. Everything we do in this section will be based on the 2-projector, its properties and evaluations based on the bracket polynomial model for the Jones polynomial. While we have outlined the general recoupling theory based on the bracket polynomial in earlier sections, one can make the calculations for the Fibonacci model completely self-contained, using only basic information about the bracket polynomial, and the essential properties of the 2-projector as shown in Fig. 3.48. See [1] for the details of this, or work them out for yourself! In Fig. 3.48, we state the definition of the 2-projector, list its two main properties (the operator is idempotent and a self-attached strand yields a zero evaluation), and give diagrammatic proofs of these properties.

In Fig. 3.49, we show the essence of the Temperley–Lieb recoupling model for the Fibonacci particle. The Fibonacci particle is, in this mathematical model, identified with the 2-projector itself. As the reader can see from

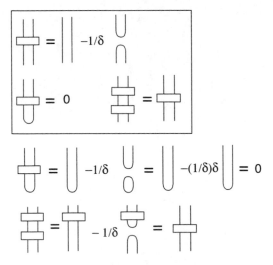

Fig. 3.48. The 2-projector

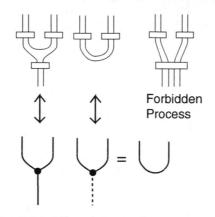

Fig. 3.49. Fibonacci particle as 2-projector

Fig. 3.49, there are two basic interactions of the 2-projector with itself: one giving a 2-projector, the other giving nothing. This is the pattern of self-interaction of the Fibonacci particle. There is a third possibility, depicted in Fig. 3.49, where two 2-projectors interact to produce a 4-projector. We could remark at the outset that the 4-projector will be zero if we choose the bracket polynomial variable $A = e^{3i\pi/5}$. If we wish, we can assume that the 4-projector is forbidden and deduce that the theory has to be at this root of unity [1].

Note that in Fig. 3.49, we have adopted a single-strand notation for the particle interactions, with a solid strand corresponding to the marked particle, a dotted strand (or nothing) corresponding to the unmarked particle. A dark vertex indicates an interaction point or may be used to indicate the single strand is shorthand for two ordinary strands. Remember that these are all shorthand expressions for underlying bracket polynomial calculations. Vertices in this theory have to be readjusted for unitarity just as we described in our general treatment of the recoupling theory in the previous sections.

In Figs. 3.50 and 3.51, we indicate the form of the recoupling matrix for this model, and the effect of braiding at a three vertex. When the three vertex has three marked lines, then the braiding operator is multiplication by $-A^4$, as in Fig. 3.51. When the three vertex has two marked lines, then the braiding operator is multiplied by A^8, as shown in Fig. 3.51.

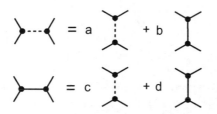

Fig. 3.50. Recoupling for 2-projectors

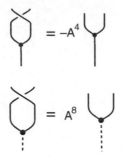

Fig. 3.51. Braiding at the three vertex

The real symmetric recoupling matrix F is given by the equation

$$F = \begin{pmatrix} a & b \\ c & d \end{pmatrix} = \begin{pmatrix} \tau & \sqrt{\tau} \\ \sqrt{\tau} & -\tau \end{pmatrix},$$

where $\tau = 1/\Delta$ and $\Delta = (1 + \sqrt{5})/2$ is the golden ratio. This gives the Fibonacci model. Using Fig. 3.51, we have that the local braiding matrix for the model is given by the formula below with $A = e^{3i\pi/5}$

$$R = \begin{pmatrix} -A^4 & 0 \\ 0 & A^8 \end{pmatrix} = \begin{pmatrix} e^{4i\pi/5} & 0 \\ 0 & -e^{2i\pi/5} \end{pmatrix}.$$

The simplest example of a braid group representation arising from this theory is the representation of the three-strand braid group generated by $S_1 = R$ and $S_2 = FRF$ (remember that $F = F^T = F^{-1}$). The matrices S_1 and S_2 are both unitary, and they generate a dense subset of the unitary group $U(2)$, supplying the first part of the transformations needed for quantum computing.

3.14 Quantum Computation of Colored Jones Polynomials and the Witten–Reshetikhin–Turaev Invariant

In this section, we make some brief comments on the quantum computation of colored Jones polynomials.

First, consider Fig. 3.52. In that figure we illustrate the calculation of the evaluation of the *(a) – colored bracket polynomial* for the *plat closure $P(B)$* of a braid B. The reader can infer the definition of the plat closure from Fig. 3.52. One takes a braid on an even number of strands and closes the top strands with each other in a row of maxima. Similarly, the bottom strands are closed with a row of minima. It is not hard to see that any knot or link can be represented as the plat closure of some braid.

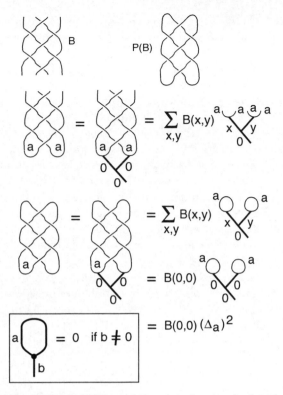

Fig. 3.52. Evolution of the plat closure of a braid

The (a) – colored bracket polynomial of a link L, denoted $< L >_a$, is the evaluation of that link where each single strand has been replaced by a parallel strands and the insertion of Jones–Wenzl projector (as discussed in Sect. 3.11). We then see that we can use our discussion of the Temperley–Lieb recoupling theory as in Sects. 3.11, 3.12, and 3.13 to compute the value of the colored bracket polynomial for the plat closure PB. As shown in Fig. 3.52, we regard the braid as acting on a process space $V_0^{a,a,\cdots,a}$ and take the case of the action on the vector v whose process space coordinates are all zero. Then the action of the braid takes the form

$$Bv(0,\cdots,0) = \Sigma_{x_1,\cdots,x_n} B(x_1,\cdots,x_n)v(x_1,\cdots,x_n),$$

where $B(x_1,\cdots,x_n)$ denotes the matrix entries for this recoupling transformation and $v(x_1,\cdots,x_n)$ runs over a basis for the space $V_0^{a,a,\cdots,a}$. Here n is even and equal to the number of braid strands. In Fig. 3.52, we illustrate with $n = 4$. Then, as the figure shows, when we close the top of the braid action to form PB, we cut the sum down to the evaluation of just one term. In the general case, we will get

$$< PB >_a = B(0,\cdots,0)\Delta_a^{n/2}.$$

The calculation simplifies to this degree because of the vanishing of loops in the recoupling graphs. The vanishing result is stated in Fig. 3.52.

The *colored Jones polynomials* are normalized versions of the colored bracket polynomials, differing just by a normalization factor.

In order to consider quantum computation of the colored bracket or colored Jones polynomials, we therefore can consider quantum computation of the matrix entries $B(0, \cdots, 0)$. These matrix entries in the case of the roots of unity $A = e^{i\pi/2r}$ and for the $a = 2$ Fibonacci model with $A = e^{3i\pi/5}$ are parts of the diagonal entries of the unitary transformation that represents the braid group on the process space $V_0^{a,a,\cdots,a}$. *We can obtain these matrix entries by using the Hadamard test as described in Sect. 3.8.* As a result we get relatively efficient quantum algorithms for the colored Jones polynomials at these roots of unity, in essentially the same framework as we described in Sect. 3.8, but for braids of arbitrary size. The computational complexity of these models is essentially the same as the models for the Jones polynomial discussed in [36].

It is worth remarking here that these algorithms give quantum algorithms for computing not only the colored bracket and Jones polynomials but also the Witten–Reshetikhin–Turaev (WRT) invariants evaluated at the above roots of unity. The reason for this is that the WRT invariant, in unnormalized form, is given as a finite sum of colored bracket polynomials:

$$WRT(L) = \Sigma_{a=0}^{r-2} \Delta_a < L >_a .$$

This means that we have, in principle, a quantum algorithm for the computation of the Witten functional integral [46] via this knot-theoretic combinatorial

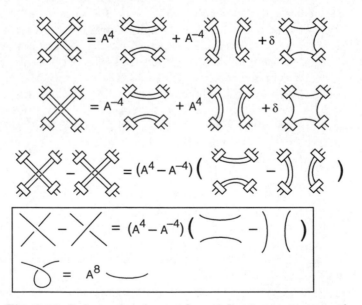

Fig. 3.53. Dubrovnik polynomial specialization at two strands

topology. It would be very interesting to understand a more direct approach to such a computation via quantum field theory and functional integration.

Finally, we note that in the case of the Fibonacci model, the (2)-colored bracket polynomial is a special case of the Dubrovnik version of the Kauffman polynomial [64]. See Fig. 3.53 for diagrammatics that resolve this fact. The skein relation for the Dubrovnik polynomial is boxed in this figure. Above the box, we show how the double strands with projectors reproduce this relation. This observation means that in the Fibonacci model, the natural underlying knot polynomial is a special evaluation of the Dubrovnik polynomial, and the Fibonacci model can be used to perform quantum computation for the values of this invariant.

Acknowledgements

The first author thanks the National Science Foundation for support of this research under NSF Grant DMS-0245588.

References

1. L. H. Kauffman and S. J. Lomonaco Jr.: *q-deformed spin networks, knot polynomials and anyonic topological quantum computation*, arXiv:quant-ph/0606114
2. L. H. Kauffman: *State models and the Jones polynomial*, Topology **26**, 395–407 (1987)
3. L. H. Kauffman: *Knots and Physics*, 3rd edn (World Scientific Publishers, Singapore 1991)
4. F. Wilczek: *Fractional Statistics and Anyon Superconductivity* (World Scientific Publishing Company, Singapore 1990)
5. E. Fradkin and P. Fendley: *Realizing non-abelian statistics in time-reversal invariant systems*, Theory Seminar, Physics Department, UIUC, 4/25/2005
6. N. E. Bonesteel, L. Hormozi, G. Zikos and S. H. Simon: *Braid topologies for quantum computation*, Phys. Rev. Lett. **95**, 140503 (2005)
7. S. H. Simon, N. E. Bonesteel, M. H. Freedman, N. Petrovic and L. Hormozi: *Topological quantum computing with only one mobile quasiparticle*, Phys. Rev. Lett. **96**, 070503 (2006)
8. C. Ernst and D. W. Sumners: *A calculus for rational tangles: Applications to DNA Recombination*, Math. Proc. Camb. Phil. Soc. **108**, 489-515 (1990)
9. L. H. Kauffman (ed): *Knots and Applications* (World Scientific, Singapore 1996)
10. L. Smolin: *Link polynomials and critical points of the Chern-Simons path integrals*, Mod. Phys. Lett. A **4**, 1091–1112 (1989)
11. C. Rovelli and L. Smolin: *Spin networks and quantum gravity*, Phys. Rev. D **52**, 5743–5759 (1995)
12. L. H. Kauffman and T. Liko: *Knot theory and a physical state of quantum gravity*, Class. Quantum Grav. **23**, R63 (2006), arXiv:hep-th/0505069
13. P. A. M. Dirac: *Principles of Quantum Mechanics* (Oxford University Press, Oxford 1958)

14. M. A. Nielsen and I. L. Chuang: *Quantum Computation and Quantum Information* (Cambrige University Press, Cambridge 2000)
15. R. J. Baxter: *Exactly Solved Models in Statistical* (Mechanics. Acad. Press, London 1982)
16. H. Dye: *Unitary solutions to the Yang-Baxter equation in dimension four*, Quant. Inf. Proc. **2**, 117–152 (2003)
17. Y. Zhang, L.H. Kauffman and M. L. Ge: *Yang-Baxterizations, universal quantum gates and Hamiltonians*, Quant. Inf. Proc. **4**, 159–197 (2005)
18. J. L. Brylinski and R. Brylinski: Universal quantum gates. In: *Mathematics of Quantum Computation*, ed by R. Brylinski and G. Chen (Chapman & Hall/CRC Press, Boca Raton 2002)
19. L. H. Kauffman and S. J. Lomonaco: *Braiding operators are universal quantum gates*, New J. Phys. **6**, 134 (2004)
20. B. Schumacher, Ph.D. Thesis.
21. P. K. Aravind: Borromean entanglement of the GHZ state. In *Potentiality, Entanglement and Passion-at-a-Distance*, ed by R. S. Cohen et al (Kluwer, Dordrecht 1997) pp. 53–59
22. L. H. Kauffman and S. J. Lomonaco Jr.: Quantum knots. In: *Quantum Information and Computation II, Proceedings of Spie, 12–14 April 2004*, ed by Donkor Pirich and Brandt, (2004) pp. 268–284.
23. V. F. R. Jones: *Hecke algebra representations of braid groups and link polynomials*, Ann. of Math. **126**, 335–338 (1987)
24. V. F. R. Jones: *On knot invariants related to some statistical mechanics models*, Pacific J. Math. **137**, 311–334 (1989)
25. L. H. Kauffman: *Statistical mechanics and the Jones polynomial*, AMS Contemp. Math. Series **78**, 263–297 (1989)
26. L. H. Kauffman: *Temperley–Lieb Recoupling Theory and Invariants of Three-Manifolds*, Annals Studies **114** (Princeton University Press, Princeton 1994)
27. L. H. Kauffman: *New invariants in the theory of knots*, Amer. Math. Monthly **95**, 195–242 (1988)
28. L. H. Kauffman (ed): *The Interface of Knots and Physics*, vol 51 AMS PSAPM (Providence, RI 1996)
29. Y. Akutsu and M. Wadati: *Knot invariants and critical statistical systems*, J. Phys. Soc. Japan **56**, 839–842 (1987)
30. L. H. Kauffman and D. E. Radford: *Invariants of 3-manifolds derived from finite-dimensional Hopf algebras*, J. Knot Theory Ramifications **4**, 131–162 (1995)
31. N. Y. Reshetikhin and V. Turaev: *Ribbon graphs and their invariants derived from quantum groups*, Comm. Math. Phys. **127**, 1–26 (1990)
32. N. Y. Reshetikhin and V. Turaev: *Invariants of three manifolds via link polynomials and quantum groups*, Invent. Math. **103**, 547–597 (1991)
33. V. G. Turaev: *The Yang-Baxter equations and invariants of links*, LOMI preprint E-3-87, Steklov Institute, Leningrad, USSR. Inventiones Math. **92**, Fasc. 3, 527–553
34. V. G. Turaev and O. Viro: *State sum invariants of 3-manifolds and quantum 6j symbols*, Topology, **31**, 865–902 (1992)
35. L. H. Kauffman: Quantum computing and the Jones polynomial. In: *Quantum Computation and Information*, ed by S. Lomonaco Jr. (AMS CONM/305, 2002) pp. 101–137, arXiv:math.QA/0105255

36. D. Aharonov, V. Jones and Z. Landau: *A polynomial quantum algorithm for approximating the Jones polynomial*, Proc. thirty-eighth Ann. ACM STC (2006), arXiv:quant-ph/0511096
37. L. H. Kauffman and S. J. Lomonaco Jr.: *Topological quantum computing and the Jones polynomial*, arXiv:quant-ph/0605004
38. D. Aharonov and I. Arad: *The BQP-hardness of approximating the Jones polynomial*, arXiv:quant-ph/0605181
39. P. Wocjan and J. Yard: *The Jones polynomial: quantum algorithms and applications in quantum complexity theory*, arXiv:quant-ph/0603069
40. G. Benkart: Commuting actions – a tale of two groups. In: *Lie algebras and Their Representations* (Seoul 1995); Contemp. Math. Series **194**, 1–46 (1996)
41. L. H. Kauffman: Knot diagrammatics. In: *Handbook of Knot Theory*, ed by Menasco and Thistlethwaite (Elsevier B. V., Amsterdam 2005) pp. 233–318, arXivmath.GN/0410329:
42. L. H. Kauffman: Teleportation topology. In: *The Proceedings of the 2004 Byelorus Conference on Quantum Optics*, Opt. Spectrosc. **9**, 227–232 (2005), arXiv:quant-ph/0407224
43. B. Coecke: *The logic of entanglement*, arXiv:quant-phy/0402014
44. S. Abramsky and B. Coecke: *A categorical semantics of quantum protocols*, arXiv:quant-ph/0402130
45. M.F. Atiyah: *The Geometry and Physics of Knots* (Cambridge University Press, Cambridge 1990)
46. E. Witten: *Quantum field Theory and the Jones Polynomial*, Commun. Math. Phys. **121**, 351–399 (1989)
47. C. N. Yang: *Some exact results for the many-body problem in one dimension with repulsive delta-function interaction*, Phys. Rev. Lett. **19**, 1312 (1967)
48. T. Kohno: *Conformal field theory and topology*, AMS Translations of Mathematical Monographs, **210** (1998)
49. L. Crane: *2-d physics and 3-d topology*, Comm. Math. Phys. **135**, 615–640 (1991)
50. G. Moore and N. Seiberg: *Classical and quantum conformal field theory*, Comm. Math. Phys. **123**, 177–254 (1989)
51. A. Ashtekar, C. Rovelli and L. Smolin: *Weaving a classical geometry with quantum threads*, Phys. Rev. Lett. **69**, 237 (1992)
52. A. Ashetekar and J. Lewandowski: *Quantum theory of geometry I: Area operators*, Class. Quant. Grav. **14**, A55–A81 (1997)
53. M. Freedman: *A magnetic model with a possible Chern-Simons phase*, Comm. Math. Phys. **234**, 129–183 (2003), arXiv:quant-ph/0110060v1
54. M. Freedman: *Topological Views on Computational Complexity*, Documenta Mathematica – Extra Volume (ICM, 1998) pp. 453–464
55. M. Freedman, M. Larsen and Z. Wang: *A modular functor which is universal for quantum computation*, Comm. Math. Phys. **227**, 605–622, arXiv:quant-ph/0001108v2
56. M. H. Freedman, A. Kitaev and Z. Wang: *Simulation of topological field theories by quantum computers*, Commun. Math. Phys. **227**, 587–603 (2002), arXiv:quant-ph/0001071
57. M. Freedman: *Quantum computation and the localization of modular functors*, Found. Comp. Math. **1**, 183–204 (2001), arXiv:quant-ph/0003128

58. A. Marzuoli and M. Rasetti: *Spin network quantum simulator*, Phys. Lett. A **306**, 79–87 (2002)
59. S. Garnerone, A. Marzuoli and M. Rasetti: *Quantum automata, braid group and link polynomials*, Quant. Inf. Comp. **7**, 479–503 (2007), arXiv:quant-ph/0601169
60. R. Penrose: Angular momentum: An approach to Combinatorial Spacetime. In: *Quantum Theory and Beyond*, ed T. Bastin (Cambridge University Press, Cambridge 1969)
61. A. Kitaev: *Anyons in an exactly solved model and beyond*, Ann. Phys. **321**, 2–111 (2006), arXiv:cond-mat/0506438
62. Preskill, J.: *Topological Computing for Beginners*, (slide presentation), Lect. Notes Phys. **219**, – Quantum Computation (Chap. 9). http://www.iqi.caltech.edu/preskill/ph219
63. G. Spencer–Brown: *Laws of Form* (George Allen and Unwin Ltd., London 1969)
64. L. H. Kauffman: *An invariant of regular isotopy*, Trans. Amer. Math. Soc. **318**, 417–471 (1990)

4 Entanglement in Phase Space

A.M. Ozorio de Almeida

Centro Brasileiro de Pesquisas Fisicas, Rua Xavier Sigaud 150, 22290-180, Rio de Janeiro, RJ, Brazil.

4.1 Introduction

The realization that quantum mechanics admits entangled states goes back to Schrödinger in 1926 [1]. He went on to coin the term *entanglement* in 1935 [2], but the dramatic example of a biological cat coupled to a decaying nucleus was never meant to be operational. Einstein, Rosen and Podolsky (EPR) [3] discussed an example of a simple entangled bipartite state in the same year. Their concern was the compatibility between Heisenberg's indeterminacy principle and the generation of strong correlations through a measurement on a member of a pair of particles, even when they could no longer be interacting. It was the formulation of Bell inequalities, starting in 1964 [4, 5] that provided a *litmus test* for nonlocal correlations in quantum mechanics. The initial concern was centred on hidden variable theories and the possibility of their emulating quantum correlations even for particles that have ceased to interact. Such violations of local causality, detected by Bell inequalities, could not have developed within any kind of classically evolved ensemble, irrespective of whether the variables are explicit or hidden.

Quantum information theory [6] has given a new boom to the study of the qualitative distinctions between classical and quantum mechanics and to establishing their quantitative measures. There is nothing so dramatic about the development of nonclassical correlations between particles that are still undergoing an interaction, but this question has acquired promising applications in future quantum computations. Necessarily, these deal with finite-dimensional (Hilbert) state spaces, for which the appropriate entanglement measures are now well established.

One of the difficulties in applying semiclassical methods to the study of entanglement is that the former have been developed for infinite-dimensional Hilbert spaces. Not only are these an extrapolation from the few qubits that have been usually considered in quantum information theory, but entanglement is most clearly exhibited through the correlations in elementary *either-or* experiments. This seems to privilege simple state spaces of a single qubit, such as spin-1/2 systems. For this reason, Bohm's version of EPR [7] has become much more popular than the original full phase space version. A way around this difficulty is to consider the measurement of special observables that have only a pair of eigenvalues, even though they operate on states

Ozorio de Almeida, A.M.: *Entanglement in Phase Space*. Lect. Notes Phys. **768**, 157–219 (2009)
DOI 10.1007/978-3-540-88169-8_4

within an infinite space. It turns out that one of the most renowned phase space representations in quantum mechanics, the Wigner–Weyl representation, is based on such operators. Usually, this representation is viewed as a way of eliciting classical features in a quantum state, but it will be used here mainly as a probe into nonclassical correlations.

The development of semiclassical theory throughout the last century allows us to trace the classical skeleton underlying many features of quantum evolution. These classical structures are the core of approximations that improve asymptotically in the limit of large classical actions, or, more formally, as Planck's constant, $\hbar \to 0$. In the case of a finite-dimensional Hilbert space, this becomes the limit of large dimensions. Even though entanglement is a subtle phenomenon, it leads to gross violation of inequalities and to quantitative measures that are not beyond the accuracy of semiclassical approximations. Therefore, it is appropriate to enquire into the manner in which classical structures can be implicated in such a very nonclassical feature of quantum mechanics.

Traditionally, semiclassical theory is concerned with the unitary quantum evolution of closed systems, which are thus described classically by Hamiltonian dynamical systems. Each point in phase space accounts completely for the state of the classical system that evolves along a trajectory. A bipartite or multipartite system is accommodated in this correspondence by a higher dimensional phase space. Each point still evolves as a single-dimensional (1-D) trajectory, but its projections onto the subspaces, which describe the succession of possible states of each component of the system, are also 1-D trajectories in their own right, as shown in Fig. 4.1.

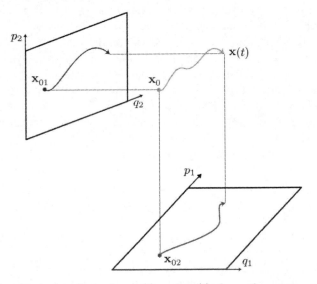

Fig. 4.1. The classical trajectories $x_1(t)$ and $x_2(t)$, for each component in its own phase space, are projections of the full trajectory for the entire system

Part of the power of Hamiltonian dynamics lies in the freedom to transform between different sets of phase space coordinates. This *canonical invariance* emphasizes the importance of the unified evolution of the full system over that of the component trajectories, which are seen to depend on the particular choice of coordinates. In contrast, it is in the separation into components that the phenomenon of quantum entanglement emerges. The particular nature of quantum measurement lies behind this difference, as is discussed in Sect. 4.2: It is only when this is combined to the preceding unitary evolution, that the unclassical correlations between the components become manifest.

Therefore, the study of the properties related to entanglement should be viewed as an objective that is imposed externally on semiclassical physics, which perhaps explains the low priority received by this goal so far. In these lectures, we will only be concerned with the most elementary kind of entanglement, i.e., that of pure bipartite states, for which the measures of entanglement are well established. Even so, it will be seen that this simple case requires the introduction of theoretical instruments of semiclassical theory that are far from elementary. Not only do we need to cope with a higher dimensional phase space for the description of a bipartite system, but it will be shown how the simplest semiclassical description of operators is achieved in a phase space with double the dimension of the one corresponding to the states on which they act. Conversely, some of the most relevant structures for entanglement, such as partial traces and probability densities, can be interpreted as projections of the Wigner function, or sections of its Fourier transform.

The following section reviews the different ways in which features of quantum mechanics, interference and entanglement are *nonclassical*. Simple examples introduce the reflection symmetries, quite familiar for classical waves, that will play a major role in the Wigner function formalism. Prior to this though, it is useful to consider classical–quantum correspondence in a more simple-minded way. This is the subject of Sect. 4.3, which introduces product spaces for both quantum states and classical probability distributions in phase space. In either case, the factorizability is broken by an interaction Hamiltonian, leading to correlations for measurements on the different components. In the classical case, these correlations are constrained by general Bell inequalities. This section also introduces the Schmidt decomposition of quantum states.

Section 4.4 reviews standard semiclassical theory for quantum states. Special emphasis is given to products and factorization of both the phase spaces themselves and the internal *Lagrangian* surfaces that support the quantum states. This product structure is then generalized in Sect. 4.5 to the representation of operators. Dyadic operators of position or momentum eigenstates form a complete linear basis for all quantum operators, which correspond to planes in double phase space. Linear canonical transformations take these into the phase space coordinates for the Weyl representation and the chord representation, its Fourier transform. These bases are associated respectively to phase space reflections and translations and to the corresponding quantum operators. In the case of the density operator, we thus obtain the Wigner function and the chord function, both presented in Sect. 4.6. Though the Wigner function cannot be interpreted as a probability distribution, because it may be negative, it coincides with the difference for probabilities of measuring either the positive or the negative eigenvalue of the reflection operator. Section 4.7 is dedicated to projections of the Wigner function and sections of the chord function, which represent the reduced density operators. The loss of purity of the latter, obtained as integrals of the square of either the reduced Wigner function or the reduced chord function, indicates that the overall state is entangled.

It may be guessed that an initially *classical* pure state, the product of Gaussian Wigner functions, would not be entangled by a simple rotation of positions and momenta. After all, this class of states, including the original EPR states, could stand in for a classical phase space distribution. However, this is not so, as shown in Sect. 4.8: The reflection correlations for such states violate Bell inequalities, even though measurements of positions and momenta can only correlate classically. The transformation to centre of mass coordinates for any number of particles, studied in Sect. 4.9, has similar features. By invoking the Central Limit Theorem for Wigner functions, we obtain features of the nonunitary evolution of the centre of mass in agreement with Markovian theory, i.e., the exact solution of the Lindblad equation for the density operator.

The final section relates double phase space geometry to the semiclassical Wigner and chord functions. These are not known in detail for eigenstates of chaotic Hamiltonians, but it has been proved that *ergodic eigenstates* are supported by the entire energy shell. In this case, the unitary transformation that factorizes the state can have no classical correspondence.

A lot of the experimental work related to entanglement has been carried out in quantum optics. Rarely is the full generality of semiclassical states employed there and one can rely mainly on states derived from the eigenstates of the harmonic oscillator, even when phase space is invoked [8]. For this reason, the initial examples of phase space structures are here chosen among states of this type. The reader who wishes to avoid the more subtle aspects of semiclassical theory can mostly skip Sects. 4.4, 4.10 and parts of 4.5.

4.2 Entanglement and Classical Physics

Entanglement is considered to be a quintessential quantum property which defies all attempts at a classical correspondence. For this reason, its description in terms of the classical concept of phase space might appear foolhardy. It could be that the semiclassical program of uncovering meaningful relationships between XIX'th and XX'th century mechanics would be overstretched.

Perhaps, though, such an endeavour would make more sense if it were recalled that the usual validity of a classical description of macroscopic phenomena can be attributed to the effect of decoherence. In its turn, this results from the entanglement of a given system with an uncontrolled environment, caused by interactions that can be minimized, but never entirely eliminated. Thus, in spite of the fact that the common working languages employed in classical and quantum mechanics are quite alien to each other, it is hard to fully comprehend why the outcome of decoherence should be the emerging appropriateness of a classical description for quantum systems, unless we can detect its traces even within entanglement itself. A simplified version of this program will be sketched in Sect. 4.10.

Before attempting to establish a bridge between some features of quantum entanglement and classical mechanics, it is worthwhile to consider the more obvious way in which interference already separates these theories. In contrast, the analogy of quantum mechanics with classical waves is much smoother: The latter may be superposed linearly and they interfere in the same way as matter waves. In a simple two-slit experiment, the initial quantum state is prepared as a coherent superposition of momentum eigenstates, with eigenvalues that can be *classically* measured: The probability for each momentum direction is the same as for a uniform ensemble of *classical* states. The evolution through a pair of slits generates classical interference, equally observable in water waves, or sound waves. Quantum strangeness only emerges if the intensity of the resulting interference pattern for the conjugate variable, the position, is interpreted as the probability for the position measurement of a single particle, moving according to classical mechanics. Even so, the particular nature of quantum measurement itself does not play a prominent role in the phenomenon of quantum interference. The subsequent quantum state is certainly redefined by the result of the measurement, but this is not a crucial feature of quantum interference, no matter how unclassical its interpretation for a single particle.

The success of the semiclassical treatment of interference phenomena is no real surprise. If we start from Feynman's path integral formalism [9, 10], quantum evolution is described by a continuum of interfering paths. Semiclassical theory merely groups these around a few particular classical trajectories with their Feynman phase. The amplitude of each of these discrete interfering terms is then given by a local integration over the continuum of paths. Classical mechanics takes its part in the theory, but there is no limitation to classical phenomena and interference is well described. Indeed, the role of classical mechanics is the same as ray optics in classical wave theory.

In contrast to interference, the unintuitive nature of entanglement is derived from that of quantum measurement itself. In no way does this tally with the common sense description of the macroscopic world. Nothing in our everyday experience prepares us for the collapse of a state that is measured into one of several possibilities. The common sense presupposition would be that the effect of the measurement on the system could and should be made negligible. Entanglement highlights this phenomenon in a specially subtle way because it involves pairs of measurements on systems with at least two degrees of freedom, or components.

If we consider classical waves, or particles, it would be indeed strange to imagine that such a collapse could result from the measurement of a subsystem, thus constraining the possible states of the complementary subsystem: It is well known that playing a note on a piano, i.e., exciting a finite string, will provoke a response on the next octave string. Here we have two nearly independent systems, stretched strings, weakly coupled by the surrounding air. Perhaps, it is better to consider the same note on two nearby pianos, so that we consider the interaction of identical systems. The wave form assumed instantaneously by the pair of strings may be used to describe the state of the whole system, or else, we may prefer the Fourier representation, in terms of the eigenstates for the discrete set of allowed frequencies of each string. These classical strings are completely analogous to the textbook example in quantum mechanics of particles moving in 1-D, each in its own box. But there is no way in which a photograph of one of the piano strings will affect the sound produced by the other string, no matter how entangled the quantum analogues happen to be! Likewise, the measurement of the frequency spectrum of the vibrations of one of the strings does not oblige it to choose among the various overtones and we would be even more surprised if this led to a correlated jump in the other string.

Yet this is just what we would expect for the analogous quantum system composed of two particles in their 1-D boxes, coupled by the same Hamiltonian that may account for the atmospheric interaction. Such a measurement would single out a discrete energy, or equivalently a discrete momentum modulus. Furthermore, in the quantum system, we could also measure the position of the particle, with a probability density that is specified by the wave intensity. No equivalent interpretation can be imputed to the classical wave, so that such a *position measurement* would then be devoid of meaning.[1]

Just as there are measurements on a quantum system that are meaningless for a classical wave, there are others that make no sense for a classical particle. Consider the excitation of a piano string by the same note, but played on a clarinet. This has only even harmonics because it is equivalent to a string that is free on one side. Then only the even harmonics will be excited in the string, which will be symmetric about its midpoint. Such an even (or odd) parity, i.e., the symmetry (or antisymmetry) of the classical stationary wave, is certainly a measurable property of the analogous quantum state. Indeed, even a classical wave, a string that is free on one side, could in principle be used as a probe to measure directly the even component of the wave, instead of exciting it. But what would it mean to measure the parity of the corresponding classical particle in a box? Generalizations of such *parity*

---

[1] It should be remembered that the classical particle analogy here is not related to the phonons that are generated by second quantization within each mode.

measurements, distinguishing the eigenvalues of *non-mechanical observables*, will play a major role in the following discussions of entanglement.

Measurement theory lies outside the scope of a semiclassical treatment. However, such experimental outcomes will be preceded by (unitary) quantum evolution, which is not so adverse to a classical description. Indeed the process by which subsystems become entangled is a *preparation* that precedes any quantum measurement. It is only in the probabilistic interpretation of the subsequent measurement on the system that the quantum and the classical viewpoints fundamentally diverge. As it happens, standard measures of entanglement require that the components of the system be completely defined, but do not pre-specify the measurements to be performed. Thus, the presence of entanglement only indicates the possibility that some subset of measurements will have nonclassical correlations. It is precisely this lack of definition with respect to future quantum measurements that allows space for a semiclassical treatment.

The study of classical waves displays many of the properties of a simple quantum system. Indeed, Rayleigh's *The Theory of Sound* [11] anticipates some results later rediscovered in semiclassical theory. However, each piano string is a system with an infinite number of degrees of freedom. Though it is not forbidden to consider coupled fields,[2] the following lectures will concern mainly systems with a finite number of degrees of freedom. In most cases, two degrees of freedom already suffice to discuss the relation between the concept of entanglement and classical mechanics. So we start with a review of classical–quantum correspondence.

4.3 Classical–Quantum Correspondence

The simplest quantum systems with a classical correspondence have a single degree of freedom, e.g., a particle constrained to move on a straight line. The classical state of the system is described by its position, q, and its momentum, p. Together they define a point in phase space, $x = (p, q)$, which is a 2-D plane. Perhaps, classical state space would be a more appropriate term because each point specifies all future motion of a classical system, once the Hamiltonian, $H(x)$, is specified, through Hamilton's equations:

$$\dot{p} = -\frac{\partial H}{\partial q}, \quad \dot{q} = \frac{\partial H}{\partial p}. \tag{4.1}$$

These equations may be compactified into the form

$$\dot{x} = \mathbf{J}\frac{\partial H}{\partial x}, \tag{4.2}$$

with the definition of the (2×2)-dimensional matrix

---

[2] Perhaps, quantum superstring theory will tackle entanglement someday.

$$\mathbf{J} = \left[\begin{array}{c|c} 0 & -1 \\ \hline 1 & 0 \end{array} \right], \qquad (4.3)$$

acting on the phase space points, $x = (p, q)$. Unless $H(x)$ is quadratic, this motion is nonlinear.

Corresponding to this 2-D plane, quantum mechanics matches the states $|\psi\rangle$ of an infinite-dimensional Hilbert space, $\mathbf{H}$, on which act the operators, $\widehat{q}$ and $\widehat{p}$. Each eigenstate of $\widehat{q}$, labeled by the eigenvalue q_0, corresponds to the vertical line, $q = q_0$, whereas the horizontal phase space lines are matched by eigenstates of $\widehat{p}$. These operators do not commute, $[\widehat{p}, \widehat{q}] = i\hbar$, but if we appropriately symmetrize the order in which p and q appear in $H(x)$, then the motion of the states $|\psi\rangle$ is also determined by the quantum Hamiltonian $H(\widehat{x})$, through the linear equation

$$i\hbar \frac{\partial}{\partial t} |\psi\rangle = H(\widehat{x}) |\psi\rangle, \qquad (4.4)$$

i.e., Schrödinger's equation.

The uncertainty principle excludes the existence of a quantum state that corresponds precisely to a phase space point. However, the unavoidable dispersion in measurements of position or momentum allows to seek an approximate correspondence with probability distributions of phase space points. This is unsatisfactory as far as interpretation is concerned because probabilities are associated to the square of a state rather than the state itself. Nonetheless, a certain intuition can be obtained through this analogy. Given a phase space probability density, $f(x)$, the expectation value of any *classical observable* $O(x)$ is given by

$$E(O) = \int \mathrm{d}x \, O(x) \, f(x) \,. \qquad (4.5)$$

Hence, the dispersions in position and momentum are $\delta q^2 = E\left((q - E(q))^2\right)$ and $\delta p^2 = E\left((p - E(p))^2\right)$. The uncertainty principle then imposes that only phase space distributions for which $\Delta' = \delta q \delta p \geq \hbar$ should be considered.

However, this quantity is not a classical invariant. The flow, $x(0) \to x(t)$, generated by the Hamiltonian is a *canonical transformation*, so that [12], for all t,

$$\oint_{\gamma_0} p(0) \cdot \mathrm{d}q(0) = \oint_{\gamma_t} p(t) \cdot \mathrm{d}q(t) \,, \qquad (4.6)$$

where γ_0 is any circuit and $\gamma_0 \to \gamma_t$. General Hamiltonian evolution will stretch and bend any closed curve that is initially compact, so that a probability distribution that is unity inside γ_0 and zero outside will not have constant Δ'. Linear canonical transformations, that is, *symplectic transformations*, are well known to be specially favourable for classical–quantum correspondence, as will be further discussed. It will be shown in Sect. 4.6 that, Δ, the determinant of the *covariance matrix*,

$$\mathbf{K} = \left[\begin{array}{c|c} \delta p & \delta pq \\ \hline \delta pq & \delta q \end{array} \right], \tag{4.7}$$

where $(\delta pq)^2 = E\,(pq - E(p)E(q))$, is invariant under symplectic transformations.

To discuss entanglement, we need more than one degree of freedom. Quantum states can then be decomposed into a basis of product states,

$$|\psi\rangle = |\psi_1\rangle \otimes \ldots |\psi_l\rangle \otimes \ldots |\psi_L\rangle, \tag{4.8}$$

which span the full Hilbert space, $\boldsymbol{H} = \boldsymbol{H}_1 \otimes \ldots \boldsymbol{H}_l \otimes \ldots \boldsymbol{H}_L$, i.e., the tensor product of the factor Hilbert spaces that describe each degree of freedom. Likewise, the full phase space is now a Cartesian product of 2-D conjugate planes, each the phase space for a particular degree of freedom,

$$x = x_1 \times \ldots x_l \times \ldots x_L \tag{4.9}$$

and thus has $2L$ dimensions. However, we must be wary of the difference between the classical and quantum geometries: a phase space strip, δq, corresponds to this range of eigenvalues for the operator $\widehat{q}$. This set of eigenstates spans an infinite-dimensional subspace of the product Hilbert space, whatever the number of degrees of freedom. On the other hand, each of these position eigenstates corresponds to one of the parallel L-D q-planes within the $2L$-D phase space strip.

The classical or quantum motion for systems with more than one degree of freedom is still defined by a Hamiltonian, $H(x)$, or $H(\widehat{x})$, but now $\partial H / \partial x$ is a $2L$-dimensional vector and $\mathbf{J}$ is a block matrix. We shall also use the *skew product*,

$$x \wedge x' = \sum_{n=1}^{L} (p_l q_l' - q_l p_l') = \mathbf{J}\, x \cdot x'. \tag{4.10}$$

This *symplectic area* of the parallelogram formed by the vectors x and x' is invariant with respect to symplectic transformations. Again, these are linear canonical transformations, with (4.6) interpreted as a line integral in the $2L$-D phase space. For higher dimensional systems, all even dimensional volumes, from 2 to $2L$, are preserved by canonical transformations [12].

If the degrees of freedom are completely decoupled, each with its own probability distribution, $f_l(x_l)$, the full probability distribution will be just the product,

$$f(x) = f_1(x_1) \ldots f_l(x_l) \ldots f_L(x_L). \tag{4.11}$$

In this case, the probability distribution for a single degree of freedom is reobtained by *tracing over* the other variables:

$$f_1(x_1) = \int f(x)\, \mathrm{d}x_2 \ldots \mathrm{d}x_L. \tag{4.12}$$

If the full probability distribution cannot be factored into a product, the above equation then defines the *marginal distribution*. This process foreshadows that of partial tracing over the density operator, to be studied in Sect. 4.7, which is central to the study of entanglement.

The product nature of the classical distribution will be retained throughout the evolution if the full classical Hamiltonian is purely additive,

$$H(x) = H_1(x_1) + \ldots H_l(x_l) + \ldots H_L(x_L), \qquad (4.13)$$

i.e., if there is no coupling between the motions of the several degrees of freedom. This follows from the decoupling of Hamilton's equations into

$$\dot{x}_l = \mathbf{J}_l \frac{\partial H_l}{\partial x_l}, \qquad (4.14)$$

for each degree of freedom. In other words, if $k \neq l$, then $x_l(t; x_{l0})$ does not depend on x_k (nor on the initial value, x_{k0}). Furthermore, we then have

$$f_l(t; x_l) = f_l(0; x_l(-t; x_l)) \text{ and } f(t; x) = f_1(t; x_1) \ldots f_L(t; x_L), \qquad (4.15)$$

where $x_l(-t; x_l)$ specifies the past location of x_l. Likewise, the volumes in each subspace will be preserved, and the conservation of the $2L$-dimensional volume is just that resulting from the conservation of the factor volumes.

For a classical system, the transition from product probabilities to general probabilities can only be generated by coupling terms in the driving Hamiltonian, containing cross products, which are at least bilinear in the different variables. A general classical observable will be a function of all the phase space variables, and its expectation is accordingly given by (4.5). For instance, this might be the *either-or observable*, $O_1 = \pm 1$ for detecting some physical properties associated with one particle, or the detection of $O_2 = \pm 1$ for a second particle. For classical particles that have been allowed to drift sufficiently far from each other after interacting, the result of the O_1 measurement will not affect the O_2 measurement and vice versa. Therefore, the correlation must be represented in the form

$$E(O_1; O_2) = E(O_1 O_2) = \int dx_1 dx_2 \, O_1(x_1) \, O_2(x_2) \, f(x_1, x_2). \qquad (4.16)$$

This equation has the same form as correlations postulated for *local hidden variable theories* [4, 13]. Indeed, one of the reasons for this choice is that (4.16) must hold for any evolution of $f(x)$ governed by classical mechanics. This form for the correlation between different components of the system is then taken as a prerequisite for theories that in all other respects should give the same results as quantum mechanics. Since this is certainly not one of the objectives of classical mechanics, such conjectures then necessarily demand extra, unknown and hence *hidden* variables.

It is due to the seminal work of Bell [4] that we are able to compare, through inequalities, the correlations predicted by quantum mechanics with a very wide range of possible local correlations. The point is that any measurement affects the entire quantum state, i.e., both its components, unless the state happens to be an eigenstate of the measured observable. So quantum measurements are not local in the sense that led to (4.16). In case of the general CHSH inequality [6, 14], involving either-or observables, O_{1a}, O_{1b}, O_{2a} and O_{2b}, (4.16) implies that

$$|E(O_{1a}; O_{2a}) + E(O_{1a}; O_{2b}) + E(O_{1b}; O_{2a}) - E(O_{1b}; O_{2b})| \leq 2. \qquad (4.17)$$

As well as constraining possible hidden variable theories, this inequality can be used as a detector of nonclassical correlations in quantum mechanics. This kind of nonclassicality, *entanglement*, is much more subtle than quantum interference effects, as will be discussed in the later sections. A dip into *Bertlmann's socks and the nature of reality* [5, 15] by Bell provides a delightful discussion of all the main points concerning classical locality versus quantum correlations. The book by Peres [13] is also recommended.

It is worthwhile to discuss some specific examples of systems with more than one degree of freedom. An obvious possibility is a collection of particles, each moving in one dimension. Another is a single particle moving in two, or three dimensions. Classical and quantum mechanics make no distinction between these alternative interpretations of the dynamical variables. All that is demanded is that the variables pertaining to different degrees of freedom commute, $[\widehat{p}_k, \widehat{q}_j] = i\hbar\delta_{kj}$, or, correspondingly, that the classical Poisson bracket $\{p_k, q_j\} = \delta_{kj}$ (see e.g., [16]). We can also use angular momentum and their conjugate angles. But are other variables, obtained through classical canonical transformations, allowed?

For example, consider our piano string, now modeled as L masses connected by harmonic springs. We can switch to the L normal modes of vibration. This is a linear canonical transformation, which substitutes the original L conjugate planes, $x_l = (p_l, q_l)$, by new conjugate planes, $x_l' = (p_l', q_l')$, that now describe collective motions of the L masses. This is also a proper phase space to be quantized, $x_l' \to \widehat{x}_l'$. Another important example of a quantizable canonical transformation follows from the description of a collection of particles in terms of the centre of mass together with internal coordinates.

Whatever the physical realization, symplectic transformations correspond exactly to unitary quantum transformations and hence to equivalent quantum systems [17]. These transformations generally redefine the components of the full system and may take an entangled state into a product state, or vice versa. Any measure of entanglement is affected by such a general transformation, so one requires only that the measure be invariant with respect to local unitary transformations, lying within each separate component. As for nonlinear canonical transformations, these are not exactly matched by quantum unitary transformations [17] and, hence, cannot be directly quantized. It might still be useful sometimes to push this correspondence through, but it must be remembered that the result is only a semiclassical approximation.

Taking again the continuum limit, $L \to \infty$, each normal mode of the finite chain converges onto one of the lower modes of the continuous string. There is no essential difference between the interaction and hence the entanglement among these modes of the continuum and that of finite modes (caused by residual non-quadratic terms in the Hamiltonian). In each case, there corresponds a plane in the phase space, which is of infinite-dimension in the case of a field. The entanglement between modes of the electromagnetic field within a finite cavity also has a similar interpretation in terms of a classical field. The unperturbed motion is now that of a quantized harmonic oscillator, corresponding to a classical oscillation in each phase plane .

Another example is that of a particle with *internal structure*. The latter may be described by an angular momentum, coupled to the translational degrees of freedom by an external field. The Stern–Gerlach experiment describes just such a system, in which the magnetic moment, tied to the spin angular momentum of the electron is coupled to its position by an inhomogeneous magnetic field. The spin is an intrinsically quantum mechanical two-level system and the interest in quantum information theory tends to emphasise such simple quantum systems. But, in principle, there is no difference between this case and a Rydberg atom, prepared in a state with a large electric dipole moment, coupled to position through an inhomogeneous electric field. Such a system can be described more naturally in classical terms. Cavity quantum optics deals with the coupling and hence the entanglement of the internal states of individual Rydberg atoms with a specific mode of the electromagnetic field.

For all these systems, coupling terms in the overall Hamiltonian will destroy the product form of an initially decoupled quantum state, or classical distribution. We should bear in mind three basic differences between classical and quantum systems: (i) the nature of the initial state; (ii) the nature of the evolution and (iii) the effect of experiments. As we have seen, the last is the most radical difference, which, indeed, gives rise to the concept of entanglement. Our objective here is to cast the quantum mechanical description of (i) and (ii) in the most classical terms possible, so as to highlight the truly innovative elements of the quantum theory when (iii) is considered.

A fundamental difference between the quantum and classical descriptions should be discussed before proceeding: The analogy between the evolution of classical probability distributions and quantum states is somewhat deceptive in as much as the latter determine only probability amplitudes that can be complex and interfere with each other. To arrive at a closer analogue of probabilities, we should, in some sense, square the quantum states. The correct procedure is to define density operators, or their phase space representation, Wigner functions, to be studied in Sect. 4.6. However, their evolution is nonclassical, unless the Hamiltonian is quadratic.

It will be only in the context of the density operator that it becomes meaningful to distinguish between pure states and mixed states. Taking an average over a set of probability distributions defines a new probability distribution.

Likewise, if we superpose the corresponding quantum states, $|\psi_j\rangle$, we obtain a new quantum pure state. But if we average over the corresponding *pure state density operators*, $\rho_j = |\psi_j\rangle\langle\psi_j|$, there results a mixed state. The latter will be discussed in Sect.4.6.

Now it is important to bring out a special form of state superposition. This is the *Schmidt decomposition*,

$$|\psi\rangle = \sum_j \lambda_j \, |\psi_1\rangle_j \otimes |\psi_2\rangle_j, \qquad (4.18)$$

which exists for any bipartite state (see Sect. 2.3.1 or e.g., [6]). It must be emphasised that both factor states in the above tensor products may themselves correspond to several degrees of freedom, but the result is only proved if there are only two of them. The product states form a particular orthonormal basis in which to describe the state, $|\psi\rangle$, so that the real, non-negative coefficients, λ_j, satisfy $\sum_j \lambda_j{}^2 = 1$. The state is entangled, unless $\lambda_j = \delta_{1,j}$. The Schmidt decomposition is often employed for the description of entangled states in finite Hilbert spaces. In this case, the number of nonzero eigenvalues, λ_j, is a relevant quantifier of entanglement, known as the *Schmidt number*. For infinite-dimensional Hilbert spaces, there may be an infinite number of nonzero Schmidt coefficients.

4.4 Semiclassical Quantum States

Consider a momentum eigenstate $|p'\rangle$ for $L = 1$. In the momentum representation, this is just

$$\langle p|p'\rangle = \delta(p' - p)\,, \qquad (4.19)$$

which is not in a good form for semiclassical extrapolation. For this purpose, it is better to use the complementary representation,

$$\langle q|p'\rangle = \exp\left(\frac{iqp'}{\hbar}\right) = \exp\left(i\frac{S_{p'}(q)}{\hbar}\right)\,. \qquad (4.20)$$

The phase in this expression can be interpreted as the area between the classical curve (the straight line $p' = p$) and the q-axis. There is also an arbitrary constant phase, which is established by the choice of the initial point for the integral,

$$S(q) = \int_{q_0}^q p(q)\,\mathrm{d}q\,. \qquad (4.21)$$

Consider now a general observable, $K(\widehat{p},\widehat{q})$. Its eigenstates correspond classically to curves, γ, in phase space: $K(p,q) = k$. These may be viewed locally as (possibly multivalued) functions, $p_j(q)$. Then the simplest semiclassical approximation is

$$\langle q|k\rangle = \sum_j A_j(q) \exp\left[i\left(\frac{S_j(q)}{\hbar} + \nu_j\right)\right], \tag{4.22}$$

see e.g., [18]. The phases, $S_j(q)$, are again obtained from (4.21). The extra constant phases, ν_j, are known as *Maslov indices* [18–20], but they will not be discussed here. The amplitudes, $A_j(q)$, are defined purely in terms of the classical structure. They are finite wherever the vertical line, $q = constant$, intersects the classical curve transversely. Where this vertical line is tangent to the classical curve, such as q_c in Fig. 4.2, the amplitude diverges. These points where the semiclassical approximation breaks down are known as *caustics*. The different branches of the function $p_j(q)$ are connected at caustic points.

In the case of bound eigenstates of $\widehat{K}$, the curves γ are closed. Then the eigenvalues are approximately obtained by the Bohr–Sommerfeld quantization condition,

$$\oint_\gamma p\,dq = (n + \frac{1}{2})\hbar. \tag{4.23}$$

The quality of the semiclassical approximation for both the states themselves and their eigenvalues improves for large quantum numbers n. Ground states, including that of the harmonic oscillator, are badly described by these approximations.

Even for large n, a closed curve, γ, must inevitably have at least a pair of caustics. The way around this is to switch to the p-representation. Then the vertical tangent at the caustic position, q_c, shown in Fig. 4.2, would correspond to the state $\langle p|q_c\rangle$, which is in a nice semiclassical form. This means that the local branch of the multivalued function $q(p)$ gives rise to a semiclassical approximation that is a superposition of terms of the form

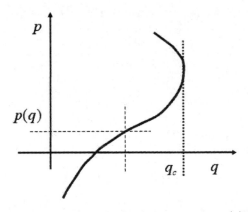

Fig. 4.2. The caustic of the semiclassical approximation to $\langle q|k\rangle$ lies in the neighbourhood of the point q_c, where the tangent to the *classical curve* is vertical. The projection of this region onto the p-axis is nonsingular, leading to a good semiclassical approximation for $\langle p|k\rangle$

$$\langle p|k\rangle = B(p) \exp\left[i\left(\frac{S(p)}{\hbar} + \mu\right)\right].$$ (4.24)

This allows us to define the correct semiclassical approximation in the q-representation through the caustic region by the Fourier transform

$$\langle q|k\rangle = \frac{1}{(2\pi\hbar)^{1/2}} \int \mathrm{d}p\, \langle p|k\rangle \exp\left(\frac{iqp}{\hbar}\right),$$ (4.25)

which leads to a more refined approximation in terms of Airy functions instead of exponentials. This is usually referred to as the Maslov method of dealing with caustics [20] (also discussed in [18]).

Let us now consider a product state for $L > 1$. Then,

$$\langle q|p'\rangle = \exp\left(\frac{iq_1 p_1'}{\hbar}\right) \ldots \exp\left(\frac{iq_L p_L'}{\hbar}\right) = \exp\left(\frac{iq \cdot p'}{\hbar}\right),$$ (4.26)

and we can generalize the definition of action,

$$S(q) = \int_{q_0}^{q} p(q) \cdot \mathrm{d}q.$$ (4.27)

This does not depend on the choice of path between q_0 and q because $p'(q)$ is a constant in this simple case. Hence, this function defines a *Lagrangian surface*, i.e., a surface such that

$$\oint p \cdot \mathrm{d}q = 0,$$ (4.28)

for any (reducible) circuit [12].

In general, the product state will involve arbitrary eigenstates of L observables, $\widehat{K} = \widehat{K}_1 \widehat{K}_2 \ldots \widehat{K}_L$, each in its own Hilbert space:

$$\langle q|k\rangle = \langle q_1|k_1\rangle \ldots \langle q_L|k_L\rangle.$$ (4.29)

The wave function will be a superposition of terms with the form

$$\langle q|k\rangle = \prod_l A_l(q_l) \exp\left[\frac{i}{\hbar}\left(S_1(q_1) + \ldots + S_L(q_L)\right)\right],$$ (4.30)

one term for each branch of the functions, $p_l(q_l)$.

Defining again $S(q)$ as the above phase, it is seen to be independent of the order in which we progress along each segment (q_{0l}, q_l), while keeping the other integration variables constant: The definition (4.21), now reinterpreted as a path integral, is independent of the path on the surface. Therefore, this more general surface, $K(p,q) = k$, is also Lagrangian.

If the surface is the product of L-quantized *circles* (closed curves), it will be an L-torus, τ. Each of the L irreducible circuits, γ_l, must then satisfy the Bohr–Sommerfeld conditions,

$$\oint_{\gamma_l} p_l \cdot \mathrm{d}q_l = (n_l + \frac{1}{2})\hbar, \tag{4.31}$$

or some suitable generalization (see e.g., [18]). Notice that the line integral here used is not restricted to plane sections of τ because all topologically equivalent circuits on a Lagrangian surface must have the same action.

Let us now evolve the product state semiclassically. The basic result, due to van Vleck [21], can be reinterpreted as the statement that classical and quantum evolutions *commute*. In other words, we can evolve classically each curve, γ_l, if there are no cross terms in the Hamiltonian, so that the different degrees of freedom are decoupled. Each evolved observable then corresponds to $K_l(x_l, t) = K_l(x_l(x_{l0}, t), 0)$ and we approximately reconstruct the classically evolved state from the evolved torus, $\tau_l(t)$, which is the product of the $\gamma_l(t) : K_l(x_l, t) = k_l$.

Notice that this classical evolution of products of curves fits into the general view concerning the evolution of product probability distributions in the previous section, by merely choosing $f_l(t; x_l) = \delta(K_l(x_l, t) = k_l)$ and running time backwards. The important distinction between classical and semiclassical evolution is that the latter contains interferences between the different branches of the evolving classical curve. Each representation exhibits these interferences in a different way.

Just as cross terms containing products of the different variables in the Hamiltonian destroy the product form of a classical probability distribution, the classically evolved L-D surface corresponding to an original product state also ceases to be a product. However, the smoothness of the evolution implies that the topology of the surface must be preserved (be it plane, torus, or, in between cylindrical). Furthermore, the classical evolution, $x_0 \rightarrow x_t$, is a canonical transformation, and hence all reducible circuits on the evolved surface have zero action, i.e., τ_t still has the Lagrangian property, which allows to define the path-independent action $S(q)$, and the irreducible circuits of τ_t still satisfy the same Bohr–Sommerfeld conditions to first order in $\hbar$.

Let us investigate further the case of two degrees of freedom. The separable torus, $\tau = \gamma_1 \otimes \gamma_2$, can be pictured through the separate γ_1 and γ_2 curves. These coincide with sections of the 2-D torus by alternative 3-D planes (the normal case for *Poincare sections*, see e.g., [18, 19]). The γ_2 curve does not depend on the choice of the $q_1 = constant$ section. The separable torus projects as a rectangle onto position space (q_1, q_2), as shown in Fig. 4.3. Within this rectangle, there are four different branches of the torus, which project onto each position, q, corresponding to the combinations of the two branches of each circle. The caustics at the side of the rectangle are *double fold lines*.

After a general canonical evolution, the sections of τ are no longer equal for different choices of $q_1 = constant$ (or $q_2 = constant$), though all the sections have the same area, S_1 (or S_2). In some cases (to do with time invariance of the Hamiltonian), the projection onto the q-plane will merely distort the rectangle, which will still have finite-angled corners connecting

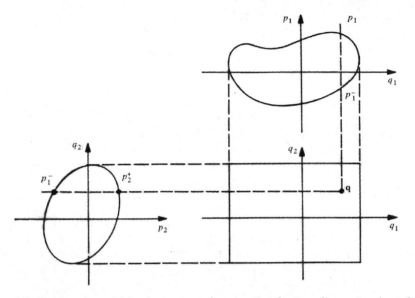

Fig. 4.3. Each point within the rectangular caustic of a two-dimensional product torus is the image of four phase space points under p-projection

double-fold lines. But in general, these corners, *hyperbolic umbilic points*, will unfold in the generic form specified by catastrophe theory, as shown in Fig. 4.4. There are four possibilities for the topology of the unfolding of the rectangle, shown in Fig. 4.5. For $L > 2$, the projection of the L-torus onto the L-D q-plane will be a solid hypercube that will be distorted, or unfolded by the motion generated by a coupling Hamiltonian. (These geometries are reviewed in [18], but are more thoroughly discussed in [22].)

The representations of quantum states in terms of orthogonal position, or, alternatively, momentum eigenstates are the best that we can do because of Heisenberg's uncertainty principle. Semiclassically, this corresponds to viewing a Lagrangian surface through a set of Lagrangian planes that *foliate*

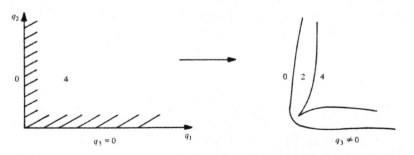

Fig. 4.4. Catastrophe theory establishes the generic form for the unfolding of the double-fold caustic at each corner of the projection of a product torus as it evolves

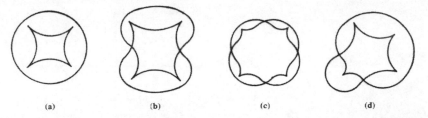

Fig. 4.5. The full topology of the full fold lines is not determined by catastrophe theory: Each of the above forms corresponds to a different symplectic evolution from an initial product torus

phase space. We switch from the q-representation to the p-representation by means of a Fourier transform of $\langle q|\psi\rangle$. This corresponds classically to taking the *Legendre transform* of $S(q)$ [12]. For $L > 1$, we may take the Fourier transform for a subset of the degrees of freedom. This corresponds to using a classical description in terms of the alternative Lagrangian planes $(p_1, \ldots, p_l, q_{l+1}, \ldots, q_L)$.

One way to achieve a full phase space description is to use the basis of *coherent states* [8, 23–26], labeled by the phase space vector, $\eta = (\eta_p, \eta_q)$,

$$\langle q|\eta\rangle = \left(\frac{\omega}{\pi\hbar}\right)^{1/4} \exp\left[-\frac{\omega}{2\hbar}(q - \eta_q)^2 + i\frac{\eta_p}{\hbar}(q - \frac{\eta_q}{2})\right] . \tag{4.32}$$

Even though the coherent state basis is overcomplete, the exact decomposition,

$$|\psi\rangle = \frac{1}{\pi} \int d\eta |\eta\rangle\langle\eta|\psi\rangle , \tag{4.33}$$

is unique. The coherent states are phase space translations of the ground state of the harmonic oscillator (with unit mass):

$$\langle q|0\rangle = \left(\frac{\omega}{\pi\hbar}\right)^{1/4} \exp\left(-\frac{\omega}{2\hbar}q^2\right) . \tag{4.34}$$

These result from the action of the *translation operator*:

$$\hat{T}_\eta = \exp\left[\frac{i}{\hbar}(\eta_p \cdot \hat{q} - \eta_q \cdot \hat{p})\right] = \exp\left(\frac{i}{\hbar}\eta \wedge \hat{x}\right) , \tag{4.35}$$

using the skew product (4.10). If either $\eta_p = 0$ or $\eta_q = 0$, we obtain the usual translation operators for momenta, or positions, respectively. The arbitrary phase due to noncommutation of $\hat{p}$ and $\hat{q}$ is here chosen in the most symmetric way, using the Baker–Hausdorff relation [27].

In quantum optics, it is customary to switch to the basis of creation and annihilation operators $(\hat{q} \pm i\hat{p})/\sqrt{2\hbar}$. In this context, the translation operator

(4.35) depends on the complex chords $(\eta_p \pm i\eta_q)/\sqrt{2\hbar}$ and is called the *displacement operator* [23]. The semiclassical limit for a complex phase space is not as transparent as the real theory treated here. However, it is quite feasible to effect phase space translations in an experimental optical context [28].

The coherent state representation is not orthogonal and is overcomplete. The alternative, to be explored in the next section, is to work directly with operators: We represent operators in *orthogonal* operator bases in analogy to the way that quantum states are commonly decomposed. This allows us to work directly with the translation operators, without having to apply them to the ground state of the harmonic oscillator.

4.5 Operator Representations and Double Phase Space

The linear operators, $\widehat{A}$, that act on the quantum Hilbert space form a vector space of their own: $|A\rangle\!\rangle$. Defining the *Hilbert–Schmidt product*,

$$\langle\!\langle A|B\rangle\!\rangle = \text{tr}\,\widehat{A}^\dagger\widehat{B}\,, \tag{4.36}$$

we find that the dyadic operators $|Q\rangle\!\rangle = |q_-\rangle\langle q_+|$ form a complete basis, i.e.,

$$\langle\!\langle Q|A\rangle\!\rangle = \langle q_+|\widehat{A}|q_-\rangle = \text{tr}\,|q_-\rangle\langle q_+|\widehat{A}\,, \tag{4.37}$$

provides a complete representation of the operator $\widehat{A}$. Here, $\widehat{A}^\dagger$ is the adjoint of $\widehat{A}$. One should note the similarity between this dyadic basis, $|q_-\rangle\langle q_+|$, in the case of $L = 1$ with the basis of product states, $|q_1\rangle\otimes|q_2\rangle$. The substitution of a bra by a ket in the former will in most cases imply no more than complex conjugation.

Thus, we may relate the vector space of quantum operators to a *double Hilbert space* with respect to that of quantum states. Since we have explored the correspondence of the state-Hilbert space with classical phase space, it is now natural to relate the double Hilbert space to a double phase space : $X = x_- \times x_+$ (see e.g., [29]). The operator $|Q\rangle\!\rangle$ should then correspond to the Lagrangian plane $Q = constant$ in the double phase space. This does hold, within a minor adaptation, analogous to the use of the adjoint operator in the definition of the Hilbert–Schmidt product. That is, we should define $Q = (q_-, q_+)$, but $P = (-p_-, p_+)$ as coordinates of the double phase space $X = (P, Q)$.

A good reason for this is that then we include among the set of Lagrangian surfaces in double phase space all the canonical transformations in single phase space, $x_- \to x_+ = \mathbb{C}(x_-)$. This also transports closed curves, $\gamma_- \to \gamma_+$, so that we may rewrite the definition of a canonical transformation as

$$\oint_{\Gamma} P\cdot dQ = 0\,, \tag{4.38}$$

where $\Gamma = (\gamma_-, \gamma_+)$. Thus, we may consider $\gamma_\pm$ as projections of the curve Γ defined on the $(2L)$-dimensional surface, $\Lambda_\mathbb{C}$, which specifies the canonical transformation, within the $(4L)$-dimensional double phase space, $X = (P, Q)$.

It is worthwhile to consider the richness of structures in double phase space. On the one hand, a canonical transformation defines a Lagrangian surface as $x_+(x_-)$, a one-to-one function. On the other hand, the product of a Lagrangian surface, λ_- in x_- with another surface λ_+ in x_+, $\Lambda = \lambda_- \otimes \lambda_+$, is also Lagrangian in double phase space, but projects singularly onto either of the factor spaces. In the case that both surfaces are tori, we obtain a double phase space torus, $\tau = \tau_- \otimes \tau_+$, as if we had doubled the number of degrees of freedom. (All Lagrangian surfaces will hereon be labeled τ, even when they are not necessarily a torus; in the case that $L = 1$, τ is just a closed curve, γ.) If $L = 1$, it will be a 2-D product torus, with the only difference that $p_- \to -p_-$ in the present construction. If each Lagrangian surface corresponds to a state, i.e., $|\psi_-\rangle$ and $|\psi_+\rangle$, then we represent $|\Psi\rangle\rangle = |\psi_+\rangle\langle\psi_-|$ in the $|Q\rangle\rangle$ representation as

$$\langle\langle Q|\Psi\rangle\rangle = \langle q_+|\psi_+\rangle\langle\psi_-|q_-\rangle . \tag{4.39}$$

Therefore, the semiclassical approximation is just a superposition of terms of the form

$$\langle\langle Q|\Psi\rangle\rangle = A_J(Q) \exp[iS_J(Q)/\hbar] \tag{4.40}$$

with

$$A_J(Q) = A_{j-}(q_-)^* \, A_{j+}(q_+) \tag{4.41}$$

and

$$S_J(Q) = \oint_0^Q P_J(Q') \cdot \mathrm{d}Q' . \tag{4.42}$$

Again this is in strict analogy to the construction of semiclassical product states of higher degrees of freedom. Note that the projection of the double Lagrangian torus onto P or Q is just the rectangle discussed previously for product states, whereas the projections onto the planes, x_- and x_+, are specially singular.

The semiclassical approximation for a unitary operator, $\widehat{U}$, that corresponds to a canonical transformation, $\mathbb{C} : x_- \to x_+$, has exactly the same form, i.e., a superposition

$$\langle\langle Q|U\rangle\rangle = \langle q_+|\widehat{U}|q_-\rangle = U_J(Q) \exp[iS_J(Q)/\hbar] , \tag{4.43}$$

for each branch of the function $P_J(Q)$ defined by the Lagrangian surface in double phase space. Note that the situation with respect to projection singularities is now reversed, as compared to $|\Psi\rangle\rangle$. The fact that the projections of the Lagrangian surface, $\Lambda_\mathbb{C}$, onto either x_-, or x_+ are both nonsingular in no way guarantees that the projections onto the P or the Q Lagrangian planes will be likewise free of caustics.

Conversely, any function, $S(Q)$, is, at least locally, the generating function of a canonical transformation through the implicit equations:

$$\frac{\partial S}{\partial Q} = P(Q) \,, \quad \text{or} \quad \frac{\partial S}{\partial q_+} = p_+ \,, \quad \frac{\partial S}{\partial q-} = -p_- \,. \tag{4.44}$$

Here we recognize the standard generating functions $S(q_-, q_+)$ in Goldstein [16]. If $S(Q)$ is quadratic, then these implicit equations will be linear, so that the explicit transformation will result from a matrix inversion (if it is nonsingular). There will be a single branch in $S(Q)$ for such a symplectic transformation, and it turns out that the semiclassical approximation is exact in this case.

The well-known alternatives to these generating functions are usually obtained by Legendre transforms. However, we can consider the $\pi/2$ rotation, $q_+ \to p_+, p_+ \to -q_+$, times the identity in x_-, as an example of canonical transformation in double phase space: $X \to X'$. Then $Q' = (q_-, p_+)$ is also a good Lagrangian plane that can be used as the new coordinate plane for the description of $\Lambda_{\mathbb{C}}$. In the new coordinates, the implicit equations for the canonical transformation are just

$$\frac{\partial S'}{\partial Q'} = P'(Q') \,, \quad \text{or} \quad \frac{\partial S'}{\partial p_+} = -q_+ \,, \quad \frac{\partial S'}{\partial q-} = -p_- \,. \tag{4.45}$$

The correspondence with a semiclassical state,

$$\langle\!\langle Q'|\Psi\rangle\!\rangle = A'_J(Q') \exp[iS'_J(Q')/\hbar] \,, \tag{4.46}$$

will be exact in the case of a symplectic transformation. Note that $|Q'\rangle\!\rangle$ is a first example of an operator basis that corresponds to a set of parallel Lagrangian planes in double phase space, which, nonetheless, have internal coordinates that can be identified with a phase space on its own.

The crucial step is now to explore other kinds of canonical transformations in double phase space [30]. In particular,

$$Q' = \mathbf{x} = \frac{x_+ + x_-}{2} \,, \quad P' = \mathbf{y} = \mathbf{J}(x_+ - x_-) = \mathbf{J}\xi \,. \tag{4.47}$$

Here, the $\mathbf{J}$ symplectic matrix in single phase space is essential to *canonize* what would be just a $\pi/4$ rotation. It accounts for the change of sign in the p_- coordinate. We will here have to bare the discomfort that the canonical coordinate in double phase space is $\mathbf{y}$, but the geometrically meaningful variable in single phase space is ξ, the trajectory *chord*, in the case of continuous evolution. The coordinate $\mathbf{x}$ will be referred to as the *centre*.

If we consider the *horizontal* Lagrangian planes $\mathbf{y} = constant$, each is identified with a uniform classical *translation*. Thus, we have departed from coordinate planes corresponding to dyadic operators to those planes in double phase that describe canonical transformations and hence correspond to unitary transformations. In this case, $x_- \to x_+ = x_- + \xi$ are the group of phase space translations, which include the the identity, i.e., the *identity plane* is defined as $\xi = 0$.

On the other hand, the *vertical* plane, $\mathbf{x} = 0$, defines the canonical *reflection* through the origin, $x_- \to x_+ = -x_-$ (or inversion), since all the chords for this transformation are centred on the origin. Other vertical planes specify reflections through other points, $x_- \to x_+ = -(x_- - 2\mathbf{x})$. The reflections do not form a group on their own (no identity), but together with the translations they form the *affine group* of geometry [31].

Since there is an exact correspondence between linear canonical transformations and unitary transformations, each plane $\mathbf{y} = constant$ corresponds precisely to the translation operator, $\widehat{T}_\xi$, previously defined as (4.35). Notice that this was written with a phase that is a skew product involving ξ, but we could also use $\widehat{T}_\xi = \exp(i\mathbf{y} \cdot \widehat{x}/\hbar)$. In terms of the previous dyadic $|Q\rangle\rangle$ basis, this is expressed as

$$\widehat{T}_\xi = \int d\mathbf{q} \left| \mathbf{q} + \frac{\xi_q}{2} \right\rangle \left\langle \mathbf{q} - \frac{\xi_q}{2} \right| e^{i\xi_p \cdot \mathbf{q}/\hbar}, \qquad (4.48)$$

a symmetrized Fourier transform (see e.g., [32]).

Just as a $\pi/2$ rotation in single phase space, $q \to p$ and $p \to -q$, corresponds to a Fourier transform, so the transformation between *horizontal* and *vertical* planes in double phase space is also achieved by a full Fourier transform (except for an annoying factor of 2^L):

$$2^L \widehat{R}_\mathbf{x} = \int \frac{d\xi}{(2\pi\hbar)^L} \, \widehat{T}_\xi \exp(\frac{i}{\hbar}\mathbf{x} \wedge \xi). \qquad (4.49)$$

In terms of the dyadic $|Q\rangle\rangle$ basis, we have

$$2^L \widehat{R}_\mathbf{x} = \int d\xi_q \left| \mathbf{q} + \frac{\xi_q}{2} \right\rangle \left\langle \mathbf{q} - \frac{\xi_q}{2} \right| e^{i\mathbf{p} \cdot \xi_q/\hbar}, \qquad (4.50)$$

the complementary symmetrized Fourier transform to (4.48).

We are now free to switch from the usual (position) dyadic basis to the unitary operator basis, $|\mathbf{y}\rangle\rangle = \widehat{T}_\xi$:

$$\langle\langle \mathbf{y}|A\rangle\rangle = \mathrm{tr}\, \widehat{T}_{-\xi}\widehat{A} = A(\xi), \qquad (4.51)$$

where we use $\widehat{T}_{-\xi} = \widehat{T}_\xi^\dagger$. $A(\xi)$ is the *chord representation* of the operator $\widehat{A}$ (also referred to as the chord symbol). Notice that the chord basis includes the identity operator, $\widehat{I} = |I\rangle\rangle = |\mathbf{y} = 0\rangle\rangle$. To verify that (4.51) is indeed the expansion coefficient for an arbitrary operator in the basis of translation chords, we use

$$\mathrm{tr}\, \widehat{T}_\xi = (2\pi\hbar)^L \delta(\xi) = \langle\langle \mathbf{y}|I\rangle\rangle \qquad (4.52)$$

(note the double phase space analogy with $\langle p'|(p = 0)\rangle = \delta(p')$), as well as the quantum version of the group of translations:

$$\widehat{T}_{\xi_2}\widehat{T}_{\xi_1} = \widehat{T}_{\xi_1+\xi_2} \exp\left[\frac{-i}{2\hbar}\xi_1 \wedge \xi_2\right] \qquad (4.53)$$

(see e.g., [33]). Then, the expansion,

$$\widehat{A} = \int \frac{\mathrm{d}\xi}{(2\pi\hbar)^L} \, A(\xi) \, \widehat{T}_\xi \,, \tag{4.54}$$

leads to

$$\begin{aligned}
\mathrm{tr}(\widehat{T}_{-\xi}\widehat{A}) &= \mathrm{tr} \int \frac{\mathrm{d}\xi'}{(2\pi\hbar)^L} \, A(\xi')\widehat{T}_{-\xi}\widehat{T}_{\xi'} \\
&= \int \frac{\mathrm{d}\xi'}{(2\pi\hbar)^L} \, A(\xi') \exp\left[\frac{i}{2\hbar} \, \xi' \wedge \xi\right] \mathrm{tr} \, \widehat{T}_{\xi'-\xi} \\
&= A(\xi) \,.
\end{aligned} \tag{4.55}$$

The chord representation is thus a second example of a representation of operators in terms of an operator basis that can be identified uniquely to a phase space. Indeed, each chord corresponds to a Lagrangian surface in double phase space and hence a particular uniform translation in single phase space.

The next representation will be based on phase space reflections, $\widehat{R}_{\mathbf{x}}$. But first, it is worthwhile to examine some characteristics of these operators. Unlike the translations, they do not form a group on their own, though they combine with the latter to form the affine group. The products are [33]

$$\widehat{R}_{\mathbf{x}}\widehat{T}_\xi = \exp[-\frac{i}{\hbar}\mathbf{x} \wedge \xi] \, \widehat{R}_{\mathbf{x}-\xi/2} \,, \tag{4.56}$$

$$\widehat{T}_\xi \widehat{R}_{\mathbf{x}} = \exp[-\frac{i}{\hbar}\mathbf{x} \wedge \xi] \, \widehat{R}_{\mathbf{x}+\xi/2} \tag{4.57}$$

and

$$\widehat{R}_{\mathbf{x}_2}\widehat{R}_{\mathbf{x}_1} = \exp[\frac{2i}{\hbar}\mathbf{x}_1 \wedge \mathbf{x}_2] \, \widehat{T}_{2(\mathbf{x}_2-\mathbf{x}_1)} \,. \tag{4.58}$$

Except for the phases, these are just the classical relations. The last one is specially interesting. Note that $\widehat{R}_{\mathbf{x}}^2 = \widehat{I}$, the identity, hence the (degenerate) eigenvalues of $\widehat{R}_{\mathbf{x}}$ must be either $+1$, or -1. Therefore, these operators are Hermitian, as well as unitary.

Are they true observables? Consider the effect of $\widehat{R}_0$ on the eigenstates of the harmonic oscillator. Taking $q \to -q$ and $p \to -p$, leads to a change of sign for all the odd states, while preserving the even states. In other words, the latter are just the $(+1)$-eigenstates, while the odd states are (-1)-eigenstates. Though it is hard to imagine measuring the parity of a particle, we saw in Sect. 4.2 that the parity decomposition of even a classical wave can certainly be effected. Measurements of the eigenvalues of this *non-mechanical observable* are currently performed for single photons in optical cavities [34]. It is true that these measurements are performed on a mode of the electromagnetic field rather than a particle, but it only makes sense to discuss the parity *within a specific mode* if it is quantized.

Reflection operators are very strange observables as far as phase space correspondence is concerned. It was discussed in Sect. 4.4 that usual observables correspond to smooth phase space functions and their eigenvalues correspond to level curves if $L = 1$. This is just not the case of reflection operators with their infinitely degenerate ± 1 eigenvalues. In their dual role as both unitary and Hermitian (observable) operators, reflections are almost schizophrenic: They are perfectly ordinary unitary operators, corresponding to Lagrangian planes in double phase space, but they do not correspond to any smooth classical function in phase space, as expected of a mechanical observable.

This should furnish sufficient motivation to investigate the representation of arbitrary operators in terms of reflection centres. The assumption that

$$\widehat{A} = \int d\mathbf{x}\, A(\mathbf{x})\, 2^L \widehat{R}_\mathbf{x} \tag{4.59}$$

leads to

$$\langle\!\langle \mathbf{x}|A \rangle\!\rangle = \mathrm{tr}\,(2^L \widehat{R}_\mathbf{x})\widehat{A} = \mathrm{tr} \int \frac{d\mathbf{x}'}{(2\pi\hbar)^L} A(\mathbf{x}')(2^L \widehat{R}_\mathbf{x})(2^L \widehat{R}_{\mathbf{x}'}) = A(\mathbf{x})\,. \tag{4.60}$$

This is the *Weyl representation* of the operator $\widehat{A}$ (also known as the Weyl symbol). Once again we use half the coordinates of double phase space, within a Lagrangian plane that is a phase space on its own, to describe a quantum operator. This perception that we are really dealing with different phase spaces for each operator representation was clearly stated in the excellent review by Balazs and Jennings [35]. What was lacking was merely the identification of each of these different phase spaces with a specific foliation of Lagrangian planes in double phase space.

As far as unitary operators, $\widehat{U}$, are concerned, the semiclassical limit of the representations, either in terms of centres or chords, has exactly the same form as for any other Lagrangian basis. For instance, the Weyl symbol will be a superposition of terms, such as

$$U(\mathbf{x}) = A(\mathbf{x}) \exp[iS(\mathbf{x})/\hbar]\,, \tag{4.61}$$

in terms of the centre action, defined as

$$S(\mathbf{x}) = \int_0^\mathbf{x} \mathbf{y}(\mathbf{x}') \cdot d\mathbf{x}' = \int_0^\mathbf{x} \xi(\mathbf{x}') \wedge d\mathbf{x}'\,. \tag{4.62}$$

For symplectic transformations, the Lagrangian surface is a plane, and so there is only a single branch of the action function $S(\mathbf{x})$, which is quadratic. Then (4.61) is an exact representation of the corresponding quantum *metaplectic transformation*. However, in the general nonlinear case, there may be caustics in the projection of the Lagrangian $\mathbf{y}(\mathbf{x}')$ surface onto the $\mathbf{x}$-plane. Recall that this is just the plane that defines the identity operator, $\widehat{I}$ (corresponding to $\mathbf{y} = 0$, or $S = 0$).

For the canonical transformation generated by a Hamiltonian, $H(x)$, it turns out that the generating function has the limit [33]

$$S_\epsilon(\mathbf{x}, t = \epsilon) \rightarrow -\epsilon H(\mathbf{x}) + \mathcal{O}(\epsilon^3). \tag{4.63}$$

There are no caustics for small times in the centre representation, since the corresponding Lagrangian surface is nearly horizontal.

The smooth real Hamiltonian itself can be equated to the Weyl symbol for the corresponding operator, $\widehat{H}$, within semiclassically small ordering terms. This is the case of the Weyl representation for any observable that corresponds classically to a smooth classical function of the points in phase space [17]. Since we can always consider classical observables as infinitesimal generators of motion through Hamilton's equations, it is appropriate to picture them as functions on the $\mathbf{y} = 0$ plane, so that the Hamiltonian vectors form a field on this plane that indicates which way it will evolve. In contrast, the chord symbol for these smooth mechanical observables is not at all smooth. This is because the chord and centre symbols are related to each other through the Fourier transform,

$$A(\xi) = \frac{1}{(2\pi\hbar)^L} \int d\mathbf{x} \, \exp\left(-\frac{i}{\hbar}\xi \wedge \mathbf{x}\right) A(\mathbf{x}), \tag{4.64}$$

just as the translation and reflection operators themselves in (4.49). This Fourier transform takes the symbol for the identity, $I(\mathbf{x}) = 1$, into $I(\xi) = \delta(\xi)$ and a Taylor series in $\mathbf{x}$ into a series of derivatives of δ-functions. However, we shall see in the next section that the chord representation of density operators has very useful properties.

It is fitting to consider here another feature which distinguishes the reflection operators from mechanical observables. Far from being represented by a smooth phase space function, their centre representation is just

$$R_\mathbf{x}(\mathbf{x}') = 2^{-L}\delta(\mathbf{x}' - \mathbf{x}). \tag{4.65}$$

These singular functions cannot be interpreted as corresponding to classical states (i.e., individual phase space points) because the $\widehat{R}_\mathbf{x}$ have the eigenvalue -1, so they are not density operators.

Probably the first to remark on the general structure of translations and reflections underlying the Weyl and the chord representations were Grossmann and Huguenin [36]. There exists an exact correspondence, between these operators of the affine quantum group, together with the unitary operators of the metaplectic group, with the classical transformations of the inhomogeneous symplectic group [17]. In other words, all linear canonical transformations, including reflections and translations, are exactly matched by quantum unitary transformations. Thus, the unitary transformation, $\widehat{U}_\mathbf{C}$, corresponding to $x \rightarrow x' = \mathbf{C}x$, where $\mathbf{C}$ is a symplectic matrix, takes

$$\widehat{R}_\mathbf{x} \rightarrow \widehat{R}'_\mathbf{x} = \widehat{U}_\mathbf{C}^\dagger \widehat{R}_\mathbf{x} \widehat{U}_\mathbf{C} = \widehat{R}_{\mathbf{x}'} \quad \text{and} \quad \widehat{T}_\xi \rightarrow \widehat{T}'_\xi = \widehat{U}_\mathbf{C}^\dagger \widehat{T}_\xi \widehat{U}_\mathbf{C} = \widehat{T}_{\xi'}. \tag{4.66}$$

This has the consequence that both the centre and the chord representations are invariant with respect to metaplectic transformations, because the transformed operator $\widehat{A} \to \widehat{A}'$ is represented by

$$A'(\mathbf{x}) = \operatorname{tr} \widehat{U}_{\mathbf{C}} \widehat{A} \widehat{U}_{\mathbf{C}}^\dagger \widehat{R}_{\mathbf{x}} = \operatorname{tr} \widehat{A} \widehat{U}_{\mathbf{C}}^\dagger \widehat{R}_{\mathbf{x}} \widehat{U}_{\mathbf{C}} = \operatorname{tr} \widehat{A} \widehat{R}'_{\mathbf{x}} = A(\mathbf{x}') \qquad (4.67)$$

and, likewise, $A'(\xi) = A(\xi')$.

This section is concluded with some general formulae concerning these representations. For the trace of an operator, we have the alternative forms:

$$\operatorname{tr} \widehat{A} = \operatorname{tr} \widehat{I} \, \widehat{A} = \langle\!\langle \widehat{T}_{\xi=0} | A \rangle\!\rangle = A(\xi = 0) = \frac{1}{(2\pi\hbar)^L} \int d\mathbf{x} \, A(\mathbf{x}) \,. \qquad (4.68)$$

The adjoint operator, $\widehat{A}^\dagger$, is represented by

$$A^\dagger(\mathbf{x}) = [A(\mathbf{x})]^* \,, \quad \text{or} \quad A^\dagger(\xi) = [A(-\xi)]^* \,, \qquad (4.69)$$

where * denotes complex conjugation. Thus, if $\widehat{A}$ is Hermitian, $A(\mathbf{x})$ is real, though $A(\xi)$ may well be complex. The Weyl or chord symbols for products of operators are not at all obvious (see e.g., [33]), but

$$\operatorname{tr} \widehat{A}_2 \widehat{A}_1 = \int \frac{d\xi}{(2\pi\hbar)^L} A_2(\xi) A_1(-\xi) = \int \frac{dx}{(2\pi\hbar)^L} A_2(\mathbf{x}) A_1(\mathbf{x}) \,. \qquad (4.70)$$

4.6 The Wigner Function and the Chord Function

It is customary to alter the normalization of the centre and the chord symbols for the density operator, $\widehat{\rho}$, so as to define

$$W(\mathbf{x}) = \frac{\rho(\mathbf{x})}{(2\pi\hbar)^L} \quad \text{and} \quad \chi(\xi) = \frac{\rho(\xi)}{(2\pi\hbar)^L} \,, \qquad (4.71)$$

respectively the *Wigner function* and the *chord function*. Combining with the general definition of the Weyl representation and the expression for the reflection operator, we obtain the original definition of $W(\mathbf{x})$, proposed by Wigner [37]. In both cases of (4.71), the representation of the trace of a product leads to the expectation of any observable, $\widehat{A}$, as

$$\langle \widehat{A} \rangle = \int d\mathbf{x} \, W(\mathbf{x}) \, A(\mathbf{x}) = \int d\xi \, \chi(-\xi) \, A(\xi) \,. \qquad (4.72)$$

The first integral is more interesting because $A(\mathbf{x})$ is at least semiclassically close to the classical variable, which tempts us to identify the Wigner function with a *nearly classical* probability distribution. However, we will see below that $W(\mathbf{x})$, though real and normalized so

$$\int d\mathbf{x}\, W(\mathbf{x}) = 1\,, \tag{4.73}$$

may well take on negative values.

The chord function behaves like a classical characteristic function, in as much as the *moments* are

$$\langle q^n \rangle = \operatorname{tr} \widehat{q}^{\,n}\, \widehat{\rho} = (i\hbar)^n\, \frac{\partial^n}{\partial \xi_p^n} (2\pi\hbar)^L\, \chi(\xi)\Big|_{\xi=0} \tag{4.74}$$

and

$$\langle p^n \rangle = \operatorname{tr} \widehat{p}^{\,n}\, \widehat{\rho} = (-i\hbar)^n\, \frac{\partial^n}{\partial \xi_q^n} (2\pi\hbar)^L\, \chi(\xi)\Big|_{\xi=0}\,. \tag{4.75}$$

Taking the zeroth moment, we obtain the normalization,

$$1 = (2\pi\hbar)^L\, \chi(0)\,, \tag{4.76}$$

because $\operatorname{tr} \widehat{\rho} = \rho(\xi = 0) = 1$.

Shifting the phase space origin to $\langle x \rangle = (\langle p \rangle, \langle q \rangle)$, we can define the *Schrödinger covariance matrix* [38] just as its classical counterpart (4.7), with $\delta p^2 = \langle \widehat{p}^2 \rangle$, $\delta q^2 = \langle \widehat{q}^2 \rangle$ and $(\delta pq)^2 = \langle (\widehat{pq} + \widehat{qp})/2 \rangle$. It is then obvious that the expansion of the chord function at the origin is given by a quadratic form

$$\chi(\xi) = (2\pi\hbar)^{-L} - \xi\, \mathbf{K}\, \xi + \dots, \tag{4.77}$$

and we can interpret the *uncertainty*,

$$\Delta_{\mathbf{K}} = \sqrt{\det \mathbf{K}}\,, \tag{4.78}$$

as proportional to the volume of the ellipsoid: $\xi\, \mathbf{K}\, \xi = 1$. Evidently, this volume is invariant with respect to symplectic transformations, so that $\Delta_{\mathbf{K}}$ is a symplectically invariant measure of the uncertainty of the state.

The projection of the Wigner function

$$\int d\mathbf{p}\, W(\mathbf{p}, \mathbf{q}) = \Pr(\mathbf{q}) \tag{4.79}$$

is a true probability for position measurements [37]. Furthermore, the invariance of the chord and the centre representations with respect to symplectic transformations then guarantees that the projection of the Wigner function along any set of Lagrangian planes $\mathbf{p}'$ supplies the probability distribution for the conjugate variable $\mathbf{q}'$. In particular, the probability $\Pr(\mathbf{p})$ results from the projection of $W(\mathbf{p}, \mathbf{q})$ with respect to $\mathbf{q}$. All these planes are Lagrangian, so it follows that the projection of the Wigner function onto any Lagrangian plane in phase space is a probability distribution for the corresponding variable. It may appear somewhat contrived, as far as measurement is concerned, to consider general linear combinations of position and momentum. However, it should be recalled that these observables will evolve from an initial position for the motion driven by any quadratic Hamiltonian, even including free motion through a laboratory. The reconstruction of the Wigner function from a suitable set of these marginal distributions is known as quantum tomography. This is achieved through the *Radon transform* (see e.g., [39]).

It is equally remarkable, but less well known, that the characteristic function corresponding to the marginal probability distribution for positions is obtained by merely taking a section of the chord function:

$$\int d\mathbf{q} \; Pr(\mathbf{q}) \exp\left(-\frac{i}{\hbar}\eta_q \cdot \mathbf{q}\right) = (2\pi\hbar)^L \; \chi(0, \eta_q). \tag{4.80}$$

Since the chord function is also symplectically invariant, it follows that the characteristic functions for all the probability distributions, which result from Wigner projections onto Lagrangian planes, are equal to the corresponding sections of the chord function.

So far, we have emphasised the seemingly classical aspects of the Wigner function. However, it must be remembered that the Weyl representation is defined in terms of a very anomalous observable, as far as classical correspondence is concerned. In order to reveal the full quantum nature of the Wigner function, let us divide the Hilbert space of quantum states into even and odd subspaces for a given reflection operator, $\hat{R}_\mathbf{x}$. This is achieved through the projection operator introduced by Grossmann [40] and Royer [41],

$$\hat{P}^\mathbf{x}_\pm = \frac{1}{2}\left(1 \pm \hat{R}_\mathbf{x}\right), \tag{4.81}$$

so that, in its turn, we can express each reflection operator as the superposition of this pair of projections onto the even and the odd subspaces:

$$\hat{R}_\mathbf{x} = \hat{P}^\mathbf{x}_+ - \hat{P}^\mathbf{x}_-. \tag{4.82}$$

But

$$\operatorname{tr} \hat{\rho} \, \hat{P}^\mathbf{x}_\pm = Pr^\mathbf{x}_\pm \tag{4.83}$$

is just the probability of measuring $\hat{R}_\mathbf{x}$ to have the eigenvalue ± 1, so it follows that [41]

$$W(\mathbf{x}) = \frac{1}{(\pi\hbar)^L}[Pr^\mathbf{x}_+ - Pr^\mathbf{x}_-] = \frac{1}{(\pi\hbar)^L}[2Pr^\mathbf{x}_+ - 1]. \tag{4.84}$$

We thus find that the Wigner function does not admit the interpretation as a probability distribution in phase space because it can certainly be negative. Even so, it is a simple linear function of a distribution of probabilities of positive eigenvalues for all possible reflection measurements. Its maximum possible value $(\pi\hbar)^{-L}$ is attained for any point, $\mathbf{x}$, such that $\hat{P}^\mathbf{x}_+ \hat{\rho} = \hat{\rho}$, whereas the commutation of the density operator with $\hat{P}^\mathbf{x}_-$ specifies a phase space point where $W(\mathbf{x}) = -(\pi\hbar)^{-L}$.

Let us now investigate the effect of reflections and translations on a density operator. Evidently, the centre and chord representations are specially suitable for this purpose. In the case of a phase space translation by the vector, η, i.e., $\widehat{\rho}_\eta = \widehat{T}_\eta \, \widehat{\rho} \, \widehat{T}_{-\eta}$, the respective Wigner and chord functions become

$$W_\eta(\mathbf{x}) = W(\mathbf{x} - \eta) \quad \text{and} \quad \chi_\eta(\xi) = e^{i\eta \wedge \xi / \hbar} \, \chi(\xi) \,, \tag{4.85}$$

which shows that, unlike the Wigner function, the chord function is not generally real. The sensitivity of a state to translations is described by the phase space correlations of a given density operator, defined as [32]

$$
\begin{aligned}
C(\xi) &= \operatorname{tr} \widehat{\rho} \, \widehat{T}_\xi \, \widehat{\rho} \, \widehat{T}_\xi^\dagger \\
&= (2\pi\hbar)^L \int d\mathbf{x} \, W(\mathbf{x}) \, W(\mathbf{x} - \xi) \\
&= (2\pi\hbar)^L \int d\eta \, e^{i\eta \wedge \xi / \hbar} \, |\chi(\eta)|^2 \,.
\end{aligned}
\tag{4.86}
$$

From the reciprocal relation that supplies the intensity of the chord function as the Fourier transform of these correlations and the normalization condition (4.76), we see that

$$\int d\xi \, C(\xi) = (2\pi\hbar)^{3L} |\chi(\eta = 0)|^2 = (2\pi\hbar)^L \,. \tag{4.87}$$

So, even though these correlations are defined in terms of classical translations in phase space, they are purely quantum and disappear in the classical limit. However, if we fix $\hbar$ and adopt this constant as our phase space scale, then we can picture $C(\xi)$ as a classical-like phase space distribution for which the characteristic function is just $|\chi(\xi)|^2$.

Specializing to the case of a pure state, $\widehat{\rho} = |\psi\rangle\langle\psi|$, we find that

$$\langle \psi | \widehat{T}_\xi | \psi \rangle = (2\pi\hbar)^L \chi(-\xi) \,, \tag{4.88}$$

so that the phase space correlations take the form [32]

$$C(\xi) = |\langle \psi | \widehat{T}_\xi | \psi \rangle|^2 = (2\pi\hbar)^{2L} |\chi(\xi)|^2 \,. \tag{4.89}$$

Thus, for instance, in the case that $\xi = (0, \xi_q)$, the phase space correlations, $C(\xi)$, are just the usual spacial correlations inferred from neutron scattering experiments. Nonetheless, we must be careful to distinguish between phase space correlations and the correlations between the quantum measurements of observables defined on the different components of a bipartite system, such as the CHSH inequality. For a pure state, (4.89) is the square modulus of the expectation for a translation, which is not a quantum observable. However, (4.86) defines the phase space correlation in the same way as for a classical distribution.

The chord function always assumes its maximum value $1/(2\pi\hbar)^L$ at the origin. But also an average of overlaps cannot exceed one, so $\chi(0)$ is the maximum even for mixed states. As for the correlations, we always have

$$\operatorname{tr} \widehat{\rho}^2 = (2\pi\hbar)^L \int \mathrm{d}\mathbf{x}\, [W(\mathbf{x})]^2 = (2\pi\hbar)^L \int \mathrm{d}\xi\, |\chi(\xi)|^2 = C(0)\,, \qquad (4.90)$$

being that $\operatorname{tr} \widehat{\rho}^2 = 1$ for pure states. But consider a mixture of orthogonal states,

$$\widehat{\rho} = \sum_j \Pr(n)\, |n\rangle\langle n|\,, \qquad (4.91)$$

then the *purity*

$$C(0) = \sum_j \Pr(n)^2 \le 1\,. \qquad (4.92)$$

Another form in which this quantity appears is the *linear entropy* : $1 - \operatorname{tr} \widehat{\rho}^2$. This may be considered as a first-order expansion of the *von Neumann entropy*:

$$S = -\operatorname{tr} \widehat{\rho} \ln \widehat{\rho} \qquad (4.93)$$

a quantum version of the classical Shannon entropy. In [32], the correlations were normalized by the purity so as to be always unity at the origin, but it is convenient to include this quantity as a special case of the correlations.

General invariance with respect to Fourier transformation characterizes the correlation in the case of pure states. Indeed, inserting the above expression in the definition of the phase space correlation , we obtain [32]

$$C(\xi) = \int \frac{\mathrm{d}\eta}{(2\pi\hbar)^L}\, e^{i\eta\wedge\xi/\hbar}\, C(\eta)\,. \qquad (4.94)$$

This is a remarkable property of all pure states and is in no way restricted by special symmetry properties that will be shown to relate certain Wigner functions to their respective chord functions. An immediate consequence is that oscillations of the phase space correlation of a pure state involving a large displacement, ξ, are necessarily bound to small ripples on the scale, $|\xi|^{-1}$, in the direction, $\mathbf{J}\xi$. Of course, these small-scale oscillations of the phase space correlations, which have been attractively described as *subplanckian* [42], show up in the pure state Wigner function because of (4.86).

The Fourier invariance condition (4.94) includes as a special case the more familiar one obtained by tracing over the full pure state condition $\widehat{\rho}^2 = \widehat{\rho}$. It follows that the difference of both sides of (4.94) for each chord ξ generalizes (4.92) as a measure of the degree of *purity* of a state. All the same, the loss of the phase information in $C(\xi)$, but contained in the chord function, would seem to imply that these are necessary conditions, whereas the full sufficient condition of purity is $\widehat{\rho}^2 = \widehat{\rho}$, which is expressed in the chord representation as [33]

$$\int \mathrm{d}\eta\, \chi(\eta)\, \chi(\xi - \eta)\, e^{i\xi\wedge\eta/2\hbar} = \int \mathrm{d}\eta\, \chi_{\xi/2}(\eta)\, \chi(\xi - \eta) = \chi(\xi) \qquad (4.95)$$

with $\chi_{\xi/2}(\eta)$ defined by (4.85).[3] However, the particular condition $C(0) = 1$ is indeed a sufficient condition, because, for any mixture of pure states, $\hat{\rho} = \sum_n \text{Pr}(n)|\psi_n\rangle\langle\psi_n|$, we obtain

$$\text{tr } \hat{\rho}^2 = 1 - \sum_{n \neq n'} \text{Pr}(n)\text{Pr}(n')[1 - |\langle\psi_n|\psi_{n'}\rangle|^2]. \tag{4.96}$$

A single phase space point does not correspond to any pure state in Hilbert space. The only pure states that are *classical-like*, i.e., have positive Wigner functions, are either coherent states, or their image by a symplectic transformation [43, 44].

Let us now consider the effect of measuring a general phase space reflection, $\hat{R}_{\mathbf{x}}$. The density operator, $\hat{\rho}$, will be projected by $\hat{P}^{\mathbf{x}}_{\pm}$, defined by (4.81), onto either the even or odd subspace for this particular reflection:

$$\hat{\rho}^{\mathbf{x}}_{\pm} = \frac{\hat{P}^{\mathbf{x}}_{\pm} \hat{\rho} \hat{P}^{\mathbf{x}}_{\pm}}{\text{tr } \hat{\rho} \hat{P}^{\mathbf{x}}_{\pm}}. \tag{4.97}$$

The Weyl symbol for $\hat{\rho}\hat{P}^{\mathbf{x}}_{\pm}$ defines the symmetric Wigner function, $W^{\pm}_{\mathbf{x}}(\mathbf{x}')$, within a normalization factor, so that, using the group relations (4.56), (4.57) and (4.58), we obtain [45]:

$$\begin{aligned}
W^{\pm}_{\mathbf{x}}(\mathbf{x}') &= (\pi\hbar)^{-L}\text{tr } \hat{R}_{\mathbf{x}'} \hat{\rho}^{\mathbf{x}}_{\pm} \\
&= \frac{W(\mathbf{x}') + W(2\mathbf{x} - \mathbf{x}') \pm \Re 2^{(L+1)} e^{2i \, \mathbf{x}' \wedge \mathbf{x}/\hbar} \chi(2(\mathbf{x}' - \mathbf{x}))}{2\left[1 \pm (\pi\hbar)^L W(\mathbf{x})\right]},
\end{aligned} \tag{4.98}$$

where $\Re$ denotes the real part of a number. It follows that the Wigner function and the chord function for a reflection symmetric density operator are trivially related. Shifting the origin of phase space to the symmetry point leads to [32]

$$W^{\pm}_0(\mathbf{x}) = \pm \, 2^L \chi^{\pm}_0(-2\mathbf{x}). \tag{4.99}$$

Thus, all pure state Wigner functions for density operators that commute with a reflection symmetry attain the largest amplitude at the symmetry point, but this will be negative in the case of odd symmetry.

Let us consider some standard examples of Wigner and chord functions. All the following cases are related to eigenstates of a harmonic oscillator with one degree of freedom and unit mass.

(i) Coherent states: the Wigner function is just a Gaussian centred on η,

$$W_\eta(\mathbf{x}) = \frac{1}{\pi\hbar} \exp\left[-\frac{\omega}{\hbar} \, (\mathbf{q} - \eta_q)^2 - \frac{1}{\hbar\omega} \, (\mathbf{p} - \eta_p)^2\right] \xrightarrow{\omega=1} \frac{1}{\pi\hbar} e^{-(\mathbf{x}-\eta)^2/\hbar}, \tag{4.100}$$

---

[3] For distributions $\text{Pr}(n)$ over eigenstates $|n\rangle$ of an observable with discrete spectrum, the condition $\text{Pr}(n)^2 = \text{Pr}(n)$ also singles out a *pure state*, $\text{Pr}(n) = \delta_{n,m}$, but this condition is not generalizable to a continuous spectrum.

whereas

$$\chi_\eta(\xi) = \frac{1}{2\pi\hbar} \exp\left(\frac{i\eta \wedge \xi}{\hbar}\right) \exp\left[-\frac{\omega}{\hbar}\left(\frac{\xi_q}{2}\right)^2 - \frac{1}{\hbar\omega}\left(\frac{\xi_p}{2}\right)^2\right]$$

$$\xrightarrow{\omega=1} \frac{1}{2\pi\hbar} e^{i\eta \wedge \xi/\hbar} e^{-\xi^2/4\hbar}.$$
(4.101)

So, any translation of the coherent state merely alters the phase of the Gaussian chord function that sits on the origin. The coherent states, or more generally all equivalent Gaussian states obtained from them by symplectic transformations, are the only examples of pure states for which the Wigner function is nowhere negative [43]. This is one of the reasons why these are sometimes considered to be the most classical of pure quantum states. Since the projection of a Gaussian is also a Gaussian, the measurement of position, or any other Lagrangian phase space coordinate, does not display interference fringes. The fact that the uncertainty, $\Delta = \delta p \delta q = \hbar$, is minimal allows us to interpret them as *quantum phase space points*.

(ii) A superposition of a pair of coherent states $|\eta\rangle \pm |-\eta\rangle$ is sometimes known as a *Schrödinger cat state*. Its Wigner function is[4]

$$W_\pm(\mathbf{x}) = \frac{1}{2\pi\hbar\left(1 \pm e^{-\eta^2/\hbar}\right)}$$
$$\times \left[e^{-(\mathbf{x}-\eta)^2/\hbar} + e^{-(\mathbf{x}+\eta)^2/\hbar} \pm 2e^{-\mathbf{x}^2/\hbar}\cos\left(\frac{2}{\hbar}\mathbf{x} \wedge \eta\right)\right].$$
(4.102)

It consists of two *classical* Gaussians centred on $\pm\eta$ and an interference pattern with a Gaussian envelope centred on their midpoint. The frequency of this oscillation increases with the separation $|2\eta|$. In Fig. 4.6, the displacements $\pm\eta$ have been chosen as $(\pm3, \pm3)$. The phase of the pair of coherent states merely shifts the phase of the interference fringes, so that the midpoint is an absolute maximum for $W_+(\mathbf{x})$ and an absolute minimum for $W_-(\mathbf{x})$. It might be supposed that for small $\eta \to 0$, we would have $W_+(\mathbf{x}) > 0$ for all $\mathbf{x}$, but it is easy to verify that there are very shallow negative regions far removed from the classical superposed Gaussians, in agreement with [43, 44]. The interference pattern of the Wigner function does not survive the projection orthogonal to η: In this direction, the interference disappears to produce a purely classical pattern. Conversely, the projection along η is marked by interference fringes.

For the chord function,

$$\chi_\pm(\xi) = \frac{1}{4\pi\hbar\left(1 \pm e^{-\eta^2/\hbar}\right)}$$
$$\times \left[e^{-(\xi/2-\eta)^2/\hbar} + e^{-(\xi/2+\eta)^2/\hbar} \pm 2e^{-\xi^2/4\hbar}\cos\left(\frac{1}{\hbar}\xi \wedge \eta\right)\right],$$
(4.103)

---

[4] Here and below we set $\omega = 1$.

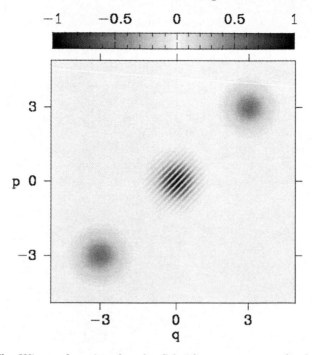

Fig. 4.6. The Wigner function for the Schrödinger cat state displays a pair of *classical* Gaussians, one for each coherent state, and a third Gaussian modulated by interference fringes halfway between them. The chord function is a mere rescaling of the Wigner function if the midpoint lies on the origin

this same configuration has to be reinterpreted. Now the local phase space correlations of the individual coherent states, as in (i), are placed in the neighbourhood of the origin, where they interfere, while their cross-correlation generates new Gaussians centred on the separation vectors $\pm 2\eta$. The general case of coherent states $|\eta_1\rangle$ and $|\eta_2\rangle$ merely leads to Gaussians centred on $\pm(\eta_1 - \eta_2)$ with addition of the phase factor $\exp[i(\eta_1 + \eta_2) \wedge \xi/2\hbar]$.

Recalling that the phase space correlations of a pure state are just the square modulus of the chord function, we can immediately verify the general relation between large- and small-scale structures in the case of Schrödinger cat states. Indeed, the spacial frequency of the oscillations of the chord function increases directly with the separation of the pair of coherent states.

The particular superpositions of coherent states, $|+\rangle$ and $|-\rangle$, are respectively even and odd eigenstates of the parity operator $\widehat{R}_0$, i.e., reflection about the origin. Therefore, they are the two possible states that could be produced by a parity measurement effected on the single coherent state $|\eta\rangle$. Thus, the parity measurement would generate a sizable probability of finding a particle near $x = -\eta$, even though this was most unlikely before the measurement.

The states $|\pm\rangle$ are orthogonal, even though the coherent states, $|\eta\rangle$ and $|-\eta\rangle$, are not. It is true that such a pair of coherent states will be nearly orthogonal if η is large enough and thus considered to form a qubit. Within this approximation, the symmetrical states would then be a mere unitary transformation of a single qubit. However, no approximation is needed in this process of carving a qubit from an infinite-dimensional system, if we use $|\pm\rangle$ as the original basis states. We would then consider a *common garden* coherent state to be the superposition of a symmetrical pair of Schrödinger cats. (Is there some approximation involved?) Indeed, this generation of a qubit by a reflection is not limited to coherent states, but could in principle be realized for any unsymmetrical initial state.

(iii) Fock states, $|n\rangle$, i.e., the excited states of the harmonic oscillator, also have reflection symmetry with respect to the origin. Thus, from the exact Wigner function, first derived by Grönewold [46],

$$W_n(\mathbf{x}) = \frac{(-1)^n}{\pi\hbar} e^{-\mathbf{x}^2/\hbar} L_n\left(\frac{2\mathbf{x}^2}{\hbar}\right), \qquad (4.104)$$

where L_n is a Laguerre polynomial, we obtain the chord function

$$\chi_n(\xi) = \frac{e^{-\xi^2/4\hbar}}{2\pi\hbar} L_n\left(\frac{\xi^2}{2\hbar}\right). \qquad (4.105)$$

It is interesting to note that the symmetry centre, which produces the maximum amplitude of the Wigner function, is nowhere near the classical manifold with energy $E_n = \left(n + \frac{1}{2}\right)\hbar\omega$. However, this point lies in a region of narrow oscillations, so that it does not affect the average of smooth observables. Figure 4.7 shows the Wigner function for the Fock state with $n = 2$; the origin is a maximum because of the positive parity. The unfolding of this peak for nonsymmetric Wigner functions is discussed in Sect. 4.10.

The Wigner function exhibits the interference fringes for the measurement of any variable $ap + bq$. In the case of the Fock state, these are always present. A simple way to see this is that any direction for the projection will be somewhere tangent to each of the continuous curves that form the Wigner function fringes. These regions dominate the projection. This example thus illustrates the *necessity* for the Wigner function to have negative regions: This is the only way that interference can result from a mere projection in phase space.

The Fock states are an example of a complete parity basis, which is even or odd according to the state label, n. Hence, if a pure state, $|\psi\rangle$, is specified in this basis, then

$$W_\psi(0) = \frac{1}{\pi\hbar} \sum_n \left[|\langle 2n|\psi\rangle|^2 - |\langle(2n+1)|\psi\rangle|^2\right]. \qquad (4.106)$$

Such a decomposition can in principle be achieved for general arguments of the Wigner function, but then it is necessary to translate the whole Fock state

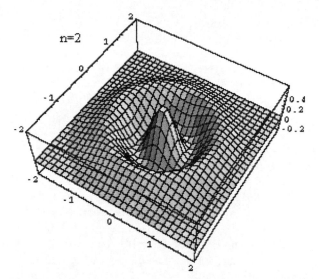

Fig. 4.7. The Wigner function for the $n = 2$ Fock state. The classical Bohr-quantized circle lies just outside the maximum of the outer fringe. The chord function is a mere rescaling of the Wigner function

basis instead of just the ground state, as in the definition of coherent states. If we similarly translate the Hamiltonian, it will commute with $\widehat{R}_\mathbf{x}$ instead of commuting with $\widehat{R}_0$. The eigenstates of all such Hamiltonians will form a good odd–even basis. The difficulty with defining a semiclassical correspondence for both these classes of eigenstates is that the alternative odd and even Bohr-quantized curves approach each other as $\hbar \to 0$.

All the above examples are singled out by some point of reflection symmetry, which needs to be chosen as the origin for the chord function to be real. The chord function must assume its maximum value $1/(2\pi\hbar)^L$ at the origin, whatever the symmetry, because of normalization. The Wigner amplitude, $|W(\mathbf{x})|$, need not have such a prominent peak in general. However we shall see in Sect. 4.10 that the large-scale features of the semiclassical forms of the Wigner function and the chord function maintain a mutual correspondence, even in the absence of a reflection symmetry.

It is important to note that the commutation, $[\widehat{\rho}, \widehat{R}_0] = 0$, guarantees that $W(\mathbf{x})$ is a symmetric function with respect to (classical) reflection at the origin. This is a consequence of the fact that if $\widehat{\rho}_R = \widehat{R}\,\widehat{\rho}\,\widehat{R}$, then $W_R(\mathbf{x}) = W(R(\mathbf{x}))$, the classical reflection of the argument. However, it is the maximum (or minimum) value at the origin which guarantees that the density operator is pure with respect to parity, i.e., it is either $\widehat{\rho}_+$, or $\widehat{\rho}_-$. Indeed, even though a mixture of an even density and an odd density (i.e., $\widehat{\rho} = c_+\widehat{\rho}_+ + c_-\widehat{\rho}_-$) will trivially satisfy $[\widehat{\rho}, \widehat{R}_0] = 0$, we see that $W(0) = (c_+ - c_-)/(\pi\hbar)^L$ will not be maximal.

Perhaps the converse property is of even more interest: If a Wigner function is symmetric about the origin, but $|W(0)| < (\pi\hbar)^{-L}$, then the state must be a mixture. After all, if it is a mixture of parities, it cannot be a pure state. It is only if the mixture is restricted to states of the same parity that it will not be detected by $W(0)$. There are many measures of degrees of mixedness, or impurity, but it is specially nice to be able to spot this property by a mere glance at the Wigner function. Furthermore, the Wigner function, i.e., the parity decomposition, is a measurable property [28, 34, 47].

The Wigner function may be considered as a field of probabilities for parity decompositions in phase space. Each reflection separates the infinite-dimensional Hilbert space into a pair of orthogonal components. If we just consider a single reflection, this goes a long way to reducing the Hilbert space to that of a single qubit, a two state system. No matter how classical the appearance of the Wigner function (i.e., it may be smooth and positive), it is always fully quantum as far as parity measurements are concerned. The situation is quite different, for instance, for position measurements. Then there is an important difference between the Wigner function for a pure Schrödinger cat state and a mixture of cats with different phases. This is revealed by the fine interference fringes between the two classical regions, but it is even more clearly displayed by the pair of correlation peaks far from the origin of the chord function. The relation between these features and entanglement is discussed in Sect. 4.8.

There is a vast literature concerning the Wigner function. Only a few topics have been mentioned here, and it has been necessary to leave out even as relevant a topic as quantum tomography. The adaptation of the Wigner function for finite Hilbert spaces is of special relevance for quantum computing and quantum information theory. Then the rule that each quantum state corresponds to a volume of $(2\pi\hbar)^L$ in classical phase space restricts the overall phase space volume. Thus, one must first face a choice of the topology in which to compactify phase space. It turns out that the simplest choice is a torus, though single qubits are more naturally displayed on a *Bloch sphere*. In spite of the intrinsic interest in many of the aspects of finite space Wigner functions [48, 49], there is no overall agreement on the choice of Wigner function properties to emphasise. Not all formalisms lead to a corresponding natural definition of a conjugate chord function as in [50], nor is there an overall preoccupation with invariance with respect to those symplectic transformations which preserve the torus topology of phase space [51]. A final difficulty concerns the appearance of ghost images and dimensionally dependent features [52].

So far nothing has been said of an alternative phase space representation, the *Husimi function* [53, 54]. Defined in terms of coherent states $|\eta\rangle$ as

$$\rho_H(\eta) = \langle\eta|\rho|\eta\rangle = \operatorname{tr} \rho \, |\eta\rangle\langle\eta| , \qquad (4.107)$$

it can be interpreted as a smoothed Wigner function,

$$\rho_H(\eta) = \int d\mathbf{x} \, W_\eta(\mathbf{x}) \, W(\mathbf{x}) \tag{4.108}$$

because of (4.70). The lack of purity of a state can be described in terms of the *Wehrl entropy*,

$$S_W = -(2\pi\hbar)^L \int d\eta \, \rho_H(\eta) \ln \rho_H(\eta) \,. \tag{4.109}$$

According to Wehrl's inequality [55, 56] (see also [57]), the Wehrl entropy is always bounded from below by the von Neumann entropy (4.93). For more recent developments concerning the Wehrl entropy, see e.g., [58].

The Husimi function is most appropriate for the study of *quantum chaos*, because it highlights the classical region. But such a downplay of the quantum interferences, achieved by coarse graining the Wigner function, is not what one would ordinarily seek in quantum information theory. In a way, this is just the opposite of the chord function, which squashes all classical structure to the neighbourhood of the origin, so as to display the purely quantum coherences. It is remarkable that both these antithetical representations are intimately related to the translation operators, since the Husimi function for a pure state, $\hat{\rho} = |\psi\rangle\langle\psi|$, can be rewritten as

$$\rho_H(\eta) = |\langle\psi|\eta\rangle|^2 = |\langle\psi|\widehat{T}_\eta|(\eta = 0)\rangle|^2 \,. \tag{4.110}$$

Hence, the basic difference with respect to C_η in (4.89) is the exchange of $|0\rangle$, the Gaussian ground state of the harmonic oscillator, for $|\psi\rangle$ itself.

A further comment is that the quantum interferences are displayed by the isolated zeroes of Husimi functions [59], in the case that $L = 1$. A uniform distribution of zeroes has been used to characterize the eigenstates of classically chaotic systems. Even though this is of great theoretical interest, these zeroes are usually located in regions where the Husimi function is already tiny, so that they may be very hard to compute. For instance, in the case of the cat state (4.102) with small η, they are found in the shallow negative regions where $|W_+(\mathbf{x})|$ is exponentially small.

4.7 The Partial Trace: Sections and Projections

Recall that the representation of operators, $\widehat{A} = |A\rangle\rangle$, in a given basis, such as $\langle\langle Q|A\rangle\rangle$, corresponds to the foliation of the double phase space, $X = (P, Q)$, by a set of Lagrangian planes, $Q = constant$. Performing linear canonical transformations in double phase space, we are free to choose the alternative coordinate planes, $Q = (q_-, q_+)$, or $Q = \mathbf{x}$, or $Q = \mathbf{y} = J\xi$ among others. In all cases, it is the fact that

$$\langle\langle Q'|Q\rangle\rangle = \delta(Q' - Q), \tag{4.111}$$

which permits us to identify the expansion coefficient in

$$\widehat{A} = |A\rangle\rangle = \int dQ \, A(Q) \, |Q\rangle\rangle \tag{4.112}$$

with $\langle\langle Q|A\rangle\rangle$.

Let us now assume that the (single) phase space is itself a product of a pair of phase spaces, $x = x_1 \times x_2$, each with $2L_j$ dimensions, and that these correspond to Hilbert spaces, $\boldsymbol{H}_j$, so that $\boldsymbol{H} = \boldsymbol{H}_1 \otimes \boldsymbol{H}_2$. Then we can always decompose the Lagrangian planes chosen as a basis for double phase space as the product $Q = Q_1 \otimes Q_2$, corresponding to operators $|Q\rangle\rangle = |Q_1\rangle\rangle \otimes |Q_2\rangle\rangle$. Thus the complete $|Q\rangle\rangle$ representation becomes

$$\widehat{A} = |A\rangle\rangle = \int dQ_1 dQ_2 \, A(Q_1, Q_2) \, |Q_1\rangle\rangle \otimes |Q_2\rangle\rangle \,. \tag{4.113}$$

The definition of the partial trace is then

$$\mathrm{tr}_2 \, \widehat{A} = \mathrm{tr}_2 \, \widehat{I}_2 \, \widehat{A} = \int dQ_1 dQ_2 \, A(Q_1, Q_2) \, |Q_1\rangle\rangle \, \langle\langle I_2|Q_2\rangle\rangle \,, \tag{4.114}$$

so that

$$A_1(Q_1) = \int dQ_2 \, A(Q_1, Q_2) \langle\langle I_2|Q_2\rangle\rangle \tag{4.115}$$

defines the $|Q_1\rangle\rangle$ representation of a reduced operator $\widehat{A}_1$, which acts on the Hilbert space $\boldsymbol{H}_1$. It is well known that in the case of the density operator $\widehat{\rho}$, the reduced operator $\widehat{\rho}_1$ describes the same probability as the full density operator for all measurements concerning the subsystem-1.[5]

The different forms of the partial trace depend essentially on the Hilbert–Schmidt product (4.36) of each basis with the identity. In the case of the position basis, we have

$$\langle\langle I|Q\rangle\rangle = \mathrm{tr} \, \widehat{I} \, |q_-\rangle\langle q_+| = \delta(q_- - q_+) \,, \tag{4.116}$$

so that

$$A_1(Q_1) = \int dq_{2-} dq_{2+} \, A(Q_1, Q_2 = (q_{2-}, q_{2+})) \, \delta(q_{2-} - q_{2+})$$
$$= \int dq_2 \, A(Q_1, (q_2, q_2)) \,. \tag{4.117}$$

Here, we should recall that,

$$A(Q_1, (q_2, q_2)) = \langle q_{1-}, q_2| \, \widehat{A} \, |q_{1+}, q_2\rangle \,, \tag{4.118}$$

in matrix notation.

---

[5] A measurement on subsystem-2 only affects $\widehat{\rho}_1$ if the information on the outcome of the measurement is made available [6].

In the centre representation, we have simply

$$\langle\!\langle I|\mathbf{x}\rangle\!\rangle = \operatorname{tr} \widehat{I}\ (2^L \widehat{R}_\mathbf{x}) = 1\,, \tag{4.119}$$

leading to the phase space projection:

$$A_1(\mathbf{x}_1) = \int \mathrm{d}\mathbf{x}_2\ A(\mathbf{x}_1,\mathbf{x}_2)\,. \tag{4.120}$$

In the case of the density operator, the corresponding reduced Wigner function, $W_1(\mathbf{x}_1)$, is thus obtained from $W(\mathbf{x})$ in the same way as a marginal probability distribution is projected out of the full distribution .

The simplest choice turns out to be the chord representation. Then, $|I\rangle\!\rangle = \widehat{T}_{\xi=0}$ is an element of the operator basis, so that

$$\langle\!\langle I|\mathbf{y}\rangle\!\rangle = \delta(\mathbf{y}) = \delta(\xi)\,. \tag{4.121}$$

Thus in this case, instead of projecting, we obtain the reduced operator merely by slicing through the chord symbol:

$$A_1(\xi_1) = A(\xi_1,\xi_2{=}0)\,. \tag{4.122}$$

Of course, the reduced operator $\widehat{A}_1$ itself is insensitive to the procedure used to obtain it within the various representations, but the ease of calculating the reduction is a special bonus of the chord representation.

It should be recalled that the partial trace is invariant with respect to unitary transformations performed internally within the factor Hilbert space $\boldsymbol{H}_2$: $\widehat{U} = \widehat{U}_2 \otimes \widehat{I}_1$ (see, e.g., [6]). In the example where the subsystems are particles that have separated by a large distance, then these are truly *local transformations*. In other words, if $\widehat{A}' = \widehat{U}\,\widehat{A}\,\widehat{U}^{-1}$, then $\operatorname{tr}_2 \widehat{A}' = \operatorname{tr}_2 \widehat{A}$. This invariance corresponds semiclassically to the freedom of performing canonical transformations that leave invariant the $\mathbf{x}_1$ variables: $(x_1, x_2) \rightarrow (x_1, x_2')$. This also implies that only the double phase space corresponding to x_2 changes: $(X_1, X_2) \rightarrow (X_1, X_2')$. If the canonical transformation is linear in the single phase space, then both the centres, $\mathbf{x}$, and the chords, ξ, are propagated in the arguments of their respective functions by this same transformation.

Another point that is worth discussing concerns the completeness of the operator representations. Notice that the restricted translation operators

$$|\mathbf{y}_1\rangle\!\rangle_1 = \widehat{T}'_{\xi_1} = \widehat{T}_{\xi_1} \otimes \widehat{I}_2 \tag{4.123}$$

are a subset of the translation operators used in the chord basis for the full Hilbert space, $\boldsymbol{H}_1 \otimes \boldsymbol{H}_2$. It follows that a representation in terms of the restricted translation operators, $\widehat{T}'_{\xi_1}$, would not be complete. Likewise, we may define the restricted unitary reflection operators,

$$|\mathbf{x}_1\rangle\!\rangle_1 = 2^{L_1} \widehat{R}'_{\mathbf{x}_1} = \widehat{R}_{\mathbf{x}_1} \otimes \widehat{I}_2\,, \tag{4.124}$$

but these do not belong to the centre basis for $H_1 \otimes H_2$. Even so, we may also define directly the reduced operator $\widehat{A}_1$ as

$$\widehat{A}_1 = \int d\mathbf{x}_1 \, A_1(\mathbf{x}_1) \, |\mathbf{x}_1\rangle\!\rangle_1 \,, \tag{4.125}$$

with

$$A_1(\mathbf{x}_1) = \mathrm{tr}\,\widehat{A}\,(2^{L_1}\widehat{R}'_{\mathbf{x}_1}) \,. \tag{4.126}$$

Let us now specialize to density operators. In the case of the chord function, we must take care of the normalization, which depends on the number of degrees of freedom. Hence, the validity of (4.122) for the chord representation of density operators $\rho(\xi)$ and $\rho_1(\xi_1)$ implies that the reduced chord function is

$$\chi_1(\xi_1) = (2\pi\hbar)^{L_2} \, \chi(\xi_1, \xi_2=0) \,. \tag{4.127}$$

Clearly, $\chi_1(\xi_1)$ is the Fourier transform of $W_1(\mathbf{x}_1)$. Since the definition of phase space correlations (4.86) is valid for the reduced system, we obtain the reduced correlations as a projection of the correlations of the entangled pure state:

$$
\begin{aligned}
C_1(\xi_1) &= \mathrm{tr}\,\widehat{\rho}_1\,\widehat{T}_{(\xi_1,0)}\,\widehat{\rho}_1\,\widehat{T}^\dagger_{(\xi_1,0)} \\
&= (2\pi\hbar)^{L_1} \int d\eta_1 \, e^{i\eta_1 \wedge \xi_1/\hbar} \, |\chi_1(\eta_1)|^2 \\
&= (2\pi\hbar)^{L_1} \int d\eta_1 d\eta_2 \, e^{i\eta_1 \wedge \xi_1/\hbar} \delta(\eta_2) \, |\chi(\eta)|^2 \\
&= \int \frac{d\xi_2}{(2\pi\hbar)^{L_2}} C(\xi) \,.
\end{aligned}
\tag{4.128}
$$

It should be recalled that the relation between the Wigner function and the chord function mimics that between a classical probability distribution and its characteristic function. The definition of correlations and the classical marginal distributions also goes through as above. Therefore, the property that the correlation of the reduced state for a given displacement, ξ_1, is just the integral over all correlations in the larger space over displacements that share this component also holds for classical probability distributions. This relation does not depend on the full density operator being a pure state.

All the representations that we have been discussing will factor in the case that $\widehat{\rho} = (|\psi_1\rangle \otimes |\psi_2\rangle)(\langle\psi_2| \otimes \langle\psi_1|)$ is a product pure state. Thus we obtain product Wigner functions, $W(\mathbf{x}) = W_1(\mathbf{x}_1)W_2(\mathbf{x}_2)$, and product chord functions, $\chi(\xi) = \chi_1(\xi_1)\chi_2(\xi_2)$. These relations may be interpreted in terms of average values of the basis operators, i.e., $\langle\widehat{R}_{\mathbf{x}}\rangle = \langle\widehat{R}'_{\mathbf{x}_1}\rangle\langle\widehat{R}'_{\mathbf{x}_2}\rangle$ and $\langle\widehat{T}_\xi\rangle = \langle\widehat{T}'_{\xi_1}\rangle\langle\widehat{T}'_{\xi_2}\rangle$. Thus, a sufficient criterion for the existence of entanglement would be that either of these equalities not hold for some centre $\mathbf{x}$ or some chord ξ.

Curiously, it is not the generation of cross-correlations that is usually taken as a measure of entanglement, but instead the loss of correlations of the reduced density operator. Its von Neumann entropy (4.93) is often referred to as *the entanglement*. Expanding this to first order, results in the linear entropy of a partial trace of the full density operator,

$$1 - \text{tr}\,\hat{\rho}_1^2 = 1 - C_1(0)\,, \tag{4.129}$$

recalling (4.86). The square root of two times this last expression is the *concurrence*, another widely used entanglement measure (see, e.g., Sect. 2.4.2 or [60]). At first sight, these are not obvious measures of overall entanglement, because we should obtain the same measure by singling out instead the reduced density operator for subsystem-2. But, it is a simple consequence of (4.94), the invariance of the quantum correlations with respect to Fourier transforms for a pure state, that

$$C_1(0) = \int \frac{\mathrm{d}\xi_2}{(2\pi\hbar)^{L_2}} C(0, \xi_2)$$

$$= \int \frac{\mathrm{d}\xi_2}{(2\pi\hbar)^{L_2}} \int \frac{\mathrm{d}\eta}{(2\pi\hbar)^L} C(\eta)\, e^{i\eta_2 \wedge \xi_2} \tag{4.130}$$

$$= \int \frac{\mathrm{d}\eta_1}{(2\pi\hbar)^{L_1}} C(\eta_1, 0) = C_2(0)\,.$$

Reinterpreted in terms of Wigner functions,

$$\int \mathrm{d}\mathbf{x}_1\,[W_1(\mathbf{x}_1)]^2 = \int \mathrm{d}\mathbf{x}_2\,[W_2(\mathbf{x}_2)]^2\,, \tag{4.131}$$

this is another remarkable property of pure quantum states, for it is highly unusual for the second moment of a pair of marginal probability distributions to display a similar equality. Indeed, it is not even generally true for product distributions.

The focus on properties of the reduced density matrix makes sense when it is recalled that the concept of entanglement involves separate measurement on each of the components. The invariance of the partial traces with respect to local transformations carries over to the above measures of entanglement. Even better, it has been shown that it is possible to concentrate the entanglement within a few elements of an ensemble of identical states, by performing local measurements [61].

In terms of Husimi functions (4.108), it is natural to describe entanglement in terms of the Wehrl entropy (4.109) for the reduced density operator. Another way of describing entanglement is through the Schmidt decomposition (4.18). The corresponding Wigner and chord functions are then

$$W(\mathbf{x}) = (\pi\hbar)^{-L} \sum_{i,j} \lambda_i \lambda_j\, \langle \psi^i{}_1 | \widehat{R}'_{\mathbf{x}_1} | \psi^j{}_1 \rangle \langle \psi^i{}_2 | \widehat{R}'_{\mathbf{x}_2} | \psi^j{}_2 \rangle \tag{4.132}$$

and

$$\chi(\xi) = (2\pi\hbar)^{-L} \sum_{i,j} \lambda_i \lambda_j \, \langle \psi^i{}_1 | \widehat{T}'_{-\xi_1} | \psi^j{}_1 \rangle \langle \psi^i{}_2 | \widehat{T}'_{-\xi_2} | \psi^j{}_2 \rangle \,, \tag{4.133}$$

recalling the definitions of the restricted reflection operators (4.124) and the restricted translation operators (4.123). In both cases, the partial trace over subsystem-2 substitutes the second Dirac bracket by $\delta_{i,j}$, so that

$$W_1(\mathbf{x}_1) = (\pi\hbar)^{-L_1} \sum_i \lambda_i{}^2 \, \langle \psi^i{}_1 | \widehat{R}'_{\mathbf{x}_1} | \psi^i{}_1 \rangle = \sum_i \lambda_i{}^2 \, W_i(\mathbf{x}_1) \tag{4.134}$$

and

$$\chi_1(\xi_1) = (2\pi\hbar)^{-L_1} \sum_i \lambda_i{}^2 \, \langle \psi^i{}_1 | \widehat{T}'_{-\xi_1} | \psi^i{}_1 \rangle = \sum_i \lambda_i{}^2 \, \chi_i(\xi_1) \,. \tag{4.135}$$

Therefore, the reduced density operator is just a mixture of the factor states in the Schmidt decomposition for subsystem-1, with probabilities specified by the square of the Schmidt coefficients. The square of the concurrence is then given by

$$1 - \text{tr} \, \widehat{\rho}_1^2 = 1 - \sum_i \lambda_i{}^4 \,, \tag{4.136}$$

in terms of the second moment of the weighing factors for the mixed state. Note that, contrary to the Schmidt number, this is a well-defined entanglement measure for systems with infinite Hilbert spaces, if the above sum converges. Clearly, the purity of subsystem-2 involves the same sum over Schmidt coefficients, in agreement with our previous calculation (4.130).

Consider now the case that a subsystem can again be split up into a pair of components. If the full original state was entangled, the reduced density operator is not pure. Hence, it is an average over pure states. Obviously, this cannot be a product state overall, but if all of the pure states are products, the mixed state is not characterized as entangled, rather it is a *separable state*. The problem with mixed states is that the decomposition into pure states is not unique, so a state is considered separable if there exists any decomposition where it is separated (see Sect. 2.2.2).

Let us now define a *classical pure state* as a δ-function in phase space. Then all pure states, $f(x) = \delta(x)$, in a higher dimensional phase space will be product states, because the higher dimensional δ-functions factor as $\delta(x) = \delta(x_1)\,\delta(x_2)$. In this sense, the expression

$$f(x) = \int f(x') \, \delta(x - x') \mathrm{d}x' \tag{4.137}$$

can be reinterpreted as a *classical separable state*: Any probability distribution in phase space can be considered as a linear combination of products of classical pure states. Thus, we can never consider a classical phase space distribution to be entangled, no matter how strong the correlations may be between variables pertaining to different subsystems.

What if a mixed Wigner function for a bipartite state is everywhere positive? Can we mimic the above reasoning to conclude that there is no entanglement? In general this is not so, because the function $\delta(\mathbf{x})$ does not represent a density operator in the Weyl representation. It represents instead the reflection operator, which has an infinitely degenerate negative eigenvalue, as discussed in Sect. 4.5. The closest that is possible is the coherent state (4.100), which approaches a δ-function as $\hbar \to 0$, but imposes an extra smoothing on the Wigner function for any combination of these pure states. Indeed, a general superposition of coherent states is defined by a weight function, known as the Glauber–Sudarshan *P-function* in quantum optics [8, 62, 63]. So, it is the positivity of a P-function that guarantees a separable state, rather than that of the Wigner function, because each coherent state can be factored.

To close this section, let us now study another kind of projection of the Wigner function. Whereas, by projecting onto a component subspace we generate a reduced Wigner function, a projection onto a Lagrangian plane (4.79) results in a probability density. All the coordinates of such a plane correspond to commuting operators. In the case of a bipartite system, we can define a Lagrangian plane by separately choosing some linear combination of the variables for each subsystem, $q_1' = \alpha_1 p_1 + \beta_1 q_1$ and $q_2' = \alpha_2 p_2 + \beta_2 q_2$, so that each coordinate, q_j', pertains to a different subsystem.

Consider now pairs of either–or measurements on both these variables, i.e., we can define observables $\widehat{O}_{1a}$, $\widehat{O}_{1b}$, $\widehat{O}_{2a}$ and $\widehat{O}_{2b}$ which take the value $+1$, for q_j' in the interval ja, and -1 outside. In terms of projection operators $\widehat{P}_{ja}$, we have $\widehat{O}_{ja} = 2\widehat{P}_{ja} - \widehat{I}$ and $\widehat{O}_{1a}\widehat{O}_{2a} = 4\widehat{P}_{1a}\widehat{P}_{2a} - 2\widehat{P}_{1a} - 2\widehat{P}_{2a} + \widehat{I}$, with similar formulae for the other products of commuting operators. Combining the expectation values for these products in the form of the CHSH inequality (4.17),

$$\langle \widehat{O}_{1a}\widehat{O}_{2a} \rangle + \langle \widehat{O}_{1a}\widehat{O}_{2b} \rangle + \langle \widehat{O}_{1b}\widehat{O}_{2a} \rangle - \langle \widehat{O}_{1b}\widehat{O}_{2b} \rangle$$
$$= 4 \left(\langle \widehat{P}_{1a}\widehat{P}_{2a} \rangle + \langle \widehat{P}_{1a}\widehat{P}_{2b} \rangle + \langle \widehat{P}_{1b}\widehat{P}_{2a} \rangle - \langle \widehat{P}_{1b}\widehat{P}_{2b} \rangle \right) - 4 \left(\langle \widehat{P}_{1a} \rangle + \langle \widehat{P}_{2a} \rangle \right) + 2 \,,$$

$$(4.138)$$

we can now evaluate each expectation value on the right-hand side as a definite integral of the probability density over some region of the (q_1', q_2') plane. This is a purely classical set-up, so that by regrouping,

$$2 - 4 \left(\langle \widehat{P}_{1a} \rangle - \langle \widehat{P}_{1a}\widehat{P}_{2b} \rangle \right) - 4 \left(\langle \widehat{P}_{2a} \rangle - \langle \widehat{P}_{1b}\widehat{P}_{2a} \rangle - \langle \widehat{P}_{1a}\widehat{P}_{2a} \rangle \right) - 4 \langle \widehat{P}_{1b}\widehat{P}_{2b} \rangle \leq 2 \,,$$

$$(4.139)$$

we rederive the CHSH inequality, because none of the terms with the factor -4 can be positive.

We thus verify that the correlations measured among commuting pairs of either–or observables of each subsystem lie within strictly classical bounds, irrespective of any possible entanglement of their combined state. It makes no difference whether, or not, the Wigner function has negative regions. The point is that we need to deal only with a single positive projection, which is a true probability distribution. To obtain a violation of the CHSH inequality, we must choose noncommuting pairs of observables for each component. The correlation for the observables from each pair may still be computed from the probabilities in the respective Lagrangian plane, but we must use different planes in each of the four CHSH correlations. Then, if the overall Wigner function that generates all these densities has negative regions, the CHSH inequality may be violated, as discussed in the following section.

Apparently, there has not been much effort to relate the intuitively appealing picture of entanglement, as the source of nonclassical correlations in Bell inequalities, to the technical entanglement measures appropriate to quantum information theory. However, a recent paper by Cirone [64] bridges this gap for finite-dimensional systems. The main point is that measurements are restricted to projection operators for the factor states in the Schmidt basis. It is then shown that the same concurrence, which was introduced in terms of the partial trace, can be expressed as a sum over conditional probabilities for measurements on either component.

4.8 Generating a *Classical* Entanglement: The EPR State

We have seen how symplectic transformations correspond exactly to unitary transformations in Hilbert space. Let us now examine how these can produce entangled states, given that the initial state, $\hat{\rho}$, is a product of states, each represented by its Wigner function, $W_j(\mathbf{x}_j)$, or its chord function, $\chi_j(\xi_j)$, so that $W(\mathbf{x}) = W_1(\mathbf{x}_1)W_2(\mathbf{x}_2)$ and $\chi(\xi) = \chi_1(\xi_1)\chi_2(\xi_2)$. For the canonical transformation to be linear, the classical interaction Hamiltonian $H(x_1, x_2)$ can only be bilinear in the phase space variables. A convenient choice is $H = p_1 q_2 - p_2 q_1$, which may be interpreted as angular momentum, L_3, if the degrees of freedom refer to Cartesian coordinates in a plane. This Hamiltonian merely rotates both p and q coordinates in the argument of $W(\mathbf{x})$ and $\chi(\xi)$. Then, after a rotation by $\pi/4$, the density operator becomes $\hat{\rho}'$, represented by

$$\chi'(\xi) = \chi_1\left(\frac{\xi_{p_1} + \xi_{p_2}}{\sqrt{2}}, \frac{\xi_{q_1} + \xi_{q_2}}{\sqrt{2}}\right) \chi_2\left(\frac{\xi_{p_1} - \xi_{p_2}}{\sqrt{2}}, \frac{\xi_{q_1} - \xi_{q_2}}{\sqrt{2}}\right). \qquad (4.140)$$

Since the partial trace is specified by (4.127), a section of the chord function, the reduced density for the first component becomes

$$\chi_1'(\xi_1) = (2\pi\hbar)\chi_1\left(\frac{\xi_{p_1}}{\sqrt{2}}, \frac{\xi_{q_1}}{\sqrt{2}}\right) \chi_2\left(\frac{\xi_{p_1}}{\sqrt{2}}, \frac{\xi_{q_1}}{\sqrt{2}}\right), \qquad (4.141)$$

in the chord representation.

So as to emphasise how *classical* an entanglement can be, let us choose for example initial Gaussian states, the product of harmonic oscillator ground states, described by

$$W_j(\mathbf{x}_j) = \frac{1}{\pi\hbar} \exp\left(-\frac{\omega_j}{\hbar}\, \mathbf{q}_j^2 - \frac{1}{\hbar\omega_j}\, \mathbf{p}_j^2\right) \tag{4.142}$$

or

$$\chi_j(\xi_j) = \frac{1}{2\pi\hbar} \exp\left(-\frac{\omega_j}{\hbar}\left(\frac{\xi_{q_j}}{2}\right)^2 - \frac{1}{\hbar\omega_j}\left(\frac{\xi_{p_j}}{2}\right)^2\right). \tag{4.143}$$

Thus, the probability distribution for positions,

$$f(\mathbf{q}) = \int \mathrm{d}\mathbf{p}\, W(\mathbf{x}), \tag{4.144}$$

is also a Gaussian with elliptic level curves that are rotated if $\omega_1 \neq \omega_2$. In this case, the effect of rotation, followed by the partial trace, is just a narrowing of the Gaussians in the chord representation:

$$\chi_1'(\xi_1) = \frac{1}{2\pi\hbar} \exp\left[-\frac{\omega_1+\omega_2}{2\hbar}\left(\frac{\xi_{q_1}}{2}\right)^2 - \frac{1}{2\hbar}\left(\frac{1}{\omega_1}+\frac{1}{\omega_2}\right)\left(\frac{\xi_{p_1}}{2}\right)^2\right]. \tag{4.145}$$

Notice that normalization is maintained, because we still have $\chi_1'(\xi_1) = (2\pi\hbar)^{-1}$ at the chord origin, but now the widths of the position Gaussian and of the momentum Gaussian are obtained through different kinds of average. The overall narrowing indicates that this is no longer a pure state.

The Wigner function presents a more intuitive picture of a mixed state. Taking the Fourier transform:

$$W_1'(\mathbf{x}_1) = \frac{1}{\pi\Delta} \exp\left[-\frac{2\omega_1\omega_2\mathbf{q}_1^2}{\hbar(\omega_1+\omega_2)} - \frac{2\mathbf{p}_1^2}{\hbar(\omega_1+\omega_2)}\right]. \tag{4.146}$$

This still integrates to one, as demanded by normalization, but the Gaussian is now broader, with the uncertainty $\Delta = (\omega_1+\omega_2)/2\sqrt{\omega_1\omega_2} > \hbar$, if $\omega_1 \neq \omega_2$. Therefore, this is not a pure state. The way that this example relates entanglement to initial states and evolution, which may both be considered *classical*, is even more extreme than those discussed in [65], which relies on projections of the Husimi function, in the approximate role of phase space probability density.

Another confirmation that this is not a pure state is that

$$\mathrm{tr}\,(\hat{\rho}_1')^2 = 2\pi\hbar \int \mathrm{d}\mathbf{x}_1\, [W_1'(\mathbf{x}_1)]^2 = 2\pi\hbar \int \mathrm{d}\xi_1\, |\chi_1'(\xi_1)|^2 < 1, \tag{4.147}$$

and yet it might seem that this is just a freak result. After all, our state has remained a smooth *classical-like* Gaussian throughout. There are none of the quantum oscillations which are supposed to be the fingerprint of nonclassicality: For a start, nothing would prevent us from identifying the original Wigner function with a classical probability distribution. We then perform a simple rotation with perfect classical correspondence and obtain a new Gaussian, which pretends to be a quintessentially quantum entangled state! Have we been led astray?

Let us go back to the full Wigner function, resulting from choice (4.142) of Gaussians for the initial product state. After the $\pi/4$ rotation, this is just

$$W'(\mathbf{x}) = \left(\frac{1}{\pi\hbar}\right)^2 \exp\left[-\frac{\omega_1}{2\hbar}(q_1+q_2)^2 - \frac{1}{2\hbar\omega_1}(p_1+p_2)^2\right]$$
$$\times \exp\left[-\frac{\omega_2}{2\hbar}(q_1-q_2)^2 - \frac{1}{2\hbar\omega_2}(p_1-p_2)^2\right]. \quad (4.148)$$

In the extreme limit where $\omega_1 \to 0$ and $\omega_2 \to \infty$, we would obtain a normalized version of

$$W'(x) = \delta(q_1-q_2)\,\delta(p_1+p_2), \quad (4.149)$$

which is just the Wigner function derived by Bell [66] for the original EPR wave function [3], namely $\langle q|\psi\rangle = \delta(q_1 - q_2)$. It so happens that the rotation that transformed the coordinates of our initial state, i.e., the ground state of an anisotropic plane harmonic oscillator, is essentially the same as the transformation from the individual coordinates for a pair of particles into a centre of mass, together with a relative *internal* coordinate. (These transformations differ only by local unitary transformations.) The EPR state is a δ-function both in the relative position and in the total momentum, which is the conjugate variable to the centre of mass.

Thus, the entanglement verified in our initial example implies that the centre of mass is likewise entangled with the relative coordinate in the EPR state. Perhaps, it is then even more surprising that the example that was picked is in some sense *classical*, if we consider that the discussion of the nonlocal and hence nonclassical nature of quantum mechanics started off with the historic EPR paper [3]. The fact that the full Wigner function is positive, not only allows us to interpret it as a classical probability distribution, but it also ensures that there is a wide range of measurements that can be performed on either component which may be considered as classical and hence local. We already found in the previous section that any measurement of pairs of either–or variables, $\widehat{O}_{1a}$, $\widehat{O}_{1b}$, $\widehat{O}_{2a}$ and $\widehat{O}_{2b}$ which take the value $+1$, for general phase space coordinates, q'_j, in the interval ja, and -1 outside, have correlations that satisfy the CHSH inequality, even if the Wigner function has negative regions. That was the case where the quantum observables, which are measured, commute. The statement for positive Wigner functions, due to Bell [5, 66], is stronger: The inequality is then satisfied even if we choose different variables for each measurement, $q'_{1a} \neq q'_{1b}$ and $q'_{2a} \neq q'_{2b}$, corresponding to different Lagrangian planes in phase space and, hence, quantum operators that do not commute. The argument is essentially the same as in the last section, except that now we can obtain all the expectation values from the full Wigner function, acting as a global probability distribution, instead of dealing with different probability distributions, each restricted to the Lagrangian plane specific to a given pair of variables.

Let us now reexamine our classically entangled states from the point of view of the reduced reflection operators, $\widehat{R}'_{\mathbf{x}j}$, defined as (4.124), that act on each component and, in particular, the parity operators, $\widehat{R}'_{0j}$. Such observables do not correspond to smooth phase space functions in classical mechanics, indeed, the Weyl representation of these operators (4.65) is singular. Nonetheless, parity, or reflection measurements can also be carried out on classical waves, as discussed in Sect. 4.2, and the question now concerns the possible correlations between measurements for different reflections carried out on both components. The fact that the full Wigner function (4.148) is symmetric with respect to the origin implies that the density operator commutes with the full reflection operator, $\widehat{R}_0$. However, $W'_1(0) < \pi\hbar$, so it does not have pure parity, i.e., $\widehat{\rho}'_1$ does not commute with $\widehat{R}'_{01}$. Hence, according to the discussion in Sect. 4.6, there is a finite probability to obtain negative (odd) parity, if such a measurement is performed on subsystem-1.

Perhaps this would not be so obvious a priori: The original state, represented by $W_0(\mathbf{x})$, is a pure state with pure positive (even) parity and this is also a property of the rotated state. This property can be verified directly, or it may be noticed that the driving Hamiltonian commutes with $\widehat{R}_0$, so that $H(\mathbf{x}) = H(R_0(\mathbf{x}))$. But now we find that a measurement of the parity of subsystem-1 has a finite probability to be negative. How is that?

Notice that the same also holds for subsystem-2: The derivation of the reduced density operator, $\widehat{\rho}'_2$, represented by $W'_2(\mathbf{x}_2)$ and $\chi'_2(\xi_2)$, goes through exactly as above. Therefore, there is also a finite probability of measuring negative parity in subsystem-2. As was shown in Sect. 4.6, the fact that, in both cases, the Wigner function is symmetric about the origin implies that all the pure states, into which the mixed reduced density operator can be decomposed, must have pure parity, but they are not all even. For this reason, the Wigner function (4.146) had to be obtained as a Fourier transform of the chord function; not a mere rescaling.

The crucial point is that the rotated state, $\widehat{\rho}'$, does not commute with either of the restricted reflections defined by (4.124), i.e., $\widehat{R}'_{01}$ or $\widehat{R}'_{02}$, even though it commutes with their product: $\widehat{R}_0 = \widehat{R}'_{01}\widehat{R}'_{02} = \widehat{R}'_{02}\widehat{R}'_{01}$. It should be recalled that the reduced Wigner functions are entirely determined by (4.125) and (4.126) in terms of the restricted reflections. Thus, to understand the results of measurements of either $\widehat{R}'_{0j}$, we need a common basis for all these operators. This is just the product of an even–odd basis for subsystem-1 and subsystem-2, for which we obtain the table:

$$even \otimes even \rightarrow even,$$
$$even \otimes odd \rightarrow odd,$$
$$odd \otimes even \rightarrow odd,$$
$$odd \otimes odd \rightarrow even.$$

$$(4.150)$$

Since $\widehat{\rho}'$ is even, it must be a superposition of the subset of basis states: $even \otimes even$ or $odd \otimes odd$. Furthermore, we now find that the evolved state has a full parity correlation: If the measurement of $\widehat{R}'_{01}$ specifies even parity, then this must be the outcome of a measurement on $\widehat{R}'_{02}$. Conversely, if one of the subsystems has odd parity, then we know this to be the parity of the other subsystem.

An initial product state of an even Schrödinger cat state with a coherent state, which is rotated by $\pi/4$, is also susceptible to the foregoing analysis. However, an odd symmetry Schrödinger cat would have perfectly anticorrelated odd–even, or even–odd subsystems. In the case of the rotated cat, the evidence for entanglement is much more obvious. The pair of Gaussians is not centred on either of the planes in the chord phase space pertaining to the pair of subsystems. The partial trace that generates the reduced chord functions is a section of the full chord function, so that it does not capture these local maxima. Therefore, there is a deficit of phase space correlations in the reduced density operators.

Returning to the original rotated squeezed state, or, equivalently, the original EPR state, we must conclude that this is truly quantum and correctly described as entangled, i.e., just as nonclassical as the spin states in the Bohm version of EPR [7] that are commonly used to exemplify entanglement. The secret lies in choosing the property to be measured: A position measurement on one of the subsystems would not distinguish between this pure quantum state and a classical distribution. However, a measurement of reflection eigenvalues evokes a *spin-like duality* of this apparent classical state.

The violation of the CHSH inequality for reflection measurements of the smoothed EPR state completes the evidence of its nonclassicality. Banaszek and Wodkiewicz [67] first pointed out that the full pure state Wigner function of a bipartite state is proportional to the correlation for reflection measurements on each subsystem: $\langle \widehat{R}'_{\mathbf{x}1} \widehat{R}'_{\mathbf{x}2} \rangle = (\pi\hbar)^2 W(\mathbf{x})$. This leads to a violation of the CHSH inequality for reflection measurements of the EPR state. They also proposed a realistic experiment for this in quantum optics [68]. We have already verified the complete correlation for parity measurements about the origin, which is in agreement with the maximal value that the full Wigner function (4.148) attains there. Its decay for large $\mathbf{x}_1$ or $\mathbf{x}_2$ signifies that $\langle \widehat{R}'_{\mathbf{x}1} \widehat{R}'_{\mathbf{x}2} \rangle \rightarrow 0$, so that

$$C_{CHSH} = \langle \widehat{R}'_{01} \widehat{R}'_{02} \rangle + \langle \widehat{R}'_{01} \widehat{R}'_{\mathbf{x}2} \rangle + \langle \widehat{R}'_{\mathbf{x}1} \widehat{R}'_{02} \rangle - \langle \widehat{R}'_{\mathbf{x}1} \widehat{R}'_{\mathbf{x}2} \rangle \qquad (4.151)$$

sinks from 2, its maximal classical value at the origin, to the limiting value 1. However, the origin is not the maximum of $C_{CHSH}(\mathbf{x}_1, \mathbf{x}_2)$, because the lowest order expansion of

$$(\pi\hbar)^2 W'(\mathbf{x}) = 1 - \frac{\omega_1}{2\hbar}(q_1 + q_2)^2 - \frac{1}{2\hbar\omega_1}(p_1 + p_2)^2$$

$$- \frac{\omega_2}{2\hbar}(q_1 - q_2)^2 - \frac{1}{2\hbar\omega_2}(p_1 - p_2)^2 + \dots \quad (4.152)$$

leads to

$$C_{CHSH}(\mathbf{x}_1, \mathbf{x}_2) = 2 + (\omega_1 - \omega_2)q_1 q_2 + \left(\frac{1}{\omega_1} - \frac{1}{\omega_2}\right) p_1 p_2 + \dots \quad (4.153)$$

Hence, the origin is a saddle point of $C_{CHSH}(\mathbf{x}_1, \mathbf{x}_2)$, which *increases* from its maximal classical value along the directions $q_2 = -q_1$ and $p_2 = p_1$, if one chooses the EPR conditions, $\omega_2 \gg \omega_1$, i.e., if reflections are chosen in the directions where the Wigner function decays rapidly.

So we find that nonlocal correlations between two subsystems can arise even if the Wigner function for the full system is non-negative everywhere. It would thus appear that there is no relation between fringes in the Wigner function, where it attains negative values, and entanglement. The former project as interference fringes for possible measurements, but this is quite a different kind of *nonclassicality* than the delicate nonclassical correlations resulting from entanglement. But even here, one must be wary! If the measurements on the different components concern mechanical observables, natural for classical particles, then there is at least one case where negativity of the Wigner function has been shown to produce nonclassical correlations. Indeed, Bell [5, 66] constructed an example where the CHSH inequality is violated for measurements on pairs of different variables,[6] $q'_j = q_j + t_j p_j$ and $q'_j = q_j + \tau_j p_j$. The state for which this is proved is a variation of our rotated state, where one of the factor Gaussians is substituted by the second excited state of the harmonic oscillator.

It should always be remembered that entanglement is not an intrinsic property, but only acquires its meaning within a specified basis, *the computational basis*, or the basis where measurements are made. In this respect, it resembles semiclassical caustics, which depend on our choice of representation. If the physical realization of the foregoing example were the ground state of a 2-D harmonic oscillator, then the rotation, which was found to produce entanglement, could be dismissed as merely an inconvenient coordinate transformation: Unless all measurements were to be restricted to the original coordinate axes, it would not be relevant, though true, to say that the rotated system became entangled, while the original system was a mere product. In contrast, for the alternative physical interpretation of one of the new coordinates as the centre of mass for a pair of particles, its entanglement with the internal coordinate can be important.

---

[6] Even though Bell refers to the transformation parameters as *times*, these should be understood as specifications of the variables and hence of the planes onto which the Wigner function is projected.

4.9 Entanglement and Decoherence

The process of decoherence also results from the interaction of a pair of systems: the (small) *open system* and a (large) system, which we call the *environment* (see, e.g., Chap. 5). In contrast to the previous example, the component over which we trace, so as to obtain the reduced density operator, is on a scale which defies anything but a statistical description. The usual picture is that the environment lies *somewhere outside*, but it may just as well consist of the internal degrees of freedom for the centre of mass (CM) of a large system of particles. Exchanges between the large-scale motion and the internal variables lead to macroscopic energy dissipation as well as decoherence of the quantum state for the CM.

Let us consider the simplest possible example of the decoherence of the CM, because of its entanglement with internal variables. The CM for a system of L identical particles, assumed to be distinguishable is $Q = (q_1 + \ldots + q_L)/L$. The conjugate variable to Q is the total momentum, $P = p_1 + \ldots + p_L$. Let us further imagine that they are each in the same single particle state, $\widehat{\rho}$, and that these are independent, i.e., both the Wigner and the chord function are products over those of the individual states. This may seem too restrictive, because we should allow for different values of each average position $\langle q_j \rangle$, but we can redefine this as the origin for each j, so that we then measure Q from $\langle Q \rangle$.

In the case of $L = 2$, $X = (P, Q)$ is obtained from the rotated coordinate in the previous section by a mere canonical rescaling of $2^{\pm 1/2}$. It has been repeatedly emphasised that all such symplectic transformations on the argument of the Wigner or the chord function correspond exactly to unitary quantum transformations. So let us now reverse this transformation in the case of general L: We define $Q' = L^{1/2}Q$ and $P' = L^{-1/2}P$. Then, if the individual Wigner functions, $W(\mathbf{x}_j)$, were classical probability distributions, the *central limit theorem* would imply that the distribution for $X' = (P', Q')$ converges to

$$W_L(\mathbf{X}') \to [\pi \mathbf{\Delta_K}]^{-1} \exp[-\mathbf{X}'\mathbf{K}^{-1}\mathbf{X}'/2], \qquad (4.154)$$

as $L \to \infty$, where we recall the definition of the Schrödinger covariance matrix, $\mathbf{K}$, in (4.7) and its determinant $\mathbf{\Delta_K}^2$. It is remarkable that positivity is not a necessary ingredient for the proof of the central limit theorem: In the case of identical square-integrable pure state Wigner functions, it is shown by Tegmark and Shapiro [69] that convergence onto a Gaussian again results. If the state for the individual particles is not represented by a pure state Gaussian, then the moments for this state will be such that $\mathbf{\Delta_K} > \hbar$. Therefore, the centre of mass Wigner function, $W_L(\mathbf{X})$, is a broader Gaussian than is permissible for a pure state and hence it must be a mixture. So, the CM of independent particles with identical Wigner functions is generally entangled with the internal phase space variables (which it has not been necessary to describe explicitly). Curiously, the potential entanglement resulting from the central limit theorem was overlooked in [69].

How does this entanglement with the internal coordinates evolve in time? It is easy to verify that free motion, generated by the Hamiltonian $H = p_1{}^2 + \ldots + p_L{}^2$, will not alter $\Delta_{\mathbf{K}}$. Thus the entanglement of the centre of mass with the environment is invariant in this simple case. Let us suppose instead that, though the particles do not interact, there is an external nonlinear field. Furthermore, the particles are sufficiently separated and the field is smooth enough so that it is legitimate to linearize the field locally around each $\langle q_j \rangle$. Then the Wigner function for each particle will evolve classically in different ways. The restrictive form of the central limit theorem in [69] cannot be applied in this case, but one can readily adapt Levy's proof [70] to allow for different Wigner functions, as long as the moments are finite and their average values converge.

The averages of the moments resulting from the different evolutions of many Wigner functions lead to a progressive loss of purity for the CM. Just as in (4.146) for the simple example of the last section, the uncertainty, $\Delta_{\mathbf{K}}$, increases. On top of that, the central limit theorem supplies the statistical ingredient for the decoherence process. It might appear strange to obtain decoherence even for a system of noninteracting particles, but it should be recalled that the CM momentum P, or P', appears linearly in each of the terms, $p_j{}^2$, in the kinetic energy, which accounts for the coupling to the internal momenta.

So as to make contact with the theory of Markovian open systems, we can now reinterpret this evolution of the reduced density matrix as a convolution of the original (Gaussian) Wigner function for the CM with a broadening Gaussian. For its Fourier transform, the chord function, this evolution is merely the product of an initial Gaussian with another Gaussian that narrows in time,[7] This is exactly the result for *quantum Markovian evolution* of an open system, in the case of quadratic internal Hamiltonian and linear coupling to the environment [71].

The deduction of the *canonical Lindblad equation* (see Sect. 5.3.2 or [72]),

$$\frac{\partial \widehat{\rho}}{\partial t} = -\frac{i}{\hbar}[\widehat{H}, \widehat{\rho}] - \frac{1}{2\hbar} \sum_j [\widehat{L}_j, [\widehat{L}_j, \widehat{\rho}]] \,, \qquad (4.155)$$

that governs the evolution of the density operator in the quantum Markovian theory does not proceed by tracing out a larger system. All the same, the mere fact that the evolution is entirely determined by a differential equation precludes any delayed participation of previous motion. The *Lindblad operators* $\widehat{L}_j$, account for the nonunitarity of the evolution, that is, they take the part of the coupling to the environment. The Markovian approximation can in principle include arbitrary (non-quadratic) internal Hamiltonians for the system.

---

[7] It must be recalled that the average CM evolution $\langle X(t) \rangle$ has been hidden by a time-dependent coordinate transformation.

The derivation of the Markovian approximation in the context of quantum optics (the damped harmonic oscillator) was carried out originally by Agarwal [73], but this is all in the language of complex phase space. The exact solution of (4.155) in [73] and that of Diosy and Kiefer [74], for the free *open* particle, are special cases of the general result in [71]: The chord representation of (4.155) is particularly simple if the Lindblad operators are linear functions of positions and momenta, $\widehat{L}_j = l_j \cdot \widehat{x}$, and if the Hamiltonian is quadratic [71]:

$$\frac{\partial \chi}{\partial t}(\xi, t) = \{H(\xi), \chi(\xi, t)\} - \frac{1}{2\hbar} \sum_j (1_j \cdot \xi)^2 \, \chi(\xi, t) \,. \qquad (4.156)$$

Here, the first term is the classical Poisson bracket. The exact solution of this equation factors into the unitary evolution of the chord function, undisturbed by the Lindblad operators, and a narrowing Gaussian factor. In the Wigner representation this becomes a Gaussian smudging of the unitarily evolving Wigner function. It is remarkable that the Wigner function becomes positive after a time that depends only on the parameters of the Lindblad equation, regardless of the initial pure state [71, 74].

In our simple example of the evolution of the CM, the Lindblad operator for its one-dimensional motion should be chosen as the total momentum $\widehat{P}$, because this is the variable that couples to the internal motion, which is hidden within the Markovian approximation. Even though the central limit theorem supplied the Gaussian factor of the evolving chord function, the overall Gaussian form for the evolving CM does not reflect the richness of other possibilities for Markovian evolution. However, by considering the entanglement of a *small system* with the CM of a *large system* and following the treatment of the example in the preceding section, we obtain qualitatively the general Markovian picture. This allows an interpretation of the Gaussian smoothing as originating in the multiple small contributions contemplated in the central limit theorem.

The standard way of going beyond the Markovian approximation, so as to include memory kept by the environment of the previous motion of the system, is to use the Feynman–Vernon functional [75], in the manner exploited by Caldeira and Leggett [76, 77]. Some standard references for dissipation, noise and decoherence from a quantum optics point of view are the books by Louisell [78], Gardiner [79] and Weiss [80] (see also Chap. 5).

4.10 A Semiclassical Picture of Entanglement

A full semiclassical theory of entanglement is still a program for the future, fascinating but difficult. Even so, several of the main elements for this construction can be sketched in this concluding section.

For a start, one should note that it is feasible to fit semiclassical torus states with Gaussian coherent states placed along the classical torus in a very satisfactory way [81]. The number of Gaussians required increases with a fractional power of $\hbar^{-1}$. The important qualitative feature is that the interference fringes of the Wigner function, near the midpoint of the pair of Gaussians composing a Schrödinger cat, have the same wavevector as the similar fringes at the centre of a geometrical chord of the classical torus. Therefore, in both cases we can describe very fine interference fringes related to long chords. It also follows that our preliminary study of entanglement and decoherence of cat states is not at all irrelevant for understanding the evolution of product semiclassical states. Refinements of the fitting procedure allow even the description of the diffraction effects near caustics [82].

Before analysing product states and their partial trace, recall that dyadic operators, $|\psi\rangle\langle\phi|$, live in a kind of squared Hilbert space, which corresponds to a double phase space. These operators were shown in Sect. 4.5 to correspond to a product Lagrangian surface in double phase space, $\tau_\psi \otimes \tau_\phi$, if each of these states corresponds to a Lagrangian surface on its own right. Thus, the projection operator, or pure state density operator, $\hat{\rho} = |\psi\rangle\langle\psi|$, is just a particular case of this general rule. If the state, $|\psi\rangle$, corresponds to a Bohr–Sommerfeld quantized torus of L dimensions in a $2L$-D phase space, then the full density operator must correspond to a $2L$-D product torus in $4L$-D double phase space. This is in exact analogy to the way that a product torus describing the state for several particles (4.30) is obtained from lower dimensional tori. Recalling that we can describe double phase space in terms of the centre coordinates, $\mathbf{x}$, and the conjugate variables, $\mathbf{y} = \mathbf{J}\xi$ (4.47), the semiclassical Wigner function, $W(\mathbf{x})$, is then a superposition of complex exponentials, such that each phase is obtained by integrating $\mathbf{y}(\mathbf{x})$ along one of the different branches of the torus. Even though this approximation breaks down along caustics, the latter provide ready indication of regions where the Wigner function has a large intensity.

The problem is then to relate the semiclassical Wigner function, defined on the centre plane, to classical structures that are also portrayed in this same single phase space. Let us consider first the semiclassical Wigner and chord functions in the simplest case where $L = 1$. The Fock states (4.104) are good examples of semiclassical torus states when the quantum number n is large. Introducing the asymptotic expression for Laguerre polynomials,

$$\lim_{n\to\infty} L_n\left(\frac{z^2}{2n}\right) = J_0\left(\sqrt{2}z\right), \qquad (4.157)$$

together with the large argument expansion,

$$J_0(y) \approx \frac{2}{\sqrt{\pi y}} \cos\left(y - \frac{\pi}{4}\right), \qquad (4.158)$$

brings the Wigner function for these states into a semiclassical form. To understand this, we must investigate the geometry of the double torus from the point of view of the simpler quantized curve, which is just a circle in this case.

Every point on the double torus represents a pair of points on the quantized curve and vice versa. A given pair of points, $x_\pm$, on the quantized curve, defines a geometric chord: $\xi = x_+ - x_-$. Hence, $\mathbf{y} = \mathbf{J}\xi$ is the chord coordinate on the double torus, which has the centre coordinate, $\mathbf{x} = (x_+ + x_-)/2$. Obviously, the exchange of x_+ with x_- produces a new chord of the quantized curve with the same centre, $\mathbf{x}$. Viewed in double phase space, there must always be pairs of chords of the double torus projecting onto each centre, $\mathbf{x}$. The symmetry of this surface with respect to the identity plane, $\mathbf{y} = 0$, leads to complex conjugate phase contributions, in line with the above cosine for the Fock state. Actually this is a general feature: Because the Wigner function is real, the chord pairs will always produce semiclassical contributions adding up to cosines.

To obtain the phase of the cosine contribution to the semiclassical Wigner function for each pair of chords, the best course is to use a result which was put in its most general form by Littlejohn [29]. This concerns the general overlap, $\langle \psi | \phi \rangle$, of quantum states associated semiclassically to curves τ_ψ and τ_ϕ: The semiclassical contributions arise from the intersections of these classical curves and the phase difference between a pair of contributions is just the area sandwiched between the corresponding pair of intersections, divided by Planck's constant, as shown in Fig. 4.8.

We can immediately apply this principle to the Wigner and chord functions for pure states, by recalling that $W_\psi(\mathbf{x}) = (\pi\hbar)^{-1}\langle \psi | (\widehat{R}_\mathbf{x} | \psi) \rangle$ and $\chi_\psi(\xi) = (2\pi\hbar)^{-1}\langle \psi | (\widehat{T}_\xi | \psi) \rangle$. The semiclassical state $\widehat{R}_\mathbf{x} | \psi \rangle$ is merely the state constructed from the reflected curve, $R_\mathbf{x}(\tau_\psi)$, whereas $\widehat{T}_\xi | \psi \rangle$ corresponds to

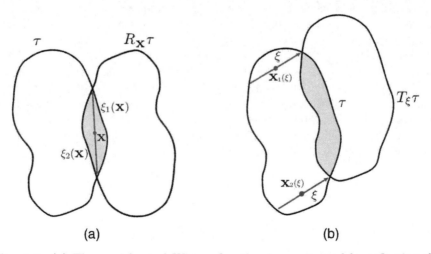

(a)   (b)

Fig. 4.8. (a) The semiclassical Wigner function is constructed by reflecting the torus τ around the point $\mathbf{x}$: The chords $\xi(\mathbf{x})$ are defined by the intersections with $R_\mathbf{x}\tau$. (b) The semiclassical chord function is constructed by translating τ by the vector ξ: The centres $\mathbf{x}(\xi)$ lie halfway back along ξ from each intersection of τ with $T_\xi\tau$

the translated curve $T_\xi(\tau_\psi)$. Therefore, in the case of the Wigner function, we obtain the phase of the cosine as half of the area sandwiched between the torus and its reflection at the centre $\mathbf{x}$ [22], which coincides with the area between the torus and the chord [83, 84] (see also [85]). Furthermore, this construction supplies, at a glance, the tips of all the chords centred on $\mathbf{x}$ as the intersections between both curves, as seen in Fig. 4.8(a). Note that once the curve has been reflected around the origin, we need only translate $R_0(\tau_\psi)$ to obtain the reflections around all other centres because of the group property, $R_\mathbf{x} = T_\mathbf{x} \circ R_0 \circ T_{-\mathbf{x}}$.

The same geometrical method can be used to study the structure of centres for a pre-specified chord ξ on the curve τ_ψ. Each intersection of τ_ψ with the translated curve $T_\xi(\tau_\psi)$ reveals one of the tips where the chord is to be placed and hence the centre of the chord, as shown in Fig. 4.8(b). In the case of open curves, the chord function may actually be simpler than the Wigner function because it is not necessary to have interference, as in the case of the parabola, for which there is only one intersection. In the case that τ_ψ has a centre of symmetry, as in the example of the Fock state, we thus find that the simple relation (4.99) between Wigner and chord functions is respected by the semiclassical approximation.

Viewed in single phase space, the caustics of the Wigner function arise from coalescing torus chords, as their centre, $\mathbf{x}$, is moved. This occurs at the tangencies of $R_\mathbf{x}(\tau_\psi)$ with the fixed curve, τ_ψ. Similarly, the caustics of the chord function are the loci of ξ such that $T_\xi(\tau_\psi)$ is tangent to τ_ψ. On the other hand, in double phase space, the Wigner caustics for a torus state are viewed as projection singularities of the double torus, $\tau_\psi \otimes \tau_\psi$, which lies *above* the area inside τ_ψ. The general geometric constructions underlying the semiclassical Wigner and chord functions are readily extended to phase space representations of dyadic operators, $|\psi\rangle\langle\phi|$, corresponding to double tori, $\tau_\psi \otimes \tau_\phi$. Their Weyl representation is known as *cross-Wigner functions* or *Moyal brackets* [86], whose semiclassical form is presented in [87].

There are many fascinating features of caustics in the phase space representations of pure states that have been studied and many more that must still be analyzed. For instance, the build up in the centre of the Wigner function for the Fock state is a caustic. Its semiclassical origin is the degeneracy of a continuum of chords conjugate to the same symmetry centre. However, this is a nongeneric feature of reflection-symmetric states. If the symmetry is broken, this supercaustic unfolds into a cusped triangle, first described by Berry [84]. The unfolding of higher dimensional caustics for rotated product tori, studied in [22], and displayed in Fig. 4.9 were also examined for the Wigner function. It turns out that the double fold surfaces of the Wigner caustic that meet along the torus do not unfold in the manner portrayed in Fig. 4.4 because of a symmetry constraint.

The limit of small chords is specially relevant for semiclassical theory. For the Wigner function, it singles out the classical torus itself as the Wigner caustic. The uniform approximation for the Wigner function throughout this

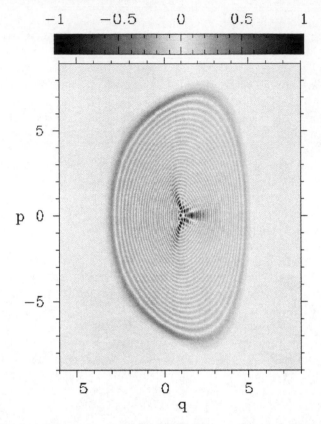

Fig. 4.9. Typical fringes for a semiclassical Wigner function. Alternating large values of the Wigner function, positive or negative, are represented by dark regions. The classical curve follows closely the border of the fringes. The *triangular structure* of interfering fringes near the centre results from the unfolding of the maximum of the Fock states due to symmetry breaking

region is presented in [84], for the case of a curve and in [88], for a two-dimensional torus. A pair of sheets of the double torus are joined on the identity plane along this curve or torus. The large amplitude of the Wigner function oscillations near the quantized curve is due to this caustic. The corresponding caustic of the chord function collapses onto the origin, whatever the geometry of the classical region. The neighbourhood of this highly non-generic chord-caustic is discussed in [32]. Once again, we find that all regions where $C(\xi) = |\chi(\xi)|^2$ is large, outside of an $\hbar^{L/2}$-neighbourhood of the origin, point to phase space correlations that are truly quantum in nature.

So far, we have only discussed static properties of the density operator. In turning to dynamics, a preliminary point is that we should distinguish the *Weyl propagator*, $U(\mathbf{x})$, that is, the Weyl representation of the evolution operator, from the propagator for Wigner functions. The former is unitary and is hence supported by its own *static* Lagrangian surface in double phase space, as discussed in Sect. 4.5, so that its semiclassical description is similar to that of the Wigner function itself. This can be deduced from a path integral in single phase space [33, 89].

In order to treat the unitary evolution of the density operator for a pure bound state, $|\psi\rangle\langle\psi|$, with L degrees of freedom, we need to consider the corresponding classical evolution of a $2L$-D Lagrangian torus, $\tau_\psi \otimes \tau_\psi$. Initially, this is separable within both single phase spaces, even though the product is not factored in the centre×chord coordinates. The classical motion must propagate the tips of each chord, x_- and x_+, in the same way. Taking account of the change of sign, $p_- \rightarrow -p_-$, in the definition of double phase space, we find that the double phase space Hamiltonian must be

$$\mathbb{H}(X) = H(x_+) - H(x_-) = H(\mathbf{x} - \mathbf{Jy}/2) - H(\mathbf{x} + \mathbf{Jy}/2)\,. \qquad (4.159)$$

This classical Hamiltonian can be verified to preserve the product form of the geometric structures in each of the phase spaces $x_\pm$, but it will not preserve initial products within each of these in the general case that the single Hamiltonian $H(x)$ has coupling terms between different degrees of freedom. It propagates Lagrangian surfaces in double phase space that correspond either to density operators (according to the Liouville–von Neumann equation) or to unitary operators (the Heisenberg equation). The explicit formulae for the semiclassical evolution of the Wigner function are given in [90, 91], whereas the evolving action of the chord function is presented in [92]. The difficulty that cannot be avoided by changing the representation lies in the caustics of the initial state, which require more sophisticated semiclassical treatment.

A promising approach lies in the definition of integral propagators for the Wigner function, or for the chord function. The former may be defined in terms of the Weyl propagator as a kind of *second-order Wigner transform* (see e. g. [90])

$$\mathbb{U}(\mathbf{x}_t, \mathbf{x}) = \int d\mu\, U_{-t}\left(\frac{\mathbf{x} + \mathbf{x}_t}{2} - \mu\right) U_t\left(\frac{\mathbf{x} + \mathbf{x}_t}{2} + \mu\right) \exp\left(\frac{2i}{\hbar}\mu \wedge (\mathbf{x}_t - \mathbf{x})\right)\,.$$
$$(4.160)$$

Their explicit semiclassical form has been developed in [93], but these propagators also have their own intrinsic caustics. More recently, caustic-free propagators, from the Wigner to the chord function and vice versa, have been defined [92]. These are constructed either in terms of the propagation of the unitary reflection operators or the translation operators, instead of directly evolving the density operator itself.

Several of the geometrical structures underlying the semiclassical theory of the Wigner function for nonseparable tori in double phase space that evolve under the action of a general Hamiltonian were analyzed in [22] for the simplest case where $L = 2$. It will be necessary to push much further this analysis, while adapting it to the chord function. The reason is again that the partial trace of the density operator is obtained immediately by a section through the chord function, $\chi(\xi_1, \xi_2 = 0)$, which is defined semiclassically by a nonseparable section of the double torus that evolved from an initial product $\tau_\psi \otimes \tau_\psi$. For instance, the quantum state and the corresponding double torus both loose their product form under the action of the simple Hamiltonian employed in Sect. 4.8. But slicing through a torus produces either a single or several lower dimensional tori.

This indicates that the semiclassical theory of reduced density operators that allow us to quantify entanglement, or to calculate the correlations on separate measurements effected on each component of the system, can still be associated to lower dimensional Lagrangian surfaces that are no longer products. The problem is to work out the actions and the amplitudes for this multidimensional geometry. This general picture agrees with initial results for nonunitary Markovian evolution of semiclassical Wigner functions [94]. It is notable that the same methods that have been used in [95] to show that the semiclassical approximation of the Wigner function satisfies the purity condition, tr $\widehat{\rho}^2 = \widehat{\rho}$, reveal the loss of purity with time due to decoherence.

A question that has deserved some attention concerns the effect of classically chaotic evolution on decoherence of an open system, or equivalently on the evolution of a system entangled to a larger system. There are indications that the reduced density matrix will loose its purity faster than in the case of regular internal motion [96], and this is verified exactly for quadratically hyperbolic Markovian systems [71], which capture a small element of chaotic behaviour. However, it must be stressed that there is no essential extra difficulty, in dealing semiclassically with the short-time evolution generated by chaotic hamiltonians, over regular motion.

In contrast, the evolution of initial pure states, which are eigenstates of chaotic Hamiltonians, is a much tougher problem. According to Shnirelman's theorem [97–99], in most cases these *ergodic states* are supported by the full energy shell, in the sense that averages of smooth observables are well approximated by classical averages over this surface. Often, such a state is thus described as having a *δ-function over the shell* for its Wigner function, but this is not true of its detailed structure: The intimacy between the Wigner function and phase space reflections implies that interference fringes generally exist halfway between classical regions of the Wigner function. According to the discussion of correlations of the Wigner function in Sect. 4.6, these are equal to their Fourier transform, for all pure states. Therefore, a Wigner function supported by an energy shell of large phase space dimensions must have oscillations in its interior of correspondingly small wavelength. These are readily derived semiclassically for mixtures of states within a narrow energy range [33, 100], but the fine features of pure states have so far eluded all theoretical efforts.

So far, our discussion has concerned a chaotic initial state for the relevant component, but nothing prevents us from defining an initial state that is ergodic for some chaotic Hamiltonian defined over the full product Hilbert space of the system with its environment. The situation is then radically different from the states that have evolved semiclassically from product states: It is possible to *disentangle* each state that has lost its original product form by merely reversing time. This corresponds classically to running Hamilton's equations backwards in the double phase space; e.g., in the example of Sect. 4.8, the product of two harmonic oscillator ground states is recovered from the EPR state. This is always possible in quantum mechanics because one can always specify a unitary transformation on the entire Hilbert space, which transforms any given state into any other state and we can choose the latter to be a product state. However, the correspondence for this disentanglement cannot exist for ergodic eigenstates of a classically chaotic Hamiltonian defined on the full Hilbert space.

In the case where $L = 2$, the energy shell is 3-D and, because it has an extra dimension, there exists no classical canonical transformation, whether linear or nonlinear, that can transform it into the product of two closed curves. Therefore, ergodic eigenstates are *essentially entangled* from the point of view of classical correspondence. The study of traces of classical chaos in quantum mechanics is known as *quantum chaology* [85]. The characterization of ergodic states as those that are not classically disentangleable establishes a bridge between entanglement theory and quantum chaology. It remains to be seen whether this special type of entangled state has any application in quantum information theory.

Acknowledgement

This text develops a series of Lectures delivered in September 2005 at the MPIPKS-Dresden, always so hospitable. Many of the ideas and results presented here arose in the context of a fruitful collaboration with Olivier Brodier in CBPF. Raul Vallejos, Marcos Saraceno and Pedro Rios also played important roles in the developments that I have here sketched. Maria Carolina Nemes and Thomas Dittrich contributed helpful criticism of the manuscript. Partial financial support from Millenium Institute of Quantum Information, PROSUL and CNPq is gratefully acknowledged.

References

1. E. Schrödinger: *Quantisierung als Eigenwertproblem*, Ann. der Physik **81**, 109 (1926)
2. E. Schrödinger: *Die gegenwärtige Situation in der Quantenmechanik*, Naturwiss. **23**, 807 (1935)
3. A. Einstein , B. Podolsky and N. Rosen: *Can quantum-mechanical description of physical reality be considered complete?*, Phys. Rev. **47**, 777 (1935)
4. J. S. Bell: *On the Einstein Podolsky Rosen paradox*, Physics **1**, 195 (1964)
5. J. S. Bell: *Speakable and Unspeakable in Quantum Mechanics* (Cambridge University Press, Cambridge 1993)

6. M. A. Nielsen and L. I. Chuang: *Quantum Computation and Quantum Information* (Cambridge University Press, Cambridge 2003)
7. D. Bohm: *Quantum Theory* (Prentice Hall, New York 1951)
8. P. W. Schleich: *Quantum Optics in Phase Space* (Wiley-VCH, Berlin 2001)
9. R. P. Feynman: *Space-time approach to non-relativistic quantum mechanics*, Rev. Mod. Phys. **20**, 367 (1948)
10. L. S. Schulman: *Techniques and Applications of Path Integration* (Wiley, New York 1981)
11. J. W. S. Rayleigh: *The Theory of Sound* (Dover Publications, New York 1976)
12. V. I. Arnold: *Mathematical Methods of Classical Mechanics* (Springer, Berlin 1978)
13. A. Peres: *Quantum Theory: Concepts and Methods* (Kluwer Academic, Dordrecht 1993)
14. J. F. Clauser, R. A. Holt, M. A. Horne and A. Shimony: *Proposed experiment to test local hidden-variable theories*, Phys. Rev. Lett. **23**, 880 (1969)
15. J. S. Bell: *Bertlmann's socks and the nature of reality*, J. de Physique **42**, C2 41 (1981).
16. H. Goldstein: *Classical Mechanics*, 2nd edn (Addison-Wesley, Reading, 1980)
17. A. Voros: *Semi-classical approximations*, Ann. Inst. Henri Poincaré **24**, 31 (1976)
18. A. M. Ozorio de Almeida: *Hamiltonian Systems: Chaos and Quantization* (Cambridge University Press, Cambridge 1988)
19. M. C. Gutzwiller: *Chaos in Classical and Quantum Mechanics* (Springer, New York 1990)
20. V. P. Maslov and M. V. Fedoriuk: *Semiclassical Approximation in Quantum Mechanics* (Reidel, Dordrecht 1981)
21. J. H. Van Vleck: *The correspondence principle in the statistical interpretation of quantum mechanics*, Proc. Natl. Acad. Sci. USA **14**, 178 (1928)
22. A. M. Ozorio de Almeida and J. Hannay: *Geometry of two dimensional tori in phase space: Projections, sections and the Wigner function*, Ann. Phys. **138**, 115 (1982)
23. R. J. Glauber: *Coherent and incoherent states of the radiation field*, Phys. Rev. **131**, 2766 (1963)
24. J. R. Klauder and B. Skagerstam: *Coherent States* (World Scientific, Singapore 1985)
25. A. Perelomov: *Generalized Coherent States and their Applications* (Springer, New York 1986)
26. C. Cohen Tannoudji, B. Diu and F. Laeoe: *Quantum Mechanics* (Wiley, New York 1977)
27. A. Messiah: *Quantum Mechanics* (North Holland, Amsterdam 1961)
28. L. G. Lutterbach and L. Davidovich: *Method for direct measurement of the Wigner function in cavity QED and ion traps*, Phys. Rev. Lett. **78**, 2547 (1997)
29. R. G. Littlejohn. In: *Quantum Chaos: Between Order and Disorder*, ed by G. Casati and B. Chirikov (Cambridge University Press, Cambridge 1995) p. 343
30. J. P. Amiet and P. Huguenin: *Generating functions of canonical maps*, Helvet. Phys. Acta **53**, 377 (1980)
31. H. S. M. Coxeter: *Introduction to Geometry* (Wiley, New York 1969)

32. A. M. Ozorio de Almeida, O. Vallejos and M. Saraceno: *Pure state correlations: chords in phase space*, J. Phys. A **38**, 1473 (2004) and arXiv:quant-ph/0410129

33. A. M. Ozorio de Almeida: *The Weyl representation in classical and quantum mechanics*, Phys. Rep. **295**, 265 (1998)

34. P. Bertet, A. Auffeves, P. Maioli, S. Ornaghi, T. Meunier, M. Brune, J. M. Raimond and S. Haroche: *Direct measurement of the Wigner function of a one-photon fock state in a cavity*, Phys. Rev. Lett. **89**, 200402 (2002)

35. N. L. Balazs and B. K. Jennings: *Wigner's function and other distribution functions in mock phase spaces*, Phys. Rep. **104**, 347 (1984)

36. A. Grossmann and P. Huguenin: *Group-theoretical aspects of the Wigner–Weyl isomorphism*, Helv. Phys. Acta **51**, 252 (1978)

37. E. P. Wigner: *On the Quantum correction for thermodynamic equilibrium*, Phys. Rev. **40**, 749 (1932)

38. E. Schrödinger: *Zum Heisenbergschen Unschärfeprinzip*, Proc. Pruss. Acad. Sci. **19**, 296 (1930)

39. S. R. Deans: *The Radon Transform and Some of Its Applications* (John Wiley & Sons, New York 1983)

40. A. Grossmann: *Parity operator and quantization of δ-functions*, Commun. Math. Phys. **48**, 191 (1976)

41. A. Royer: *Wigner function as the expectation value of a parity operator*, Phys. Rev. A **15**, 449 (1977)

42. W. H. Zurek: *Sub-Planck structure in phase space and its relevance for quantum decoherence*, Nature **412**, 712 (2001)

43. R. L. Hudson: *When is the wigner quasi-probability density non-negative?*, Rep. Math. Phys. **6**, 249 (1974)

44. V. I. Tatarskii: *The Wigner representation of quantum mechanics*, Sov. phys. Usp. **26**, 311 (1983)

45. A. M. Ozorio de Almeida and O. Brodier: *Parity measurements, decoherence and spiky Wigner functions*, J. Phys. A **37**, L245 (2004)

46. H. J. Grönewold: *On principles of quantum mechanics*, Physica **12**, 405 (1946)

47. B.-G. Englert, N. Sterpi and H. Walther: *Parity states in the one-atom maser*, Opt. Commun. **100**, 526 (1993)

48. W. K. Wootters: *A Wigner-function formulation of finite-state quantum mechanics*, Ann. Phys. NY **176**, 1 (1987)

49. C. Miquel, J. P. Paz and M. Saraceno: *Quantum computers in phase space*, Phys. Rev. A **65**, 062309 (2002)

50. A. M. F. Rivas and A. M. Ozorio de Almeida: *The Weyl representation on the torus*, Ann. Phys. NY **276**, 223 (1999)

51. A. Rivas, M. Saraceno and A. M. Ozorio de Almeida: *Quantization of multidimensional cat maps*, Nonlinearity **13**, 341 (2000)

52. A. Arguelles and T. Dittrich: *Wigner function for discrete phase space: Exorcising ghost images*, Physica A **356**, 72 (2005)

53. K. Husimi: *Some formal properties of the density matrix*, Proc. Phys. Math. Soc. Jpn. **22**, 264 (1940)

54. K. Takashi: *Wigner and Husimi functions in quantum mechanics*, J. Phys. Soc. Jpn. **55**, 762 (1986)

55. A. Wehrl: *General properties of entropy*, Rev. Mod. Phys. **50**, 221 (1978)

56. A. Wehrl: *On the relation between classical and quantum-mechanical entropy*, Rep. Math. Phys. **16**, 353 (1979)
57. I. Bengtsson and K. Życzkowski: *Geometry of Quantum States* (Cambridge University Press, Cambridge 2006)
58. F. Mintert and K. Życzkowski: *Wehrl entropy, Lieb conjecture, and entanglement monotones*, Phys. Rev. A **69**, 022317 (2004)
59. P. Leboeuf and A. Voros. In: *Quantum Chaos: Between Order and Disorder*, ed by G. Casati and B. Chirikov (Cambridge University Press, Cambridge 1995) p. 507
60. F. Mintert, A. R. R. Carvalho, M. Kus and A. Buchleitner: *Measures and dynamics of entangled states*, Phys. Rep. **415**, 207 (2005)
61. H. B. Bennet, H. J. Bernstein, S. Popescu and B. Schumacher: *Concentrating partial entanglement by local operations*, Phys. Rev. A **53**, 2046 (1996)
62. R. J. Glauber: *Photon correlations*, Phys. Rev. Lett. **10**, 84 (1963)
63. E. C. G. Sudarshan: *Equivalence of semiclassical and quantum mechanical descriptions of statistical light beams*, Phys. Rev. Lett. **10**, 277 (1963)
64. M. A. Cirone: *Entanglement correlations, Bell inequalities and the concurrence*, Phys. Lett. A **339**, 269 (2005)
65. R. M. Angelo, S. A. Vitiello, M. A. M. de Aguiar and K. Furuya: *Quantum linear mutual information and classical correlations in globally pure bipartite systems*, Physica A **338**, 458 (2004)
66. J. S. Bell. In: *New Techniques and Ideas in Quantum Measurement Theory* (New York Academy of Sciences, New York 1986)
67. K. Banaszek and K. Wodkiewicz: *Nonlocality of the Einstein-Podolsky-Rosen state in the Wigner representation*, Phys. Rev. A **58**, 4345 (1988)
68. K. Banaszek and K. Wodkiewicz: *Testing quantum nonlocality in phase space*, Phys. Rev. Lett. **82**, 2009 (1999)
69. M. Tegmark and H. S. Shapiro: *Decoherence produces coherent states: An explicit proof for harmonic chains*, Phys. Rev. E **50**, 2538 (1994)
70. P. Levy: *Theorie de l'Adition des Variables Aleatoires* (Gauthier-Villars, Paris 1954)
71. O. Brodier and A. M. Ozorio de Almeida: *Symplectic evolution of Wigner functions in Markovian open systems*, Phys. Rev. E **69**, 016204 (2004)
72. D. Giulini, E. Joos, C. Kiefer, J. Kupsch, I.-O. Stamatescu and H. D. Zeh: *Decoherence and the Appearance of a Classical World in Quantum Theory* (Springer, Berlin 1996)
73. G. S. Agarwal: *Brownian motion of a quantum oscillator*, Phys. Rev. A **4**, 739 (1971)
74. L. Diosi and C. Kiefer: *Exact positivity of the Wigner and P-functions of a Markovian open system*, J. Phys. A **35**, 2675 (2002)
75. R. P. Feynman and F. L. Vernon: *The theory of a general quantum system interacting with a linear dissipative system*, Ann. Phys. (NY) **24**, 118 (1963)
76. A. O. Caldeira and A. J. Leggett: *Quantum tunnelling in a dissipative system*, Ann. Phys. (NY) **149**, 374 (1983)
77. A.O. Caldeira and A.J. Leggett: Ann. Phys. (NY) **153**, 445 (1984)
78. W. H. Louisell: *Quantum Properties of Radiation* (Wiley, New York 1973)
79. C. W. Gardiner: *Quantum Noise* (Springer, Berlin 1991)
80. U. Weiss: *Quantum Dissipative Systems*, Series in Modern Condensed Matter Physics, vol 2 (World Scientific, Singapore 1993)

81. A. Kenfack, J. M. Rost and A. M. Ozorio de Almeida: *Optimal representations of quantum states by Gaussians in phase space*, J. Phys. B **37**, 1645 (2004)
82. F. Toscano, A. Kenfack, A. R. R. Carvalho, J. M. Rost and A. M. Ozorio de Almeida: *Husimi-Wigner representation of chaotic eigenstates*, Proc. R. Soc. Lond. A **464**, 1503 (2008)
83. F. A. Berezin and M. A. Shubin. In: *Colloquia Mathematica Societatis Janos Bolyiai* (North-Holland, Amsterdam 1972) p. 21
84. M. V. Berry: *Semi-classical mechanics in phase space: a study of Wigner's function*, Phil. Trans. Roy. Soc. A **287**, 237 (1977)
85. M. V. Berry. In: *Chaos and Quantum Physics; Les Houches LII*, ed by M.-J. Giannoni, A. Voros and J. Zinn-Justin (North Holland, 1991) p. 251
86. E. J. Moyal: *Quantum Mechanics as a Statistical Theory*, Proc. Cambridge Phil. Soc. **45**, 99 (1949)
87. A. M. Ozorio de Almeida: *Semiclassical Matrix Elements*, Rev. Bras. Fis. **14**, 62 (1984)
88. A. M. Ozorio de Almeida: *The Wigner function for two-dimensional tori: uniform approximation and projections*, Ann. Phys. (NY) **145**, 100 (1983)
89. A. M. Ozorio de Almeida: *Phase space path integral for the Weyl propagator*, Proc. R. Soc. Lond. A **439**, 139 (1992)
90. P. P. M. Rios and A. M. Ozorio de Almeida: *On the propagation of semiclassical Wigner functions*, J. Phys. A **35**, 2609 (2002)
91. T. A. Osborn and M. F. Kondratieva: *Heisenberg evolution WKB and symplectic area phases*, J. Phys. A **35**, 5279 (2002)
92. A. M. Ozorio de Almeida and O. Brodier: *Phase space propagators for quantum operators*, Ann. Phys. (NY) **321**, 1790 (2006)
93. T. Dittrich, C. Viviescas and L. Sandoval: *Semiclassical propagator of the Wigner function*, Phys. Rev. Lett. **96**, 070403 (2006)
94. A. M. Ozorio de Almeida: *Decoherence of semiclassical Wigner functions*, J. Phys. A **36**, 67 (2003)
95. A. M. Ozorio de Almeida: *Pure state condition for the semiclassical Wigner function*, Physica A **110**, 501 (1982)
96. W. H. Zurek and J. P. Paz: *Decoherence, Chaos, and the second law*, Phys. Rev. Lett. **72**, 2508 (1994)
97. A. I. Shnirelman: *Ergodic properties of eigenfunctions*, (Russian) Uspehi. Mat. Nauk. **29**, 181 (1974)
98. Y. Colin de Verdière: *Ergodicit et fonctions propres du laplacien*, Comm. Math. Phys. **102**, 497 (1985)
99. S. Zelditch: *Uniform distribution of eigenfunctions on compact hyperbolic surfaces*, Duke Math. J. **55**, 919 (1987)
100. M. V. Berry: *Fringes decorating anticaustics in ergodic wavefunctions*, Proc. R. Soc. A **424**, 279 (1989)

5 Introduction to Decoherence Theory

K. Hornberger

Arnold Sommerfeld Center for Theoretical Physics,
Ludwig–Maximilians–Universität München, Theresienstraße 37, D-80333 Munich,
Germany

5.1 The Concept of Decoherence

This introduction to the theory of decoherence is aimed at readers with an
interest in the science of quantum information. In that field, one is usually
content with simple, abstract descriptions of non-unitary "quantum channels" to account for imperfections in quantum processing tasks. However, in
order to justify such models of non-unitary evolution and to understand their
limits of applicability it is important to know their physical basis. I will therefore emphasize the dynamic and microscopic origins of the phenomenon of
decoherence, and will relate it to concepts from quantum information where
applicable, in particular to the theory of quantum measurement.

The study of decoherence, though based at the heart of quantum theory,
is a relatively young subject. It was initiated in the 1970s and 1980s with
the work of H. D. Zeh and W. Zurek on the emergence of classicality in the
quantum framework. Until that time the orthodox interpretation of quantum
mechanics dominated, with its strict distinction between the classical macroscopic world and the microscopic quantum realm. The mainstream attitude
concerning the boundary between the quantum and the classical was that
this was a purely philosophical problem, intangible by any physical analysis.
This changed with the understanding that there is no need for denying quantum mechanics to hold even macroscopically, if one is only able to understand
within the framework of quantum mechanics why the macro-world *appears*
to be classical. For instance, macroscopic objects are found in approximate
position eigenstates of their center of mass, but never in superpositions of
macroscopically distinct positions. The original motivation for the study of
decoherence was to explain these effective *super-selection rules* and the apparent emergence of classicality within quantum theory by appreciating the
crucial role played by the environment of a quantum system.

Hence, the relevant theoretical framework for the study of decoherence
is the theory of *open quantum systems*, which treats the effects of an uncontrollable environment on the quantum evolution. Originally developed to
incorporate the phenomena of friction and thermalization in the quantum
formulation, it has of course a much longer history than decoherence theory. However, we will see that the intuition and approximations developed
in the traditional treatments of open quantum systems are not necessarily

Hornberger, K.: *Introduction to Decoherence Theory*. Lect. Notes Phys. **768**, 221–276 (2009)
DOI 10.1007/978-3-540-88169-8_5   © Springer-Verlag Berlin Heidelberg 2009

appropriate to yield a correct description of decoherence effects, which may take place on a time scale much shorter than typical relaxation phenomena. In a sense, one may say that while the traditional treatments of open quantum systems focus on how an environmental "bath" affects the system, the emphasis in decoherence is more on the contrary question, namely how the system affects and disturbs environmental degrees of freedom, thereby revealing information about its state.

The physics of decoherence became very popular in the last decade, mainly due to advances in experimental technology. In a number of experiments the gradual emergence of classical properties in a quantum system could be observed, in agreement with the predictions of decoherence theory. Needles to say, a second important reason for the popularity of decoherence is its relevance for quantum information processing tasks, where the coherence of a relatively large quantum system has to be maintained over a long time.

Parts of these lecture notes are based on the books on decoherence by E. Joos et al. [1], on open quantum systems by H.-P. Breuer & F. Petruccione [2], and on the lecture notes of W. Strunz [3]. Interpretational aspects, which are not covered here, are discussed in [4, 5] and useful reviews by W. H. Zurek and J. P. Paz can be found in [6, 7]. This chapter deals exclusively with conventional, i.e., environmental decoherence, as opposed to spontaneous reduction theories [8], which aim at "solving the measurement problem of quantum mechanics" by modifying the Schrödinger equation. These models are conceptually very different from environmental decoherence, though their predictions of super-selection rules are often qualitatively similar.

5.1.1 Decoherence in a Nutshell

Let us start by discussing the basic decoherence effect in a rather general framework. As just mentioned, we need to account for the unavoidable coupling of the quantum system to its environment. Although these environmental degrees of freedom are to be treated quantum mechanically, their state must be taken unobservable for all practical purposes, be it due to their large number or uncontrollable nature. In general, the detailed temporal dynamics induced by the environmental interaction will be very complicated, but one can get an idea of the basic effects by assuming that the interaction is sufficiently short ranged to admit a description in terms of scattering theory. In this case, only the map between the asymptotically free states before and after the interaction needs to be discussed, thus avoiding a temporal description of the collision dynamics.

Let the quantum state of the system be described by the density operator ρ on the Hilbert space $\mathcal{H}$. We take the system to interact with a single environmental degree of freedom at a time – think of a phonon, a polaron, or a gas particle scattering off your favorite implementation of a quantum register. Moreover, let us assume, for the time being, that this environmental "particle" is in a pure state $\rho_E = |\psi_{in}\rangle\langle\psi_{in}|_E$, with $|\psi_{in}\rangle_E \in \mathcal{H}_E$. The scattering operator S_{tot} maps between the in- and out-asymptotes in the total Hilbert space $\mathcal{H}_{tot} = \mathcal{H} \otimes \mathcal{H}_E$, and for sufficiently short-ranged interaction potentials we may identify those with the states before and after the collision. The initially uncorrelated system and environment turn into a joint state,

$$[\text{before collision}] \quad \rho_{tot} = \rho \otimes |\psi_{in}\rangle\langle\psi_{in}|_E , \tag{5.1}$$

$$[\text{after collision}] \quad \rho'_{tot} = \mathsf{S}_{tot}[\rho \otimes |\psi_{in}\rangle\langle\psi_{in}|_E]\mathsf{S}^\dagger_{tot} . \tag{5.2}$$

Now let us assume, in addition, that the interaction is *non-invasive* with respect to a certain system property. This means that there is a number of distinct system states, such that the environmental scattering off these states causes no transitions in the system. For instance, if these distinguished states correspond to the system being localized at well-defined sites then the environmental particle should induce no hopping between the locations. In the case of elastic scattering, on the other hand, they will be the energy eigenstates. Denoting the set of these mutually orthogonal system states by $\{|n\rangle\} \in \mathcal{H}$, the requirement of non-invasiveness means that S_{tot} commutes with those states, that is, it has the form

$$\mathsf{S}_{tot} = \sum_n |n\rangle\langle n| \otimes \mathsf{S}_n , \tag{5.3}$$

where the S_n are scattering operators acting in the environmental Hilbert space. The insertion into (5.2) yields

$$\rho'_{tot} = \sum_{m,n}\langle m|\rho|n\rangle|m\rangle\langle n| \otimes \mathsf{S}_m|\psi_{in}\rangle\langle\psi_{in}|_E\mathsf{S}^\dagger_n$$

$$\equiv \sum_{m,n}\rho_{mn}|m\rangle\langle n| \otimes |\psi^{(m)}_{out}\rangle\langle\psi^{(n)}_{out}|_E , \tag{5.4}$$

and disregarding the environmental state by performing a partial trace we get the *system* state after the interaction:

$$\rho' = \text{tr}_E\left(\rho'_{tot}\right) = \sum_{m,n}|m\rangle\langle n|\rho_{mn}\underbrace{\langle\psi_{in}|\,\mathsf{S}^\dagger_n\mathsf{S}_m|\psi_{in}\rangle_E}_{\langle\psi^{(n)}_{out}|\psi^{(m)}_{out}\rangle_E} . \tag{5.5}$$

Since the S_n are unitary the diagonal elements, or *populations*, are indeed unaffected,

$$\rho'_{mm} = \rho_{mm} , \tag{5.6}$$

while the off-diagonal elements, or *coherences*, get multiplied by the overlap of the environmental states scattered off the system states m and n,

$$\rho'_{mn} = \rho_{mn}\langle\psi^{(n)}_{out}|\psi^{(m)}_{out}\rangle. \tag{5.7}$$

This factor has a modulus of less than one so that the coherences, which characterize the ability of the system state to display a superposition between $|m\rangle$ and $|n\rangle$, get suppressed.[1] It is important to note that this loss of coherence occurs in a *special basis*, which is determined only by the scattering operator, i.e., by the type of environmental interaction, and to a degree that is determined by both the environmental state and the interaction.

This loss of the ability to show quantum behavior due to the interaction with an environmental quantum degree of freedom is the basic effect of decoherence. One may view it as due to the arising correlation between the system with the environment. After the interaction the joint quantum state of system and environment is no longer separable, and part of the coherence initially located in the system now resides in the non-local correlation between system and the environmental particle; it is lost once the environment is disregarded. A complementary point of view argues that the interaction constitutes an information transfer from the system to the environment. The more the overlap in (5.7) differs in magnitude from unity, the more an observer could in principle learn about the system state by measuring the environmental particle. Even though this measurement is never made, the complementarity principle then explains that the wave-like interference phenomenon characterized by the coherences vanishes as more information discriminating the distinct, "particle-like" system states is revealed.

To finish the introduction, here is a collection of other characteristics and popular statements about the decoherence phenomenon. One often hears that decoherence (i) can be extremely fast as compared to all other relevant time scales, (ii) can be interpreted as an indirect measurement process, a monitoring of the system by the environment, (iii) creates dynamically a set of preferred states ("robust states" or "pointer states") which seemingly do not obey the superposition principle, thus providing the microscopic mechanism for the emergence of effective super-selection rules, and (iv) helps to understand the emergence of classicality in a quantum framework. These points will be illustrated in the following, though (iii) has been demonstrated only for very simple model systems and (iv) depends to a fair extent on your favored interpretation of quantum mechanics.

5.1.2 General Scattering Interaction

In the above demonstration of the decoherence effect the choice of the interaction and the environmental state was rather special. Let us therefore now take S_{tot} and ρ_E to be arbitrary and carry out the same analysis. Performing the trace in (5.5) in the eigenbasis of the environmental state, $\rho_E = \sum_\ell p_\ell |\psi_\ell\rangle\langle\psi_\ell|_E$, we have

---

[1] The value $|\rho_{mn}|$ determines the maximal fringe visibility in a general interference experiment involving the states $|m\rangle$ and $|n\rangle$, as described by the projection on a general superposition $|\psi_{\theta,\varphi}\rangle = \cos(\theta)|m\rangle + e^{i\varphi}\sin(\theta)|n\rangle$.

$$\rho' = \mathrm{tr}_E\left(S_{\mathrm{tot}}[\rho \otimes \rho_E]S_{\mathrm{tot}}^\dagger\right) = \sum_{j,\ell} p_\ell \langle \psi_j | S_{\mathrm{tot}} | \psi_\ell \rangle_E \rho \langle \psi_\ell | S_{\mathrm{tot}}^\dagger | \psi_j \rangle_E$$

$$= \sum_k W_k \rho W_k^\dagger, \tag{5.8}$$

where the $\langle \psi_j | S_{\mathrm{tot}} | \psi_\ell \rangle$ are operators in $\mathcal{H}$. After subsuming the two indices j, ℓ into a single one, we get the second line with the *Kraus operators* W_k given by

$$W_k = \sqrt{p_{\ell_k}} \langle \psi_{j_k} | S_{\mathrm{tot}} | \psi_{\ell_k} \rangle. \tag{5.9}$$

It follows from the unitarity of S_{tot} that they satisfy

$$\sum_k W_k^\dagger W_k = \mathbb{I}. \tag{5.10}$$

This implies that (5.8) is the operator-sum representation of a completely positive map $\Phi : \rho \mapsto \rho'$ (see Sect. 5.3.1). In other words, the scattering transformation has the form of the most general evolution of a quantum state that is compatible with the rules of quantum theory. In the operational formulation of quantum mechanics this transformation is usually called a *quantum operation* [9], the quantum information community likes to call it a *quantum channel*. Conversely, given an arbitrary quantum channel, one can also construct a scattering operator S_{tot} and an environmental state ρ_E giving rise to the transformation, though it is usually not very helpful from a physical point of view to picture the action of a general, dissipative quantum channel as due to a single scattering event.

5.1.3 Decoherence as an Environmental Monitoring Process

We are now in a position to relate the decoherence of a quantum system to the information it reveals to the environment. Since the formulation is based on the notion of an *indirect measurement* it is necessary to first collect some aspects of measurement theory [10, 11].

Elements of General Measurement Theory

Projective Measurements

This is the type of measurement discussed in standard textbooks of quantum mechanics. A projective operator $|\alpha\rangle\langle\alpha| \equiv P_\alpha = P_\alpha^2 = P_\alpha^\dagger$ is attributed to each possible outcome α of an idealized measurement apparatus. The probability of the outcome α is obtained by the Born rule

$$\mathrm{Prob}(\alpha|\rho) = \mathrm{tr}\left(P_\alpha \rho\right) = \langle \alpha | \rho | \alpha \rangle, \tag{5.11}$$

and after the measurement of α the state of the quantum system is given by the normalized projection

$$\mathcal{M} : \rho \mapsto \mathcal{M}(\rho|\alpha) = \frac{\mathsf{P}_\alpha \rho \mathsf{P}_\alpha}{\mathrm{tr}(\mathsf{P}_\alpha \rho)} . \tag{5.12}$$

The basic requirement that the projectors form a resolution of the identity operator,

$$\sum_\alpha \mathsf{P}_\alpha = \mathbb{I}, \tag{5.13}$$

ensures the normalization of the corresponding probability distribution Prob $(\alpha|\rho)$.

If the measured system property corresponds to a self-adjoint operator A the P_α are the projectors into its eigenspaces, so that its expectation value is

$$\langle \mathsf{A} \rangle = \mathrm{tr}(\mathsf{A}\rho) .$$

If A has a continuous spectrum the outcomes are characterized by intervals of a real parameter, and the sum in (5.13) should be replaced by a projector-valued Stieltjes integral $\int d\mathsf{P}(\alpha) = \mathbb{I}$, or equivalently by a Lebesgue integral over a *projector-valued measure* (PVM) [10, 11].

It is important to note that projective measurements are not the most general type of measurement compatible with the rules of quantum mechanics. In fact, non-destructive measurements of a quantum system are usually not of the projective kind.

Generalized Measurements

In the most general measurement situation, a positive (and therefore hermitian) operator $\mathsf{F}_\alpha > 0$ is attributed to each outcome α. Again, the collection of operators corresponding to all possible outcomes must form a resolution of the identity operator,

$$\sum_\alpha \mathsf{F}_\alpha = \mathbb{I}. \tag{5.14}$$

In particular, one speaks of a *positive operator-valued measure* (POVM) in the case of a continous outcome parameter, $\int d\mathsf{F}(\alpha) = \mathbb{I}$, and the probability (or probability density in the continuous case) of outcome α is given by

$$\mathrm{Prob}(\alpha|\rho) = \mathrm{tr}(\mathsf{F}_\alpha \rho). \tag{5.15}$$

The effect on the system state of a generalized measurement is described by a nonlinear transformation,

$$\mathcal{M} : \rho \mapsto \mathcal{M}(\rho|\alpha) = \frac{\sum_k \mathsf{M}_{\alpha,k} \rho \mathsf{M}_{\alpha,k}^\dagger}{\mathrm{tr}(\mathsf{F}_\alpha \rho)}, \tag{5.16}$$

involving a norm-decreasing completely positive map in the numerator (see Sect. 5.3.1), and a normalization which is subject to the *consistency requirement*

$$\sum_k M_{\alpha,k}^\dagger M_{\alpha,k} = F_\alpha \,. \tag{5.17}$$

The operators $M_{\alpha,k}$ appearing in (5.16) are called *measurement operators*, and they serve to characterize the measurement process completely. The F_α are sometimes called "effects" or "measurement elements". Note that different measurement operators $M_{\alpha,k}$ can lead to the same measurement element F_α.

A simple class of generalized measurements are *unsharp measurements*, where a number of projective operators are lumped together with probabilistic weights in order to account for the finite resolution of a measurement device or for classical noise in its signal processing. However, generalized measurements schemes may also perform tasks which are seemingly impossible with a projective measurement, such as the error-free discrimination of two non-orthogonal states [12, 13].

Efficient Measurements

A generalized measurement is called *efficient* if there is only a single summand in (5.16) for each outcome α,

$$\mathcal{M}(\rho|\alpha) = \frac{M_\alpha \rho M_\alpha^\dagger}{\mathrm{tr}\left(M_\alpha^\dagger M_\alpha \rho\right)}\,, \tag{5.18}$$

implying that pure states are mapped to pure states. In a sense, these are measurements where no unnecessary, that is no classical, uncertainty is introduced during the measurement process, see below. By means of a (left) polar decomposition and the consistency requirement (5.17) efficient measurement operators have the form

$$M_\alpha = U_\alpha \sqrt{F_\alpha}\,, \tag{5.19}$$

with an unitary operator U_α. This way the state after efficient measurement can be expressed in a form which decomposes the transformation into a "raw measurement" described by the F_α and a "measurement back-action" given by the U_α:

$$\mathcal{M}(\rho|\alpha) = \underbrace{U_\alpha}_{\text{back-action}} \underbrace{\frac{\sqrt{F_\alpha}\rho\sqrt{F_\alpha}}{\mathrm{tr}(F_\alpha\rho)}}_{\substack{\text{raw} \\ \text{measurement}}} \underbrace{U_\alpha^\dagger}_{\text{back-action}}\,. \tag{5.20}$$

In this transformation the positive operators $\sqrt{F_\alpha}$ "squeeze" the state along the measured property and expand it along the other, complementary ones, similar to what a projector would do, while the back-action operators U_α "kick" the state by transforming it in a way that is reversible, in principle, provided the outcome α is known. Note that the projective measurements (5.12) are a subclass in the set of back-action-free efficient measurements.

Indirect Measurements

In an *indirect measurement* one tries to obtain information about the system in a way that disturbs it as little as possible. This is done by letting a well-prepared microscopic quantum probe interact with the system. This probe is then measured by projection, i.e., destructively, so that one can infer properties of the system without having it brought into contact with a macroscopic measurement device. Let ρ_{probe} be the prepared state of the probe, S_{tot} describe the interaction between system and probe, and P_α be the projectors corresponding to the various outcomes of the probe measurement. The probability of measuring α is determined by the reduced state of the probe after interaction, i.e.,

$$\text{Prob}(\alpha|\rho) = \text{tr}_{\text{probe}}\left(P_\alpha \rho'_{\text{probe}}\right) = \text{tr}_{\text{probe}}\left(P_\alpha \text{tr}_{\text{sys}}(S_{\text{tot}}[\rho \otimes \rho_{\text{probe}}]S_{\text{tot}}^\dagger)\right).$$
(5.21)

By pulling out the system trace (extending the projectors to $\mathcal{H}_{\text{tot}} = \mathcal{H} \otimes \mathcal{H}_{\text{p}}$) and using the cyclic permutability of operators under the trace we have

$$\text{Prob}(\alpha|\rho) = \text{tr}\left(S_{\text{tot}}^\dagger[\mathbb{I} \otimes P_\alpha]S_{\text{tot}}[\rho \otimes \rho_{\text{probe}}]\right) = \text{tr}(F_\alpha\rho),$$
(5.22)

with microscopically defined measurement elements

$$F_\alpha = \text{tr}_{\text{probe}}\left(S_{\text{tot}}^\dagger[\mathbb{I} \otimes P_\alpha]S_{\text{tot}}[\mathbb{I} \otimes \rho_{\text{probe}}]\right) > 0$$
(5.23)

satisfying $\sum_\alpha F_\alpha = \mathbb{I}$. Since the probe measurement is projective, we can also specify the new system state conditioned on the click at α of the probe detector,

$$\begin{aligned}
\mathcal{M}(\rho|\alpha) &= \text{tr}_{\text{probe}}\left(\mathcal{M}_{\text{tot}}(\rho_{\text{tot}}|\alpha)\right) \\
&= \text{tr}_{\text{probe}}\left(\frac{[\mathbb{I} \otimes P_\alpha]S_{\text{tot}}[\rho \otimes \rho_{\text{probe}}]S_{\text{tot}}^\dagger[\mathbb{I} \otimes P_\alpha]}{\text{tr}(F_\alpha\rho)}\right) \\
&= \sum_k \frac{M_{\alpha,k}\rho M_{\alpha,k}^\dagger}{\text{tr}(F_\alpha\rho)}.
\end{aligned}$$
(5.24)

In the last step a convex decomposition of the initial probe state into pure states was inserted, $\rho_{\text{probe}} = \sum_k w_k|\psi_k\rangle\langle\psi_k|$. Taking $P_\alpha = |\alpha\rangle\langle\alpha|$ we thus get a microscopic description also of the measurement operators,

$$M_{\alpha,k} = \sqrt{w_k}\langle\alpha|S_{\text{tot}}|\psi_k\rangle.$$
(5.25)

This shows that an indirect measurement is efficient (as defined above) if the probe is initially in a pure state, i.e., if there is no uncertainty introduced in the measurement process, apart from the one imposed by the uncertainty relations on ρ_{probe}.

If we know that an indirect measurement has taken place, but do not know its outcome α we have to resort to a probabilistic (Bayesian) description of the new system state. It is given by the sum over all possible outcomes weighted by their respective probabilities,

$$\rho' = \sum_\alpha \text{Prob}(\alpha|\rho)\mathcal{M}(\rho|\alpha) = \sum_{\alpha,k} \mathsf{M}_{\alpha,k}\rho\mathsf{M}^\dagger_{\alpha,k}. \qquad (5.26)$$

This form is the same as above in (5.8) and (5.9), where the basic effect of decoherence has been described. This indicates that the decoherence process can be legitimately viewed as a consequence of the information transfer from the system to the environment. The complementarity principle can then be invoked to understand which particular system properties lose their quantum behavior, namely those complementary to the ones revealed to the environment. This "monitoring interpretation" of the decoherence process will help us below to derive microscopic master equations.

5.1.4 A Few Words on Nomenclature

Since decoherence phenomena show up in quite different sub communities of physics, a certain confusion and lack of uniformity developed in the terminology. This is exacerbated by the fact that decoherence often reveals itself as a loss of fringe visibility in interference experiments – a phenomenon, though, which may have other causes than decoherence proper. Here is an attempt of clarification:

- *decoherence:* In the original sense, an environmental quantum effect affecting macroscopically distinct states. The term is nowadays applied to mesoscopically different states as well, and even for microscopic states, as long as it refers to the *quantum* effect of environmental, i.e., in practice unobservable, degrees of freedom.
  However, the term is often (ab-)used for any other process reducing the purity of a micro-state.
- *dephasing:* In a narrow sense, this describes the phenomenon that coherences, i.e., the off-diagonal elements of the density matrix, get reduced in a particular basis, namely the energy eigenbasis of the system. It is a statement about the effect and not the cause. In particular, dephasing may be *reversible* if it is not due to decoherence, as revealed, e.g., in spin-echo experiments.
  This phrase should be treated with great care since it is used differently in various sub communities. It is taken as a synonym to "*dispersion*" in molecular physics and in nonlinear optics, as a synonym to "*decoherence*" in condensed matter, and often as a synonym to "*phase averaging*" in matter wave optics. It is also called a T_2-*process* in NMR and in condensed matter physics (see below).

– *phase averaging:* A classical noise phenomenon entering through the dependence of the unitary system evolution on external control parameters which fluctuate (parametric average over unitary evolutions). A typical example are the vibrations of an interferometer grating or the fluctuations of the classical magnetic field in an electron interferometer due to technical noise. Empirically, phase averaging is often hard to distinguish from decoherence proper.

– *dispersion:* Coherent broadening of wave packets during the unitary evolution, e.g., due to a velocity dependent group velocity or non-harmonic energy spacings. This unitary effect may lead to a reduction of signal oscillations, for instance, in molecular pump-probe experiments.

– *dissipation:* Energy exchange with the environment leading to thermalization. Usually accompanied by decoherence, but see Sect. 5.3.4 for a counterexample.

5.2 Case Study: Dephasing of Qubits

So far, the discussion of the temporal dynamics of the decoherence process was circumvented by using a scattering description. Before going to the general treatment of open quantum systems in Sect. 5.3, it is helpful to take a closer look on the time evolution of a special system where the interaction with a model environment can be treated exactly [2, 14].

5.2.1 An Exactly Solvable Model

Let us take a two-level system, or qubit, described by the Pauli spin operator σ_z, and model the environment as a collection of bosonic field modes. In practice, such fields can yield an appropriate effective description even if the actual environment looks quite differently, in particular if the environmental coupling is a sum of many small contributions.[2] What is fairly non-generic in the present model is the type of coupling between system and environment, which is taken to commute with the system Hamiltonian.

The total Hamiltonian thus reads

$$\mathsf{H}_{\mathrm{tot}} = \underbrace{\frac{\hbar\omega}{2}\sigma_z + \sum_k \hbar\omega_k \mathsf{b}_k^\dagger \mathsf{b}_k}_{\mathsf{H}_0} + \underbrace{\sigma_z \sum_k \left(g_k \mathsf{b}_k^\dagger + g_k^* \mathsf{b}_k\right)}_{\mathsf{H}_{\mathrm{int}}} , \qquad (5.27)$$

with the usual commutation relation for the mode operators of the bosonic field modes, $[\mathsf{b}_i, \mathsf{b}_k^\dagger] = \delta_{ik}$, and coupling constants g_k. The fact that the system Hamiltonian commutes with the interaction, guarantees that there is

---

[2] A counterexample would be the presence of a degenerate environmental degree of freedom, such as a bistable fluctuator.

no energy exchange between system and environment so that we expect pure dephasing.

By going into the interaction picture one transfers the trivial time evolution generated by H_0 to the operators (and indicates this with a tilde). In particular,

$$\widetilde{H}_{int}(t) = e^{iH_0 t/\hbar} H_{int} e^{-iH_0 t/\hbar} = \sigma_z \sum_k (g_k e^{i\omega_k t} b_k^\dagger + g_k^* e^{-i\omega_k t} b_k) , \quad (5.28)$$

where the second equality is granted by the commutation $[\sigma_z, H_{int}] = 0$. The time evolution due to this Hamiltonian can be formally expressed as a Dyson series,

$$\widetilde{U}(t) = \mathcal{T}_\leftarrow \exp\left(-\frac{i}{\hbar} \int_0^t dt' \widetilde{H}_{int}(t')\right)$$

$$= \sum_{n=0}^\infty \frac{1}{n!}\left(\frac{1}{i\hbar}\right)^n \int_0^t dt_1 \cdots dt_n \mathcal{T}_\leftarrow \left[\widetilde{H}_{int}(t_1) \cdots \widetilde{H}_{int}(t_n)\right] , \quad (5.29)$$

where $\mathcal{T}_\leftarrow$ is the time ordering operator (putting the operators with larger time arguments to the left). Due to this time ordering requirement the series usually cannot be evaluated exactly (if it converges at all). However, in the present case the commutator of $\widetilde{H}_{int}$ at different times is not an operator, but just a c-number,

$$[\widetilde{H}_{int}(t), \widetilde{H}_{int}(t')] = 2i \sum_k |g_k|^2 \sin(\omega_k(t'-t)) . \quad (5.30)$$

As a consequence, the time evolution differs only by a time-dependent phase from the one obtained by casting the operators in their natural order,[3]

---

[3] To obtain the time-evolution $\widetilde{U}(t)$ for the case $[\widetilde{H}(t), \widetilde{H}(t')] = c\mathbb{I}$ define the operators

$$\Phi(t) = \frac{1}{\hbar} \int_0^t dt' \, \widetilde{H}(t')$$

and $\overline{U}(t) = \exp[i\Phi(t)]\widetilde{U}(t)$. This way $\overline{U}(t)$ describes the "additional" motion due to the time ordering requirement. It satisfies

$$\partial_t \overline{U}(t) = \left(\left[\frac{d}{dt} e^{i\Phi(t)}\right] e^{-i\Phi(t)} + \frac{1}{i\hbar} e^{i\Phi(t)} \widetilde{H}(t) e^{-i\Phi(t)}\right) \overline{U}(t) .$$

The derivative in square brackets has to be evaluated with care since the $\widetilde{H}(t)$ do not commute at different times. By first showing that $[A, \partial_t A] = c \in \mathbb{C}$ implies $\partial_t A^n = nA^{n-1}\partial_t A - \frac{1}{2}n(n-1)cA^{n-2}$ one finds

$$\frac{d}{dt} e^{i\Phi(t)} = \frac{-1}{i\hbar} e^{i\Phi(t)} \widetilde{H}(t) + \frac{1}{2\hbar^2} e^{i\Phi(t)} \left[\int_0^t dt' \widetilde{H}(t'), \widetilde{H}(t)\right] .$$

$$\tilde{U}(t) = e^{i\varphi(t)} \exp\left(-\frac{i}{\hbar} \int_0^t dt' \tilde{H}_{int}(t')\right), \tag{5.31}$$

where the phase is given by

$$\varphi(t) = \frac{i}{2\hbar^2} \int_0^t dt_1 \int_0^t dt_2 \Theta(t_1 - t_2) \left[\tilde{H}_{int}(t_1), \tilde{H}_{int}(t_2)\right]. \tag{5.32}$$

One can now perform the integral over the interaction Hamiltonian to get

$$\tilde{U}(t) = e^{i\varphi(t)} \exp\left(\frac{1}{2}\sigma_z \sum_k \left(\alpha_k(t)b_k^\dagger - \alpha_k^*(t)b_k\right)\right), \tag{5.33}$$

with complex, time-dependent functions

$$\alpha_k(t) := 2g_k \frac{1 - e^{i\omega_k t}}{\hbar\omega_k}. \tag{5.34}$$

The operator $\tilde{U}(t)$ is diagonal in the eigenbasis of the system, and it describes how the environmental dynamics depend on the state of the system. In particular, if the system is initially in the upper level, $|\psi\rangle = |\uparrow\rangle$, one has

$$\tilde{U}(t)|\uparrow\rangle|\xi_0\rangle_E = e^{i\varphi(t)}|\uparrow\rangle \prod_k D_k\left(\frac{\alpha_k(t)}{2}\right)|\xi_0\rangle =: e^{i\varphi(t)}|\uparrow\rangle|\xi_\uparrow(t)\rangle_E, \tag{5.35}$$

and for the lower state

$$\tilde{U}(t)|\downarrow\rangle|\xi_0\rangle_E = e^{i\varphi(t)}|\downarrow\rangle \prod_k D_k\left(-\frac{\alpha_k(t)}{2}\right)|\xi_0\rangle =: e^{i\varphi(t)}|\downarrow\rangle|\xi_\downarrow(t)\rangle_E. \tag{5.36}$$

Here we introduced the unitary *displacement operators* for the kth field mode,

$$D_k(\alpha) = \exp(\alpha b_k^\dagger - \alpha^* b_k), \tag{5.37}$$

which effect a translation of the field state in its attributed phase space. In particular, the coherent state $|\alpha\rangle_k$ of the field mode k is obtained from its ground state $|0\rangle_k$ by $|\alpha\rangle_k := D_k(\alpha)|0\rangle_k$ [15].

Equations (5.35) and (5.36) show that the collective state of the field modes gets displaced by the interaction with the system and that the sense of the displacement is determined by the system state.

Therefore, we have $\partial_t \overline{U}(t) = (2\hbar^2)^{-1} \int_0^t dt' [\tilde{H}(t'), \tilde{H}(t)] \overline{U}(t)$, which can be integrated to yield finally

$$\tilde{U}(t) = \exp\left(-\frac{1}{2\hbar^2} \int_0^t dt_1 \int_0^{t_1} dt_2 \left[\tilde{H}(t_1), \tilde{H}(t_2)\right]\right) e^{-i\Phi(t)}.$$

Assuming that the states of system and environment are initially uncorrelated, $\rho_{\text{tot}}(0) = \rho \otimes \rho_{\text{E}}$, the time-evolved system state reads[4]

$$\tilde{\rho}(t) = \text{tr}_{\text{E}}\left(\tilde{\mathsf{U}}(t)[\rho \otimes \rho_{\text{E}}]\tilde{\mathsf{U}}^{\dagger}(t)\right) . \tag{5.38}$$

It follows from (5.35) and (5.36) that the populations are unaffected,

$$\langle \uparrow |\tilde{\rho}(t)| \uparrow \rangle = \langle \uparrow |\tilde{\rho}(0)| \uparrow \rangle ,$$
$$\langle \downarrow |\tilde{\rho}(t)| \downarrow \rangle = \langle \downarrow |\tilde{\rho}(0)| \downarrow \rangle ,$$

while the coherences are suppressed by a factor which is given by the trace over the displaced initial field state,

$$\langle \uparrow |\tilde{\rho}(t)| \downarrow \rangle = \langle \uparrow |\tilde{\rho}(0)| \downarrow \rangle \underbrace{\text{tr}_{\text{E}}\left(\prod_k \mathsf{D}_k(\alpha_k(t))\rho_{\text{E}}\right)}_{\chi(t)} . \tag{5.39}$$

Incidentally, the complex suppression factor $\chi(t)$ is equal to the *Wigner characteristic function* of the original environmental state at the points $\alpha_k(t)$, i.e., it is given by the Fourier transform of its Wigner function [16].

Initial Vacuum State

If the environment is initially in its vacuum state, $\rho_{\text{E}} = \bigotimes_k |0\rangle\langle 0|_k$, the $|\xi_{\uparrow}(t)\rangle_{\text{E}}$ and $|\xi_{\downarrow}(t)\rangle_{\text{E}}$ defined in (5.35), (5.36) turn into multi mode coherent states, and the suppression factor can be calculated immediately to yield:

$$\chi_{\text{vac}}(t) = \prod_k \langle 0|\mathsf{D}_k(\alpha_k(t))|0\rangle_k = \prod_k \exp\left(-\frac{|\alpha_k(t)|^2}{2}\right)$$
$$= \exp\left(-\sum_k 4|g_k|^2 \frac{1 - \cos(\omega_k t)}{\hbar^2\omega_k^2}\right) . \tag{5.40}$$

For times that are short compared to the field dynamics, $t \ll \omega_k^{-1}$, one observes a Gaussian decay of the coherences. Modifications to this become relevant at $\omega_k t \cong 1$, provided $\chi_{\text{vac}}(t)$ is then still appreciable, i.e., for $4|g_k|^2/\hbar^2\omega_k^2 \ll 1$. Being a sum over periodic functions, $\chi_{\text{vac}}(t)$ is quasi-periodic, that is, it will come back arbitrarily close to unity after a large period (which increases exponentially with the number of modes). These somewhat artificial Poincaré recurrences vanish if we replace the sum over the discrete modes by an integral over a continuum with mode density μ,

---

[4] In fact, the assumption $\rho_{\text{tot}}(0) = \rho \otimes \rho_{\text{E}}$ is quite unrealistic if the coupling is strong, as discussed below. Nonetheless, it certainly represents a valid initial state.

$$\sum_k f(\omega_k) \longrightarrow \int_0^\infty d\omega \mu(\omega) f(\omega) \,, \tag{5.41}$$

for any function f. This way the coupling constants g_k get replaced by the *spectral density* of the environment,

$$J(\omega) = 4\mu(\omega) |g(\omega)|^2 \,. \tag{5.42}$$

This function characterizes the environment by telling how effective the coupling is at a certain frequency.

Thermal State

If the environment is in a thermal state with temperature T,

$$\rho_E = \rho_{th} = \frac{e^{-H_E/k_B T}}{\text{tr}\left(e^{-H_E/k_B T}\right)} = \bigotimes_k \underbrace{\left(1 - e^{-\hbar\omega_k/k_B T}\right) \sum_{n=0}^\infty e^{-\hbar\omega_k n/k_B T} |n\rangle\langle n|_k}_{=\rho_{th}^{(k)}} \,, \tag{5.43}$$

the suppression factor reads[5]

$$\chi(t) = \prod_k \text{tr}\left(D_k(\alpha_k(t))\rho_{th}^{(k)}\right) = \prod_k \exp\left(-\frac{|\alpha_k(t)|^2}{2} \coth\left(\frac{\hbar\omega_k}{2k_B T}\right)\right) \,. \tag{5.44}$$

This factor can be separated into its vacuum component (5.40) and a thermal component, $\chi(t) = e^{-F_{vac}(t)}e^{-F_{th}(t)}$, with the following definitions of the vacuum and the thermal decay functions:

$$F_{vac}(t) := \sum_k 4 |g_k|^2 \frac{1 - \cos(\omega_k t)}{\hbar^2 \omega_k^2} \,, \tag{5.45}$$

$$F_{th}(t) := \sum_k 4 |g_k|^2 \frac{1 - \cos(\omega_k t)}{\hbar^2 \omega_k^2} \left(\coth\left(\frac{\hbar\omega_k}{2k_B T}\right) - 1\right) \,. \tag{5.46}$$

5.2.2 The Continuum Limit

Assuming that the field modes are sufficiently dense we replace their sum by an integration. Noting (5.41), (5.42) we have

---

[5] This can be found in a small exercise by using the Baker–Hausdorff relation with $\exp\left(\alpha b^\dagger - \alpha^* b\right) = \exp(-|\alpha|^2/2) \exp\left(\alpha b^\dagger\right) \exp\left(-\alpha^* b\right)$, and the fact that coherent states satisfy the eigenvalue equation $b|\beta\rangle = \beta|\beta\rangle$, have the number representation $\langle n|\beta\rangle = \exp(-|\beta|^2/2)\beta^n/\sqrt{n!}$, and form an over-complete set with $\mathbb{I} = \pi^{-1} \int d^2\beta |\beta\rangle\langle\beta|$.

$$F_{\text{vac}}(t) \longrightarrow \int_0^\infty d\omega\, J(\omega)\, \frac{1 - \cos(\omega t)}{\hbar^2 \omega^2} \tag{5.47}$$

$$F_{\text{th}}(t) \longrightarrow \int_0^\infty d\omega\, J(\omega)\, \frac{1 - \cos(\omega t)}{\hbar^2 \omega^2}\left(\coth\left(\frac{\hbar\omega}{2k_B T}\right) - 1\right). \tag{5.48}$$

So far, the treatment was exact. To continue we have to specify the spectral density in the continuum limit. A typical model takes $g \propto \sqrt{\omega}$, so that the spectral density of a d-dimensional field can be written as [17]

$$J(\omega) = a\omega \left(\frac{\omega}{\omega_c}\right)^{d-1} e^{-\omega/\omega_c} \tag{5.49}$$

with "damping strength" $a > 0$. Here, ω_c is a characteristic frequency "cutoff" where the coupling decreases rapidly, such as the Debye frequency in the case of phonons.

Ohmic Coupling

For $d = 1$ the spectral density (5.49) increases linearly at small ω ("Ohmic coupling"). One finds

$$F_{\text{vac}}(t) = \frac{a}{2\hbar^2}\log(1 + \omega_c^2 t^2), \tag{5.50}$$

which bears a strong ω_c dependence. Evaluating the second integral requires to assume that the cutoff ω_c is large compared to the thermal energy, $kT \ll \hbar\omega_c$:

$$F_{\text{th}}(t) \simeq \frac{a}{\hbar^2}\log\left(\frac{\sinh(t/t_T)}{t/t_T}\right). \tag{5.51}$$

Here $t_T = \hbar/(\pi k_B T)$ is a thermal quantum time scale. The corresponding frequency $\omega_1 = 2/t_T$ is called the (first) *Matsubara frequency*, which also shows up if imaginary time path integral techniques are used to treat the influence of bosonic field couplings [17]. For large times the decay function $F_{\text{th}}(t)$ shows the asymptotic behavior

$$F_{\text{th}}(t) \sim \frac{a}{\hbar^2}\frac{t}{t_T} \quad [\text{as } t \to \infty]. \tag{5.52}$$

It follows that the decay of coherence is characterized by rather different regimes. In the short-time regime ($t < \omega_c^{-1}$) we have the perturbative behavior

$$F(t) \simeq \frac{a}{2\hbar^2}\omega_c^2 t^2 \quad [\text{for } t \ll \omega_c^{-1}], \tag{5.53}$$

which can also be obtained from the short-time expansion of the time-evolution operator. Note that the decay is here determined by the overall width ω_c of the spectral density. The intermediate region, $\omega_c^{-1} < t < \omega_1^{-1}$, is dominated by $F_{\text{vac}}(t)$ and called the *vacuum regime*,

$$F(t) \simeq \frac{a}{\hbar^2} \log\left(\omega_c t\right) \qquad [\text{for } \omega_c^{-1} \ll t \ll \omega_1^{-1}].$$

Beyond that, for large times the decay is dominated by the thermal suppression factor,

$$F(t) \simeq \frac{a}{\hbar^2} \frac{t}{t_T} \qquad [\text{for } \omega_1^{-1} \ll t]. \tag{5.54}$$

In this *thermal regime* the decay shows the exponential behavior typical for the Markovian master equations discussed below. Note that the decay rate for this long-time behavior is determined by the *low-frequency behavior* of the spectral density, characterized by the damping strength a in (5.49), and is proportional to the temperature T.

Super-Ohmic Coupling

For $d = 3$, the case of a "super-Ohmic" bath, the integrals (5.47), (5.48) can be calculated without approximation. We note only the long-time behavior of the decay,

$$\lim_{t \to \infty} F(t) = 2a \left(\frac{k_B T}{\hbar \omega_c}\right)^2 \psi'\left(1 + \frac{k_B T}{\hbar \omega_c}\right) < \infty. \tag{5.55}$$

Here $\psi(z)$ stands for the Digamma function, the logarithmic derivative of the gamma function. Somewhat surprisingly, the coherences do not get completely reduced as $t \to \infty$, even at a finite temperature. This is due to the suppressed influence of the important low-frequency contributions to the spectral density in three dimensions (as compared to lower dimensions). While such a suppression of decoherence is plausible for intermediate times, the limiting behavior (5.55) is clearly a result of our simplified model assumptions. It will be absent if there is a small anharmonic coupling between the bath modes [18] or if there is a small admixture of different couplings to H_{int}.

Decoherence by "Vacuum Fluctuations"?

The foregoing discussion seems to indicate that the "vacuum fluctuations" attributed to the quantized field modes are responsible for a general decoherence process, which occurs at short-time scales even if the field is in its ground state. This ground state is non-degenerate and the only way to change it is to increase the energy of the field. But in our model the interaction Hamiltonian H_{int} commutes with the system Hamiltonian, so that it cannot describe energy exchange between qubit and field. One would therefore expect that after the interaction the field has the same energy as before, so that an initial vacuum state remains unchanged and decoherence cannot take place.

   This puzzle is resolved by noting that the initial state $\rho_E = \bigotimes_k |0\rangle\langle 0|_k$ is an eigenstate only in the absence of the coupling H_{int}, but not of the total Hamiltonian. By starting with the product state $\rho_{\text{tot}}(0) = \rho \otimes \rho_E$ we do not account for this possibly strong coupling. At an infinitesimally small time later, system and field thus suddenly feel that they are coupled to each other, which leads to a renormalization of their energies (as described by the Lamb shift discussed in Sect. 5.4.1). The factor χ_{vac} in (5.40) describes the "initial jolt" produced by this sudden switching on of the coupling.

It follows that the above treatment of the short-time dynamics, though formally correct, does not give a physically reasonable picture if the system state is prepared in the presence of the coupling. In this case, one should rather work with the eigenstates of the total Hamiltonian, often denoted as "dressed states". If we start with a superposition of those two dressed states, which correspond in the limit of vanishing coupling to the two system states and the vacuum field, the resulting dynamics will show no further loss of coherence. This is consistent with the above notion that at zero temperature elastic processes cannot lead to decoherence [19].

5.2.3 Dephasing of N Qubits

Let us now discuss the generalization to the case of N qubits which do not interact directly among each other. Each qubit may have a different coupling to the bath modes. The system Hamiltonian is then

$$
\mathsf{H}_{\text{tot}} = \underbrace{\sum_{j=0}^{N-1} \frac{\hbar\omega_j}{2}\sigma_z^{(j)} + \sum_k \hbar\omega_k \mathsf{b}_k^\dagger \mathsf{b}_k}_{\mathsf{H}_0} + \sum_{j=0}^{N-1} \sigma_z^{(j)} \sum_k \left(g_k^{(j)} \mathsf{b}_k^\dagger + [g_k^{(j)}]^* \mathsf{b}_k \right) .
$$

$$(5.56)$$

Similar to above, the time evolution in the interaction picture reads

$$
\tilde{\mathsf{U}}(t) = e^{i\varphi(t)} \exp\left(\frac{1}{2} \sum_{j=0}^{N-1} \sigma_z^{(j)} \sum_k \left(\alpha_k^{(j)} \mathsf{b}_k^\dagger - [\alpha_k^{(j)}]^* \mathsf{b}_k \right) \right),
$$

where the displacement of the field modes now depends on the N-qubit state.

As an example, we take $N = 2$ qubits and only a single vacuum mode. For the initial qubit states

$$
|\phi\rangle = c_{11}|\uparrow\uparrow\rangle + c_{00}|\downarrow\downarrow\rangle \tag{5.57}
$$

and

$$
|\psi\rangle = c_{10}|\uparrow\downarrow\rangle + c_{01}|\downarrow\uparrow\rangle , \tag{5.58}
$$

we obtain, respectively,

$$
\tilde{\mathsf{U}}|\phi\rangle|0\rangle_{\text{E}} = c_{11}|\uparrow\uparrow\rangle\left|\frac{\alpha^{(1)}(t) + \alpha^{(2)}(t)}{2}\right\rangle_{\text{E}} + c_{00}|\downarrow\downarrow\rangle\left|\frac{-\alpha^{(1)}(t) - \alpha^{(2)}(t)}{2}\right\rangle_{\text{E}},
$$

and

$$\tilde{U}|\psi\rangle|0\rangle_{\mathrm{E}} = c_{10}|\uparrow\downarrow\rangle|\frac{\alpha^{(1)}(t) - \alpha^{(2)}(t)}{2}\rangle_{\mathrm{E}} + c_{01}|\downarrow\uparrow\rangle|\frac{-\alpha^{(1)}(t) + \alpha^{(2)}(t)}{2}\rangle_{\mathrm{E}},$$

where the $\alpha^{(1)}(t)$ and $\alpha^{(2)}(t)$ are the field displacements (5.34) due to the first and the second qubits.

If the couplings to the environment are equal for both qubits, say, because they are all sitting in the same place and seeing the same field, we have $\alpha^{(1)}(t) = \alpha^{(2)}(t) \equiv \alpha(t)$. In this case, states of the form $|\phi\rangle$ are decohered once the factor $\langle \alpha(t)| - \alpha(t)\rangle_{\mathrm{E}} = \exp(-2|\alpha(t)|^2)$ is approximately zero. States of the form $|\psi\rangle$, on the other hand, are not affected at all, and one says that the $\{|\psi\rangle\}$ span a (two-dimensional) *decoherence-free subspace*. It shows up because the environment cannot tell the difference between the states $|\uparrow\downarrow\rangle$ and $|\downarrow\uparrow\rangle$ if it couples only to the sum of the excitations.

For an arbitrary number of qubits, using an N-digit binary notation, e.g., $|\uparrow\downarrow\uparrow\rangle \equiv |101_2\rangle = |5\rangle$, one has

$$\langle m|\tilde{\rho}(t)|n\rangle = \langle m|\tilde{\rho}(0)|n\rangle$$

$$\times \mathrm{tr}\left(\exp\left[\sum_{j=0}^{N-1}(m_j - n_j)\sum_k \left(\alpha_k^{(j)}(t)\mathsf{b}_k^\dagger - [\alpha_k^{(j)}(t)]^*\mathsf{b}_k\right)\right]\rho_{\mathrm{th}}\right),$$

$$(5.59)$$

where $m_j \in \{0,1\}$ indicates the jth digit in the binary representation of the number m.

We can distinguish different limiting cases:

Qubits Feel the Same Reservoir

If the separation of the qubits is small compared to the wave lengths of the field modes they are effectively interacting with the same reservoir, $\alpha_k^{(j)} = \alpha_k$. One can push the j-summation to the α's in this case, so that, compared to the single qubit, one merely has to replace α_k by $\sum(m_j - n_j)\alpha_k$. We find

$$\chi_{mn}(t) = \exp\left(-\left|\sum_{j=0}^{N-1}(m_j - n_j)\right|^2 (F_{\mathrm{vac}}(t) + F_{\mathrm{th}}(t))\right) \qquad (5.60)$$

with $F_{\mathrm{vac}}(t)$ and $F_{\mathrm{th}}(t)$ given by (5.47) and (5.48).

Hence, in the worst case, one observes an increase of the decay rate by N^2 compared to the single qubit rate. This is the case for the coherence between the states $|0\rangle$ and $|2^N - 1\rangle$, which have the maximum difference in the number of excitations. On the other hand, the states with an equal number of excitations form a *decoherence-free subspace* in the present model, with a maximal dimension of $\binom{N}{N/2}$.

Qubits See Different Reservoirs

In the other extreme, the qubits are so far apart from each other that each field mode couples only to a single qubit. This suggests a renumbering of the field modes,

$$\alpha_k^{(j)} \longrightarrow \alpha_{k_j} \,,$$

and leads, after transforming the j-summation into a tensor product, to

$$
\begin{aligned}
\chi_{mn}(t) &= \prod_{j=0}^{N-1} \mathrm{tr} \left(\bigotimes_{k_j} \mathsf{D}_{k_j} \left((m_j - n_j) \, \alpha_{k_j}(t) \right) \rho_{\mathrm{th}}^{(k_j)} \right) \\
&= \prod_{j=0}^{N-1} \exp\left(-\underbrace{|m_j - n_j|^2}_{=|m_j - n_j|} \left(F_{\mathrm{vac}}(t) + F_{\mathrm{th}}(t) \right) \right) \\
&= \exp\left(-\underbrace{\sum_{j=0}^{N-1} |m_j - n_j|}_{\text{Hamming distance}} \left(F_{\mathrm{vac}}(t) + F_{\mathrm{th}}(t) \right) \right).
\end{aligned}
\tag{5.61}
$$

Hence, the decay of coherence is the same for all pairs of states with the same Hamming distance. In the worst case, we have an increase by a factor of N compared to the single qubit case, and there are no decoherence-free subspaces.

An intermediate case is obtained if the coupling depends on the position $\boldsymbol{r}_j$ of the qubits. A reasonable model, corresponding to point scatterings of fields with wave vector $\boldsymbol{k}$, is given by $g_k^{(j)} = g_k \exp\left(\mathrm{i}\boldsymbol{k} \cdot \boldsymbol{r}_j\right)$, and its implications are studied in [20].

The model for decoherence discussed in this section is rather exceptional in that the dynamics of the system can be calculated exactly for some choices of the environmental spectral density. In general, one has to resort to approximate descriptions for the dynamical effect of the environment; we turn to this problem in the following section.

5.3 Markovian Dynamics of Open Quantum Systems

Isolated systems evolve, in the Schrödinger picture and for the general case of mixed states, according to the von Neumann equation,

$$\partial_t \rho = \frac{1}{\mathrm{i}\hbar} [\mathsf{H}, \rho] \,. \tag{5.62}$$

One would like to have a similar differential equation for the reduced dynamics of an "open" quantum system, which is in contact with its environment. If we extend the description to include the entire environment $\mathcal{H}_{\mathrm{E}}$ and its coupling to the system, then the total state in $\mathcal{H}_{\text{tot}} = \mathcal{H} \otimes \mathcal{H}_{\text{E}}$ evolves unitarily. The partial trace over $\mathcal{H}_{\text{E}}$ gives the evolved system state, and its time derivative reads

$$\partial_t \rho = \frac{\mathrm{d}}{\mathrm{d}t} \operatorname{tr}_{\text{E}} \left(\mathsf{U}_{\text{tot}}(t) \rho_{\text{tot}}(0) \mathsf{U}^{\dagger}_{\text{tot}}(t) \right) = \frac{1}{\mathrm{i}\hbar} \operatorname{tr}_{\text{E}} \left([\mathsf{H}_{\text{tot}}, \rho_{\text{tot}}] \right) . \qquad (5.63)$$

This exact equation is not closed and therefore not particularly helpful as it stands. However, it can be used as the starting point to derive approximate time-evolution equations for ρ, in particular, if it is permissible to take the initial system state to be uncorrelated with the environment.

These equations are often non-local in time, though, in agreement with causality, the change of the state at each point in time depends only on the state evolution in the past. In this case, the evolution equation is called a *generalized master equation*. It can be specified in terms of superoperator functionals, i.e., linear operators which take the density operator ρ with its past time evolution until time t and map it to the differential change of the operator at that time,

$$\partial_t \rho = \mathcal{K} \left[\{ \rho_\tau : \tau < t \} \right] . \qquad (5.64)$$

An interpretation of this dependence on the system's past is that the environment has a memory, since it affects the system in a way which depends on the history of the system environment interaction. One may hope that on a coarse-grained time scale, which is large compared to the inter-environmental correlation times, these memory effects might become irrelevant. In this case, a proper master equation might be appropriate, where the infinitesimal change of ρ depends only on the instantaneous system state, through a *Liouville* super operator $\mathcal{L}$,

$$\partial_t \rho = \mathcal{L} \rho . \qquad (5.65)$$

Master equations of this type are also called *Markovian*, because of their resemblance to the differential Chapman–Kolmogorov equation for a classical Markov process. However, since a number of approximations are involved in their derivation, it is not clear whether the corresponding time evolution respects important properties of a quantum state, such as its positivity. We will see that these constraints restrict the possible form of $\mathcal{L}$ rather strongly.

5.3.1 Quantum Dynamical Semigroups

The notion of a *quantum dynamical semigroup* permits a rigorous formulation of the Markov assumption in quantum theory. To introduce it we first need a number of concepts from the theory of open quantum systems [11, 21–23].

Dynamical Maps

A dynamical map is a one-parameter family of trace-preserving, convex linear, and completely positive maps (CPM)

$$\mathcal{W}_t : \rho_0 \mapsto \rho_t, \qquad \text{for } t \in \mathbb{R}_0^+, \tag{5.66}$$

satisfying $\mathcal{W}_0 = \text{id}$. As such, it yields the most general description of a time evolution which maps an arbitrary initial state ρ_0 to valid states at later times.

Specifically, the condition of trace preservation guarantees the normalization of the state,

$$\text{tr}(\rho_t) = 1,$$

and the convex linearity, i.e.,

$$\mathcal{W}_t\left(\lambda\rho_0 + (1-\lambda)\rho_0'\right) = \lambda\mathcal{W}_t(\rho_0) + (1-\lambda)\mathcal{W}_t(\rho_0') \qquad \text{for all } 0 \leqslant \lambda \leqslant 1,$$

ensures that the transformation of mixed states is consistent with the classical notion of ignorance. The final requirement of *complete positivity* is stronger than mere positivity of $\mathcal{W}_t(\rho_0)$. It means that in addition all the tensor product extensions of $\mathcal{W}_t$ to spaces of higher dimension, defined with the identity map id_{ext}, are positive,

$$\mathcal{W}_t \otimes \text{id}_{\text{ext}} > 0,$$

that is, the image of any positive operator in the higher dimensional space is again a positive operator. This guarantees that the system state remains positive even if it is the reduced part of a non-separable state evolving in a higher dimensional space.

Kraus Representation

Any dynamical map admits an *operator-sum representation* of the form (5.8) [23],

$$\mathcal{W}_t(\rho) = \sum_{k=1}^{N} \mathsf{W}_k(t)\rho\mathsf{W}_k^\dagger(t) \tag{5.67}$$

with the completeness relation[6]

$$\sum_{k=1}^{N} \mathsf{W}_k^\dagger(t)\mathsf{W}_k(t) = \mathbb{I}. \tag{5.68}$$

The number of the required Kraus operators $\mathsf{W}_k(t)$ is limited by the dimension of the system Hilbert space, $N \leqslant \dim(\mathcal{H})^2$ (and confined to a countable set in case of an infinite-dimensional, separable Hilbert space), but their choice is not unique.

---

[6] In case of a *trace-decreasing*, convex linear, completely positive map the condition (5.68) is replaced by $\sum_k \mathsf{W}_k^\dagger(t)\mathsf{W}_k(t) < \mathbb{I}$, i.e., the operator $\mathbb{I} - \sum_k \mathsf{W}_k^\dagger(t)\mathsf{W}_k(t)$ must be positive.

Semigroup Assumption

We can now formulate the assumption that the $\{\mathcal{W}_t : t \in \mathbb{R}_0^+\}$ form a continuous *dynamical semigroup*[7] [21, 23]

$$\mathcal{W}_{t_2}\left(\mathcal{W}_{t_1}(\cdot)\right) \overset{!}{=} \mathcal{W}_{t_1+t_2}(\cdot) \qquad \text{for all } t_1, t_2 > 0 \tag{5.69}$$

and $\mathcal{W}_0 = \text{id}$. This statement is rather strong, and it is certainly violated for truly microscopic times. But it seems not unreasonable on the level of a coarse-grained time scale, which is long compared to the time it takes for the environment to "forget" the past interactions with the system due to the dispersion of correlations into the many environmental degrees of freedom.

For a given dynamical semigroup there exists, under rather weak conditions, a generator, i.e., a superoperator $\mathcal{L}$ satisfying

$$\mathcal{W}_t = e^{\mathcal{L}t} \qquad \text{for } t > 0. \tag{5.70}$$

In this case $\mathcal{W}_t(\rho)$ is the formal solution of the Markovian master equation (5.65).

Dual Maps

So far we used the Schrödinger picture, i.e., the notion that the state of an open quantum system evolves in time, $\rho_t = \mathcal{W}_t(\rho_0)$. Like in the description of closed quantum systems, one can also take the Heisenberg point of view, where the state does not evolve, while the operators A describing observables acquire a time dependence. The corresponding map $\mathcal{W}_t^\sharp : A_0 \mapsto A_t$ is called the *dual map*, and it is related to $\mathcal{W}_t$ by the requirement $\text{tr}(A\mathcal{W}_t(\rho)) = \text{tr}(\rho\mathcal{W}_t^\sharp(A))$. In case of a dynamical semigroup, $\mathcal{W}_t^\sharp = \exp\left(\mathcal{L}^\sharp t\right)$, the equation of motion takes the form $\partial_t A = \mathcal{L}^\sharp A$, with the dual Liouville operator determined by $\text{tr}(A\mathcal{L}(\rho)) = \text{tr}(\rho\mathcal{L}^\sharp(A))$. From a mathematical point of view, the Heisenberg picture is much more convenient since the observables form an algebra, and it is therefore preferred in the mathematical literature.

5.3.2 The Lindblad Form

We can now derive the general form of the generator of a dynamical semigroup, taking $\dim(\mathcal{H}) = d < \infty$ for simplicity [2, 23]. The bounded operators on $\mathcal{H}$ then form a d^2-dimensional vector space which turns into a Hilbert space, if equipped with the Hilbert–Schmidt scalar product $(A, B) := \text{tr}(A^\dagger B)$.

Given an orthonormal basis of operators $\{E_j : 1 \leqslant j \leqslant d^2\} \subset L(\mathcal{H})$,

$$(E_i, E_j) := \text{tr}(E_i^\dagger E_j) = \delta_{ij}, \tag{5.71}$$

---

[7] The inverse element required for a *group* structure is missing for general, irreversible CPMs.

any Hilbert–Schmidt operator W_k can be expanded as

$$\mathsf{W}_k = \sum_{j=1}^{d^2} (\mathsf{E}_j, \mathsf{W}_k) \mathsf{E}_j \,. \tag{5.72}$$

We can choose one of the basis operators, say the d^2th, to be proportional to the identity operator,

$$\mathsf{E}_{d^2} = \frac{1}{\sqrt{d}} \mathbb{I} \,, \tag{5.73}$$

so that all other basis elements are traceless,

$$\mathrm{tr}(\mathsf{E}_j) = \begin{cases} 0 & \text{for } j = 1, \ldots, d^2 - 1 \,, \\ \sqrt{d} & \text{for } j = d^2 \,. \end{cases} \tag{5.74}$$

Representing the superoperator of the dynamical map (5.67) in the $\{\mathsf{E}_j\}$ basis we have

$$\mathcal{W}_t(\rho) = \sum_{i,j=1}^{d^2} c_{ij}(t) \mathsf{E}_i \rho \mathsf{E}_j^\dagger \tag{5.75}$$

with a time-dependent, hermitian, and positive coefficient matrix,

$$c_{ij}(t) = \sum_{k=1}^{N} (\mathsf{E}_i, \mathsf{W}_k(t))(\mathsf{E}_j, \mathsf{W}_k(t))^* \tag{5.76}$$

(positivity can be checked in a small calculation). We can now calculate the semigroup generator in terms of the differential quotient by writing the terms including the element E_{d^2} separately:

$$\mathcal{L}\rho = \lim_{\tau \to 0} \frac{\mathcal{W}_\tau(\rho) - \rho}{\tau}$$

$$= \underbrace{\lim_{\tau \to 0} \frac{\frac{1}{d} c_{d^2 d^2}(\tau) - 1}{\tau}}_{c_0 \in \mathbb{R}} \rho + \underbrace{\lim_{\tau \to 0} \sum_{j=1}^{d^2-1} \frac{c_{j d^2}(\tau)}{\sqrt{d}\tau} \mathsf{E}_j}_{B \in L(\mathcal{H})} \rho + \rho \underbrace{\lim_{\tau \to 0} \sum_{j=1}^{d^2-1} \frac{c_{d^2 j}(\tau)}{\sqrt{d}\tau} \mathsf{E}_j^\dagger}_{B^\dagger \in L(\mathcal{H})}$$

$$+ \sum_{i,j=1}^{d^2-1} \underbrace{\lim_{\tau \to 0} \frac{c_{ij}(\tau)}{\tau}}_{\alpha_{ij} \in \mathbb{R}} \mathsf{E}_i \rho \mathsf{E}_j^\dagger$$

$$= c_0 \rho + B\rho + \rho B^\dagger + \sum_{i,j=1}^{d^2-1} \alpha_{ij} \mathsf{E}_i \rho \mathsf{E}_j^\dagger$$

$$= \frac{1}{i\hbar}[\mathsf{H}, \rho] + \frac{1}{\hbar}(\mathsf{G}\rho + \rho\mathsf{G}) + \sum_{i,j=1}^{d^2-1} \alpha_{ij} \mathsf{E}_i \rho \mathsf{E}_j^\dagger \,. \tag{5.77}$$

In the last equality the following hermitian operators with the dimension of an energy were introduced:

$$G = \frac{\hbar}{2}(B + B^\dagger + c_0),$$

$$H = \frac{\hbar}{2i}(B - B^\dagger).$$

By observing that the conservation of the trace implies $\text{tr}(\mathcal{L}\rho) = 0$, one can relate the operator G to the matrix $\boldsymbol{\alpha} = (\alpha_{ij})$, since

$$0 = \text{tr}(\mathcal{L}\rho) = 0 + \text{tr}\left[\left(\frac{2G}{\hbar} + \sum_{i,j=1}^{d^2-1} \alpha_{ij}E_j^\dagger E_i\right)\rho\right]$$

must hold for all ρ. It follows that

$$G = -\frac{\hbar}{2}\sum_{i,j=1}^{d^2-1} \alpha_{ij}E_j^\dagger E_i.$$

This leads to the *first standard form* for the generator of a dynamical semi-group:

$$\mathcal{L}\rho = \underbrace{\frac{1}{i\hbar}[H,\rho]}_{\text{unitary part}} + \underbrace{\sum_{i,j=1}^{d^2-1} \alpha_{ij}\left(E_i\rho E_j^\dagger - \frac{1}{2}E_j^\dagger E_i\rho - \frac{1}{2}\rho E_j^\dagger E_i\right)}_{\text{incoherent part}}. \qquad (5.78)$$

The complex coefficients α_{ij} have dimensions of frequency and constitute a positive matrix $\boldsymbol{\alpha}$.

The *second standard form* or *Lindblad form* is obtained by diagonalizing the coefficient matrix $\boldsymbol{\alpha}$. The corresponding unitary matrix U satisfying $U\boldsymbol{\alpha}U^\dagger = \text{diag}(\gamma_1,\ldots,\gamma_{d^2-1})$ allows to define the dimensionless operators $L_k := \sum_{j=1}^{d^2-1} E_j U_{jk}^\dagger$ so that $E_j = \sum_{k=1}^{d^2-1} L_k U_{kj}$ and therefore[8]

$$\mathcal{L}\rho = \frac{1}{i\hbar}[H,\rho] + \sum_{k=1}^{N} \gamma_k\left(L_k\rho L_k^\dagger - \frac{1}{2}L_k^\dagger L_k\rho - \frac{1}{2}\rho L_k^\dagger L_k\right) \qquad (5.79)$$

---

[8] It is easy to see that the dual Liouville operator discussed in Sect. 5.3.1 reads, in Lindblad form,

$$\mathcal{L}^\sharp(A) = \frac{1}{i\hbar}[A,H] + \sum_{k=1}^{N} \gamma_k\left(L_k^\dagger A L_k - \frac{1}{2}L_k^\dagger L_k A - \frac{1}{2}A L_k^\dagger L_k\right).$$

Note that this implies $\mathcal{L}^\sharp(\mathbb{I}) = 0$, while $\mathcal{L}(\mathbb{I}) = \sum_k \gamma_k[L_k, L_k^\dagger]$, and $\text{tr}(\mathcal{L}X) = 0$.

with $N \leqslant d^2 - 1$. This shows that the general form of a generator of a dynamical semigroup is specified by a single hermitian operator H, which is not necessarily equal to the Hamiltonian of the isolated system, see below, and at most $d^2 - 1$ arbitrary operators L_k with attributed positive rates γ_k. These are called *Lindblad operators*[9] or *jump operators*, a name motivated in the following section.

It is important to note that a given generator $\mathcal{L}$ does not determine the jump operators uniquely. In fact, the equation is invariant under the transformation

$$\mathsf{L}_k \rightarrow \mathsf{L}_k + c_k \,, \tag{5.80}$$

$$\mathsf{H} \rightarrow \mathsf{H} + \frac{\hbar}{2i} \sum_j \gamma_j \left(c_j^* \mathsf{L}_j - c_j \mathsf{L}_j^\dagger \right) , \tag{5.81}$$

with $c_k \in \mathbb{C}$, so that the L_k can be chosen traceless. In this case, the only remaining freedom is a unitary mixing,

$$\sqrt{\gamma_i} \mathsf{L}_i \longrightarrow \sum_j U'_{ij} \sqrt{\gamma_j} \mathsf{L}_j \,. \tag{5.82}$$

If $\mathcal{L}$ shows an additional invariance, e.g., with respect to rotations or translations, the form of the Lindblad operators will be further restricted, see, e.g., [24].

5.3.3 Quantum Trajectories

Generally, if we write the Liouville superoperator $\mathcal{L}$ as the sum of two parts, $\mathcal{L}_0$ and $\mathcal{S}$, then the formal solution (5.70) of the master equation (5.65) can be expressed as

$$\mathcal{W}_t = \mathrm{e}^{(\mathcal{L}_0 + \mathcal{S})t} = \sum_{n=0}^{\infty} \frac{t^n}{n!} (\mathcal{L}_0 + \mathcal{S})^n$$

$$= \sum_{n=0}^{\infty} \sum_{k_0,\ldots,k_n=0}^{\infty} \frac{t^{n+\sum_j k_j}}{\left(n + \sum_j k_j\right)!} \underbrace{\mathcal{L}_0^{k_n} \mathcal{S} \mathcal{L}_0^{k_{n-1}} \mathcal{S} \cdots \mathcal{S} \mathcal{L}_0^{k_1} \mathcal{S} \mathcal{L}_0^{k_0}}_{n\,\text{times}}$$

$$= \sum_{n=0}^{\infty} \int_0^t \mathrm{d}t_n \int_0^{t_n} \mathrm{d}t_{n-1} \cdots \int_0^{t_2} \mathrm{d}t_1$$

$$\times \sum_{k_0,\ldots,k_n=0}^{\infty} \frac{(t - t_n)^{k_n}}{k_n!} \frac{(t_n - t_{n-1})^{k_{n-1}}}{k_{n-1}!} \cdots \frac{(t_1 - 0)^{k_0}}{k_0!}$$

---

[9] Lindblad showed in 1976 that the form (5.79) is obtained even for infinite-dimensional systems provided the generator $\mathcal{L}$ is bounded (which is usually not the case).

$$\times \mathcal{L}_0^{k_n} \mathcal{S} \mathcal{L}_0^{k_{n-1}} \mathcal{S} \cdots \mathcal{S} \mathcal{L}_0^{k_1} \mathcal{S} \mathcal{L}_0^{k_0}$$

$$= e^{\mathcal{L}_0 t} + \sum_{n=1}^{\infty} \int_0^t dt_n \int_0^{t_n} dt_{n-1} \cdots \int_0^{t_2} dt_1$$

$$\times e^{\mathcal{L}_0(t-t_n)} \mathcal{S} e^{\mathcal{L}_0(t_n-t_{n-1})} \mathcal{S} \cdots e^{\mathcal{L}_0(t_2-t_1)} \mathcal{S} e^{\mathcal{L}_0 t_1} . \tag{5.83}$$

The step from the second to the third line, where $t^{n+\sum_j k_j}/(n + \sum_j k_j)!$ is replaced by n time integrals, can be checked by induction.

The form (5.83) is a generalized Dyson expansion, and the comparison with the Dyson series for unitary evolutions suggests to view $\exp(\mathcal{L}_0 \tau)$ as an "unperturbed" evolution and $\mathcal{S}$ as a "perturbation", such that the exact time-evolution $\mathcal{W}_t$ is obtained by an integration over all iterations of the perturbation, separated by the unperturbed evolutions.

The particular Lindblad form (5.79) of the generator suggests to introduce the completely positive *jump superoperators*

$$\mathcal{L}_k \rho = \gamma_k \mathsf{L}_k \rho \mathsf{L}_k^\dagger , \tag{5.84}$$

along with the *non-hermitian* operator

$$\mathsf{H}_{\mathbb{C}} = \mathsf{H} - \frac{i\hbar}{2} \sum_{k=1}^N \gamma_k \mathsf{L}_k^\dagger \mathsf{L}_k . \tag{5.85}$$

The latter has a negative imaginary part, $\mathrm{Im}\,(\mathsf{H}_{\mathbb{C}}) < 0$, and can be used to construct

$$\mathcal{L}_0 \rho = \frac{1}{i\hbar} \left(\mathsf{H}_{\mathbb{C}} \rho - \rho \mathsf{H}_{\mathbb{C}}^\dagger \right) . \tag{5.86}$$

It follows that the sum of these superoperators yields the Liouville operator (5.79)

$$\mathcal{L} = \mathcal{L}_0 + \sum_{k=1}^N \mathcal{L}_k . \tag{5.87}$$

Of course, neither $\mathcal{L}_0$ nor $\mathcal{S} = \sum_{k=1}^N \mathcal{L}_k$ generates a dynamical semigroup. Nonetheless, they are useful since the interpretation of (5.83) can now be taken one step further. We can take the point of view that the $\mathcal{L}_k$ with $k \geqslant 1$ describe elementary transformation events due to the environment ("jumps"), which occur at random times with a rate γ_k. A particular realization of n such events is specified by a sequence of the form

$$R_n^t = (t_1, k_1; t_2, k_2; \ldots; t_n, k_n) . \tag{5.88}$$

The attributed times satisfy $0 < t_1 < \ldots < t_n < t$, and the $k_j \in \{1, \ldots, N\}$ indicates which kind of event "took place". We call R_n^t a *record* of length n.

The general time-evolution $\mathcal{W}_t$ can thus be written as an integration over all possible realizations of the jumps, with the "free" evolution $\exp(\mathcal{L}_0 \tau)$ in between,

$$
\mathcal{W}_t = e^{\mathcal{L}_0 t} + \sum_{n=1}^{\infty} \int_0^t dt_n \int_0^{t_n} dt_{n-1} \cdots \int_0^{t_2} dt_1
$$

$$
\times \sum_{\{R_n\}} \underbrace{e^{\mathcal{L}_0(t-t_n)} \mathcal{L}_{k_n} e^{\mathcal{L}_0(t_n-t_{n-1})} \mathcal{L}_{k_{n-1}} \cdots e^{\mathcal{L}_0(t_2-t_1)} \mathcal{L}_{k_1} e^{\mathcal{L}_0 t_1}}_{\mathcal{K}_{R_n^t}} \quad (5.89)
$$

As a result of the negative imaginary part in (5.85) the $\exp(\mathcal{L}_0 \tau)$ are *trace decreasing*[10] completely positive maps,

$$
e^{\mathcal{L}_0 \tau} \rho = \exp\left(-\frac{i\tau}{\hbar} \mathsf{H}_{\mathbb{C}}\right) \rho \exp\left(\frac{i\tau}{\hbar} \mathsf{H}_{\mathbb{C}}^{\dagger}\right) > 0, \quad (5.90)
$$

$$
\frac{d}{d\tau} \operatorname{tr}\left(e^{\mathcal{L}_0 \tau} \rho\right) = \operatorname{tr}\left(\mathcal{L}_0 e^{\mathcal{L}_0 \tau} \rho\right) = -\sum_{k=1}^{N} \operatorname{tr}(\underbrace{\mathcal{L}_k e^{\mathcal{L}_0 \tau} \rho}_{>0}) < 0. \quad (5.91)
$$

It is now natural to interpret $\operatorname{tr}\left(e^{\mathcal{L}_0 t} \rho\right)$ as the probability that no jump occurs during the time interval t,

$$
\operatorname{Prob}\left(R_0^t | \rho\right) := \operatorname{tr}\left(e^{\mathcal{L}_0 t} \rho\right). \quad (5.92)
$$

To see that this makes sense, we attribute to each record of length n a n-time probability density. For a given record R_n^t we define

$$
\operatorname{prob}\left(R_n^t | \rho\right) := \operatorname{tr}\left(\mathcal{K}_{R_n^t} \rho\right), \quad (5.93)
$$

in terms of the superoperators from the second line in (5.89),

$$
\mathcal{K}_{R_n^t} := e^{\mathcal{L}_0(t-t_n)} \mathcal{L}_{k_n} e^{\mathcal{L}_0(t_n-t_{n-1})} \mathcal{L}_{k_{n-1}} \cdots e^{\mathcal{L}_0(t_2-t_1)} \mathcal{L}_{k_1} e^{\mathcal{L}_0 t_1}. \quad (5.94)
$$

This is reasonable since the $\mathcal{K}_{R_n^t}$ are completely positive maps that do not preserve the trace. Indeed, the probability density for a record is thus determined both by the corresponding jump operators, which involve the rates γ_k, and by the $e^{\mathcal{L}_0 \tau} \rho$, which account for the fact that the likelihood for the absence of a jump decreases with the length of the time interval.

This notion of probabilities is consistent, as can be seen by adding the probability (5.92) for no jump to occur during the interval $(0; t)$ to the integral over the probability densities (5.93) of all possible jump sequences. As required, the result is unity,

$$
\operatorname{Prob}\left(R_0^t | \rho\right) + \sum_{n=1}^{\infty} \int_0^t dt_n \int_0^{t_n} dt_{n-1} \cdots \int_0^{t_2} dt_1 \sum_{\{R_n^t\}} \operatorname{prob}\left(R_n^t | \rho\right) = 1,
$$

---

[10] See the note 6.

for all ρ and $t \geqslant 0$. This follows immediately from the trace preservation of the map (5.89).

It is now natural to normalize the transformation defined by the $\mathcal{K}_{R_n^t}$. Formally, this yields the state transformation conditioned to a certain record R_n^t. It is called a *quantum trajectory*,

$$\mathcal{T}\left(\rho|R_n^t\right) := \frac{\mathcal{K}_{R_n^t}\rho}{\mathrm{tr}\left(\mathcal{K}_{R_n^t}\rho\right)} . \tag{5.95}$$

Note that this definition comprises the trajectory corresponding to a null-record R_0^t, where $\mathcal{K}_{R_0^t} = \exp\left(\mathcal{L}_0 t\right)$. These completely positive, trace-preserving, *non*linear maps $\rho \mapsto \mathcal{T}\left(\rho|R_n^t\right)$ are defined for all states ρ that yield a finite probability (density) for the given record R_n^t, i.e., if the denominator in (5.95) does not vanish.

Using these notions, the exact solution of a general Lindblad master equation (5.83) may thus be rewritten in the form

$$\rho_t = \mathrm{Prob}\left(R_0^t\right) \mathcal{T}\left(\rho|R_0^t\right)$$
$$+ \sum_{n=1}^{\infty} \int_0^t \mathrm{d}t_n \int_0^{t_n} \mathrm{d}t_{n-1} \cdots \int_0^{t_2} \mathrm{d}t_1 \sum_{\{R_n\}} \mathrm{prob}\left(R_n^t|\rho\right) \mathcal{T}\left(\rho|R_n^t\right) . \tag{5.96}$$

It shows that the general Markovian quantum dynamics can be understood as a summation over all quantum trajectories $\mathcal{T}\left(\rho|R_n^t\right)$ weighted by their probability (density). This is called a *stochastic unraveling* of the master equation. The set of trajectories and their weights are labeled by the possible records (5.88) and determined by the Lindblad operators of the master equation (5.79).

The semigroup property described by the master equation shows up if a record is formed by joining the records of adjoining time intervals, $(0; t')$ and $(t'; t)$,

$$R_{n+m}^{(0;t)} := (R_n^{(0;t')}; R_m^{(t';t)}) . \tag{5.97}$$

As one expects, the probabilities and trajectories satisfy

$$\mathrm{prob}(R_{n+m}^{(0;t)}|\rho) = \mathrm{prob}(R_n^{(0;t')}|\rho)\,\mathrm{prob}(R_m^{(t';t)}|\mathcal{T}(\rho|R_n^{(0;t')})) , \tag{5.98}$$

and

$$\mathcal{T}(\rho|R_{n+m}^{(0;t)}) = \mathcal{T}(\mathcal{T}(\rho|R_n^{(0;t')})|R_m^{(t';t)}) . \tag{5.99}$$

Note finally that the concept of quantum trajectories fits seamlessly into the framework of generalized measurements discussed in Sect. 5.1.3. In particular, the conditioned state transformation $\mathcal{T}\left(\rho|R_n^t\right)$ has the form (5.18) of an efficient measurement transformation,

$$\mathcal{T}\left(\rho | R_n^t\right) = \frac{\mathsf{M}_{R_n^t} \rho \mathsf{M}_{R_n^t}^\dagger}{\mathrm{tr}\left(\mathsf{M}_{R_n^t}^\dagger \mathsf{M}_{R_n^t} \rho\right)} \tag{5.100}$$

with compound measurement operators

$$\mathsf{M}_{R_n^t} := e^{-i\mathsf{H}_{\mathbb{C}}(t-t_n)/\hbar} \mathsf{L}_{k_n} \cdots \mathsf{L}_{k_2} e^{-i\mathsf{H}_{\mathbb{C}}(t_2-t_1)/\hbar} \mathsf{L}_{k_1} e^{-i\mathsf{H}_{\mathbb{C}} t_1/\hbar}. \tag{5.101}$$

This shows that we can legitimately view the open quantum dynamics generated by $\mathcal{L}$ as due to the *continuous monitoring* of the system by the environment. We just have to identify the (aptly named) record R_n^t with the total outcome of a hypothetical, continuous measurement during the interval $(0;t)$. The jump operators $\mathcal{L}_k$ then describe the effects of the corresponding elementary measurement events[11] ("clicks of counter k"). Since the absence of any click during the "waiting time" τ may also confer information about the system, this lack of an event constitutes a measurement as well, which is described by the non-unitary operators $\exp\left(-i\mathsf{H}_{\mathbb{C}}\tau/\hbar\right)$. A hypothetical demon, who has the full record R_n^t available, would then be able to predict the final state $\mathcal{T}\left(\rho | R_n^t\right)$. In the absence of this information we have to resort to the probabilistic description (5.96) weighting each quantum trajectory with its (Bayesian) probability.

We can thus conclude that the dynamics of open quantum dynamics, and therefore decoherence, can in principle be understood in terms of an information transfer to the environment. Apart from this conceptual insight, the unraveling of a master equation provides also an efficient stochastic simulation method for its numerical integration. In these *quantum jump* approaches [25–27], which are based on the observation that the quantum trajectory (5.95) of a pure state remains pure, one generates a finite ensemble of trajectories such that the ensemble mean approximates the solution of the master equation.

5.3.4 Exemplary Master Equations

Let us take a look at a number of very simple Markovian master equations,[12] which are characterized by a single Lindblad operator L (together with a hermitian operator H). The first example gives a general description of dephasing, while the others are empirically known to describe dissipative phenomena realistically. We may then ask what they predict about decoherence.

---

[11] For all these appealing notions, it should be kept in mind that the $\mathcal{L}_k$ are not uniquely specified by a given generator $\mathcal{L}$, see (5.80), (5.81) and (5.82). Different choices of the Lindblad operators lead to different unravelings of the master equation, so that these hypothetical measurement events must not be viewed as "real" processes.

[12] See also Sect. 6.2.2 in Cord Müller's contribution for a discussion of the master equation describing spin relaxation.

Dephasing

The simplest choice is to take the Lindblad operator to be proportional[13] to the Hamiltonian of a discrete quantum system, i.e., to the generator of the unitary dynamics, $\mathsf{L} = \sqrt{\gamma}\mathsf{H}$. The Lindblad equation

$$\partial_t \rho_t = \frac{1}{i\hbar}[\mathsf{H}, \rho_t] + \gamma \left(\mathsf{H}\rho_t\mathsf{H} - \frac{1}{2}\mathsf{H}^2\rho_t - \frac{1}{2}\rho_t\mathsf{H}^2 \right) \qquad (5.102)$$

is immediately solved in the energy eigenbasis, $\mathsf{H} = \sum_m E_m |m\rangle\langle m|$:

$$\rho_{mn}(t) \equiv \langle m|\rho_t|n\rangle = \rho_{mn}(0) \exp\left(-\frac{i}{\hbar}(E_m - E_n)t - \frac{\gamma}{2}(E_m - E_n)^2 t \right) . \qquad (5.103)$$

As we expect from the discussion of qubit dephasing in Sect. 5.2, the energy eigenstates are unaffected by the non-unitary dynamics if the environmental effect commutes with H. The coherences show the exponential decay that we found in the "thermal regime" (of times t which are long compared to the inverse Matsubara frequency). The comparison with (5.54) indicates that γ should be proportional to the temperature of the environment.

Amplitude Damping of the Harmonic Oscillator

Next, we choose H to be the Hamiltonian of a harmonic oscillator, $\mathsf{H} = \hbar\omega\mathsf{a}^\dagger\mathsf{a}$, and take as Lindblad operator the ladder operator, $\mathsf{L} = \mathsf{a}$. The resulting Lindblad equation is known empirically to describe the quantum dynamics of a damped harmonic oscillator.

Choosing as initial state a coherent state, see (5.37) and (5.114) below,

$$\rho_0 = |\alpha_0\rangle\langle\alpha_0| \equiv \mathsf{D}(\alpha_0)|0\rangle\langle 0|\mathsf{D}^\dagger(\alpha_0)$$
$$= e^{-|\alpha_0|^2} \exp(\alpha_0\mathsf{a}^\dagger)|0\rangle\langle 0| \exp(\alpha_0^*\mathsf{a}) , \qquad (5.104)$$

we are faced with the exceptional fact that the state *remains pure* during the Lindblad time evolution. Indeed, the solution of the Lindblad equation reads,

$$\rho_t = |\alpha_t\rangle\langle\alpha_t| \qquad (5.105)$$

with

$$\alpha_t = \alpha_0 \exp\left(-i\omega t - \frac{\gamma}{2}t \right) , \qquad (5.106)$$

as can be verified easily using (5.104). It describes how the coherent state spirals in phase space toward the origin, approaching the ground state as $t \to \infty$. The rate γ is the *dissipation rate* since it quantifies the energy loss, as shown by the time dependence of the energy expectation value,

$$\langle\alpha_t|\mathsf{H}|\alpha_t\rangle = e^{-\gamma t}\langle\alpha_0|\mathsf{H}|\alpha_0\rangle . \qquad (5.107)$$

---

[13] As an exception, γ does not have the dimensions of a rate here (to avoid clumsy notation).

Superposition of Coherent States

What happens if we start out with a superposition of coherent states,

$$|\psi_0\rangle = \frac{1}{\sqrt{\mathcal{N}}} \left(|\alpha_0\rangle + |\beta_0\rangle \right) \tag{5.108}$$

with $\mathcal{N} = 2 + 2\operatorname{Re}\langle\alpha_0|\beta_0\rangle$, in particular, if the separation in phase space is large compared to the quantum uncertainties, $|\alpha_0 - \beta_0| \gg 1$? The initial density operator corresponding to (5.108) reads

$$\rho_0 = \frac{1}{\mathcal{N}} \left(|\alpha_0\rangle\langle\alpha_0| + |\beta_0\rangle\langle\beta_0| + c_0|\alpha_0\rangle\langle\beta_0| + c_0^*|\beta_0\rangle\langle\alpha_0| \right) \tag{5.109}$$

with $c_0 = 1$. One finds that the ansatz,

$$\rho_t = \frac{1}{\mathcal{N}} \left(|\alpha_t\rangle\langle\alpha_t| + |\beta_t\rangle\langle\beta_t| + c_t|\alpha_t\rangle\langle\beta_t| + c_t^*|\beta_t\rangle\langle\alpha_t| \right), \tag{5.110}$$

solves the Lindblad equation with (5.106), provided

$$c_t = c_0 \exp\left(\left[-\frac{1}{2}|\alpha_0 - \beta_0|^2 + i\operatorname{Im}(\alpha_0\beta_0^*) \right] \left(1 - e^{-\gamma t} \right) \right). \tag{5.111}$$

That is, while the coherent "basis" states have the same time evolution as in (5.105), the initial coherence c_0 gets additionally suppressed. For times that are short compared to the dissipative time scale, $t \ll \gamma^{-1}$, we have an exponential decay

$$|c_t| = |c_0| \exp\left(-\underbrace{\frac{\gamma}{2}|\alpha_0 - \beta_0|^2}_{\gamma_{\mathrm{deco}}} t \right), \tag{5.112}$$

with a rate γ_{deco}. For macroscopically distinct superpositions, where the phase space distance of the quantum states is much larger than their uncertainties, $|\alpha_0 - \beta_0| \gg 1$, the decoherence rate γ_{deco} can be much greater than the dissipation rate,

$$\frac{\gamma_{\mathrm{deco}}}{\gamma} = \frac{1}{2}|\alpha_0 - \beta_0|^2 \gg 1. \tag{5.113}$$

This quadratic increase of the decoherence rate with the separation between the coherent states has been confirmed experimentally in a series of beautiful cavity QED experiments in Paris, using field states with an average of 5–9 photons [28, 29].

Given this empiric support we can ask about the prediction for a material, macroscopic oscillator. As an example, we take a pendulum with a mass of $m = 100\,\mathrm{g}$ and a period of $2\pi/\omega = 1\,\mathrm{s}$ and assume that we can prepare it in a superposition of coherent states with a separation of $x = 1\,\mathrm{cm}$. The mode variable α is related to position and momentum by

$$\alpha = \sqrt{\frac{m\omega}{2\hbar}} \left(x + i\frac{p}{m\omega} \right), \tag{5.114}$$

so that we get the prediction

$$\gamma_{\text{deco}} \simeq 10^{30}\gamma.$$

This purports that even with an oscillator of enormously low friction corresponding to a dissipation rate of $\gamma = 1/\text{year}$ the coherence is lost on a time scale of 10^{-22} s – in which light travels the distance of about a nuclear diameter.

This observation is often evoked to explain the absence of macroscopic superpositions. However, it seems unreasonable to assume that anything physically relevant takes place on a time scale at which a signal travels at most by the diameter of an atomic nucleus. Rather, one expects that the decoherence rate should saturate at a finite value if one increases the phase space distance between the superposed states.

Quantum Brownian Motion

Next, let us consider a particle in one dimension. A possible choice for the Lindblad operator is a linear combination of its position and momentum operators,

$$\mathsf{L} = \frac{p_{\text{th}}}{\hbar}\mathsf{x} + \frac{i}{p_{\text{th}}}\mathsf{p}. \tag{5.115}$$

Here p_{th} is a momentum scale, which will be related to the temperature of the environment below. The hermitian operator is taken to be the Hamiltonian of a particle in a potential $V(x)$, plus a term due to the environmental coupling,

$$\mathsf{H} = \frac{\mathsf{p}^2}{2m} + V(\mathsf{x}) + \frac{\gamma}{2}(\mathsf{xp} + \mathsf{px}). \tag{5.116}$$

This additional term will be justified by the fact that the resulting Lindblad equation is almost equal to the *Caldeira–Leggett master equation*. The latter is the high-temperature limit of the exact evolution equation following from a harmonic bath model of the environment [30, 31], see Sect. 5.4.1. It is empirically known to describe the frictional quantum dynamics of a Brownian particle, and, in particular, for $t \to \infty$ it leads to the canonical Gibbs state in case of quadratic potentials.

The choices (5.115) and (5.116) yield the following Lindblad equation:

$$\partial_t \rho_t = \overbrace{\frac{1}{i\hbar}[\frac{\mathsf{p}^2}{2m} + V(\mathsf{x}), \rho_t] + \underbrace{\frac{\gamma}{i\hbar}[\mathsf{x}, \mathsf{p}\rho_t + \rho_t\mathsf{p}]}_{\text{dissipation}} \underbrace{- \frac{\gamma}{2}\frac{p_{\text{th}}^2}{\hbar^2}[\mathsf{x}, [\mathsf{x}, \rho_t]]}_{\text{position localization}}}^{\text{Caldeira–Leggett master equation}}$$

$$- \frac{\gamma}{2}\frac{1}{p_{\text{th}}^2}[\mathsf{p}, [\mathsf{p}, \rho_t]]. \tag{5.117}$$

The three terms in the upper line [with p_{th} from (5.123)] constitute the Caldeira–Leggett master equation. It is a Markovian, but not a completely positive master equation. In a sense, the last term in (5.117) adds the minimal modification required to bring the Caldeira–Leggett master equation into Lindblad form [2, 32].

To see the most important properties of (5.117) let us take a look at the time evolution of the relevant observables in the Heisenberg picture. As discussed in Sect. 5.3.1, the Heisenberg equations of motion are determined by the dual Liouville operator $\mathcal{L}^{\sharp}$. In the present case, it takes the form

$$\mathcal{L}^{\sharp}(A) = \frac{1}{i\hbar}\left[A, \frac{p^2}{2m} + V(x)\right] - \frac{\gamma}{i\hbar}\left(p\,[x, A] + [x, A]\,p\right) - \frac{\gamma}{2}\frac{p_{th}^2}{\hbar^2}\,[x, [x, A]]$$
$$- \frac{\gamma}{2}\frac{1}{p_{th}^2}\,[p, [p, A]]\,.$$
(5.118)

It is now easy to see that

$$\mathcal{L}^{\sharp}(x) = \frac{p}{m},$$
$$\mathcal{L}^{\sharp}(p) = -V'(x) - 2\gamma p.$$
(5.119)

Hence, the force arising from the potential is complemented by a frictional force which will drive the particle into thermal equilibrium. The fact that this frictional component stems from the second term in (5.117) indicates that the latter describes the *dissipative* effect of the environment.

In the absence of an external potential, $V = 0$, the time evolution determined by (5.119) is easily obtained, since $\left(\mathcal{L}^{\sharp}\right)^n(p) = (-2\gamma)^n\,p$ for $n \in \mathbb{N}$:

$$p_t = e^{\mathcal{L}^{\sharp}t}p = \sum_{n=0}^{\infty} \frac{(-2\gamma t)^n}{n!}p = e^{-2\gamma t}p$$
$$x_t = e^{\mathcal{L}^{\sharp}t}x = x + \frac{1}{m}\sum_{n=1}^{\infty}\frac{t^n}{n!}\left(\mathcal{L}^{\sharp}\right)^{n-1}(p) = x + \frac{p - p_t}{2\gamma m} \qquad [\text{for } V = 0].$$
(5.120)

Note that, unlike in closed systems, the Heisenberg operators do not retain their commutator, $[x_t, p_t] \neq i\hbar$ for $t > 0$ (since the map $\mathcal{W}_t^{\sharp} = \exp\left(\mathcal{L}^{\sharp}t\right)$ is non-unitary). Similarly, $(p^2)_t \neq (p_t)^2$ for $t > 0$, so that the kinetic energy operator $T = p^2/2m$ has to be calculated separately. Noting

$$\mathcal{L}^{\sharp}(T) = \gamma\frac{p_{th}^2}{2m} - 4\gamma T \quad [\text{for } V = 0],$$
(5.121)

we find

$$T_t = \frac{p_{th}^2}{8m} + \left(T - \frac{p_{th}^2}{8m}\right)e^{-4\gamma t} \quad [\text{for } V = 0].$$
(5.122)

This shows how the kinetic energy approaches a constant determined by the momentum scale p_{th}. We can now relate p_{th} to a temperature by equating the stationary expectation value $\mathrm{tr}\,(\rho \mathsf{T}_\infty) = p_{th}^2/8m$ with the average kinetic energy $\frac{1}{2}k_B T$ in a one-dimensional thermal distribution. This leads to

$$p_{th} = 2\sqrt{mk_B T} \tag{5.123}$$

for the momentum scale in (5.115). Usually, one is not able to state the operator evolution in closed form. In those cases it may be helpful to take a look at the Ehrenfest equations for their expectation values. For example, given $\langle \mathsf{p}^2 \rangle_t = 2m\langle \mathsf{T} \rangle_t$, the other second moments, $\langle \mathsf{x}^2 \rangle_t$ and $\langle \mathsf{px} + \mathsf{xp} \rangle_t$ form a closed set of differential equations. Their solutions, given in [2], yield the time evolution of the position variance $\sigma_x^2(t) = \langle \mathsf{x}^2 \rangle_t - \langle \mathsf{x} \rangle_t^2$. It has the asymptotic form

$$\sigma_x^2(t) \sim \frac{k_B T}{m\gamma} t \quad \text{as } t \to \infty, \tag{5.124}$$

which shows the *diffusive* behavior expected of a (classical) Brownian parti-cle.[14]

Let us finally take a closer look at the physical meaning of the third term in (5.117), which is dominant if the state is in a superposition of spatially separated states. Back in the Schrödinger picture we have in position repre-sentation, $\rho_t(x, x') = \langle x|\rho_t|x' \rangle$,

$$\partial_t \rho_t(x, x') = \underbrace{-\frac{\gamma}{2} \frac{p_{th}^2}{\hbar^2} (x - x')^2 \rho_t(x, x')}_{\gamma_{deco}} \quad + \text{[the other terms]} . \tag{5.125}$$

The "diagonal elements" $\rho(x, x)$ are unaffected by this term, so that it leaves the particle density invariant. The coherences in position representation, on the other hand, get exponentially suppressed,

$$\rho_t(x, x') = \exp\left(-\gamma_{deco} t\right) \rho_0(x, x') . \tag{5.126}$$

Again the decoherence rate is determined by the square of the relevant dis-tance $|x - x'|$,

$$\frac{\gamma_{deco}}{\gamma} = 4\pi \frac{(x - x')^2}{\Lambda_{th}^2} . \tag{5.127}$$

Like in Sect. 5.3.4, the rate γ_{deco} will be much larger than the dissipative rate provided the distance is large on the quantum scale, here given by the *thermal de Broglie wavelength*

$$\Lambda_{th}^2 = \frac{2\pi\hbar^2}{mk_B T} . \tag{5.128}$$

---

[14] Note that the definition of γ differs by a factor of 2 in part of the literature.

In particular, one finds $\gamma_{\text{deco}} \ggg \gamma$ if the separation is truly macroscopic. Again, it seems unphysical that the decoherence rate does not saturate as $|x - x'| \to \infty$, but grows above all bounds. One might conclude from this that non-Markovian master equations are more appropriate on these short time scales. However, I will argue that (unless the environment has very special properties) Markovian master equations are well suited to study decoherence processes, provided they involve an appropriate description of the microscopic dynamics.

5.4 Microscopic Derivations

In this section we discuss two important and rather different strategies to obtain Markovian master equations based on microscopic considerations.

5.4.1 The Weak Coupling Formulation

The most widely used form of incorporating the environment is the weak coupling approach. Here one assumes that the total Hamiltonian is "known" microscopically, usually in terms of a simplified model,

$$\mathsf{H}_{\text{tot}} = \mathsf{H} + \mathsf{H}_{\text{E}} + \mathsf{H}_{\text{int}}$$

and takes the interaction part H_{int} to be "weak" so that a perturbative treatment of the interaction is permissible.

The main assumption, called the *Born approximation*, states that H_{int} is sufficiently small so that we can take the total state as factorized, both initially, $\rho_{\text{tot}}(0) = \rho(0) \otimes \rho_{\text{E}}$, and also at $t > 0$ in those terms which involve H_{int} to second order.

$$\text{Assumption 1}: \quad \rho_{\text{tot}}(t) \simeq \rho(t) \otimes \rho_{\text{E}} \quad [\text{to second order in } \mathsf{H}_{\text{int}}]. \quad (5.129)$$

Here ρ_{E} is the stationary state of the environment, $[\mathsf{H}_{\text{E}}, \rho_{\text{E}}] = 0$. Like above, the use of the interaction picture is indicated with a tilde, cf. (5.28), so that the von Neumann equation for the total system reads

$$\begin{aligned} \partial_t \tilde{\rho}_{\text{tot}} &= \frac{1}{i\hbar} [\tilde{\mathsf{H}}_{\text{int}}(t), \tilde{\rho}_{\text{tot}}(t)] \\ &= \frac{1}{i\hbar} [\tilde{\mathsf{H}}_{\text{int}}(t), \tilde{\rho}_{\text{tot}}(0)] + \frac{1}{(i\hbar)^2} \int_0^t ds \, [\tilde{\mathsf{H}}_{\text{int}}(t), [\tilde{\mathsf{H}}_{\text{int}}(s), \tilde{\rho}_{\text{tot}}(s)]] \, . \end{aligned}$$

$$(5.130)$$

In the second equation, which is still exact, the von Neumann equation in its integral version was inserted into the differential equation version. Using a basis of Hilbert–Schmidt operators of the product Hilbert space, see Sect. 5.3.2, one can decompose the general $\tilde{\mathsf{H}}_{\text{int}}$ into the form

$$\tilde{H}_{\text{int}}(t) = \sum_k \tilde{A}_k(t) \otimes \tilde{B}_k(t) \tag{5.131}$$

with $\tilde{A}_k = \tilde{A}_k^\dagger, \tilde{B}_k = \tilde{B}_k^\dagger$. The first approximation is now to replace $\tilde{\rho}_{\text{tot}}(s)$ by $\tilde{\rho}(s) \otimes \rho_{\text{E}}$ in the double commutator of (5.130), where $\tilde{H}_{\text{int}}$ appears to second order. Performing the trace over the environment one gets

$$\partial_t \tilde{\rho}(t) = \text{tr}_{\text{E}}(\partial_t \tilde{\rho}_{\text{tot}})$$

$$\cong \frac{1}{i\hbar} \sum_k \langle \tilde{B}_k(t) \rangle_{\rho_{\text{E}}} [\tilde{A}_k(t), \tilde{\rho}(0)]$$

$$+ \frac{1}{(i\hbar)^2} \sum_{k\ell} \left\{ \int_0^t ds \underbrace{\langle \tilde{B}_k(t)\tilde{B}_\ell(s) \rangle_{\rho_{\text{E}}}}_{C_{k\ell}(t-s)} \{\tilde{A}_k(t)\tilde{A}_\ell(s)\tilde{\rho}(s) - \tilde{A}_\ell(s)\tilde{\rho}(s)\tilde{A}_k(t)\} \right.$$

$$\left. + \text{h.c.} \right\}. \tag{5.132}$$

All the relevant properties of the environment are now expressed in terms of the (complex) *bath correlation functions* $C_{k\ell}(t - s)$. Since $[H_{\text{E}}, \rho_{\text{E}}] = 0$, they depend only on the time difference $t - s$,

$$C_{k\ell}(\tau) = \text{tr}\left(e^{iH_{\text{E}}\tau} B_k e^{-iH_{\text{E}}\tau} B_\ell \rho_{\text{E}}\right) \equiv \left\langle e^{iH_{\text{E}}\tau} B_k e^{-iH_{\text{E}}\tau} B_\ell \right\rangle_{\rho_{\text{E}}}. \tag{5.133}$$

This function is determined by the environmental state alone, and it is typically appreciable only for a small range of τ around $\tau = 0$.

Equation (5.132) has the closed form of a generalized master equation, but it is non-local in time, i.e., non-Markovian. Viewing the second term as a superoperator $\mathcal{K}$, which depends essentially on $t - s$ we have

$$\partial_t \tilde{\rho}(t) = \underbrace{\frac{1}{i\hbar}[\langle \tilde{H}_{\text{int}}(t) \rangle_{\rho_{\text{E}}}, \tilde{\rho}(0)]}_{\text{disregarded}} + \int_0^t ds \mathcal{K}(t - s)\tilde{\rho}(s) , \tag{5.134}$$

where $\mathcal{K}$ is a superoperator memory kernel of the form (5.64). We may disregard the first term since the model Hamiltonian H_{E} can always be reformulated such that $\langle \tilde{B}_k(t) \rangle_{\rho_{\text{E}}} = 0$.

A naive application of second order of perturbation theory would now replace $\tilde{\rho}(s)$ by the initial $\tilde{\rho}(0)$. However, since the memory kernel is dominant at the origin it is much more reasonable to replace $\tilde{\rho}(s)$ by $\tilde{\rho}(t)$. The resulting master equation is local in time,

$$\partial_t \tilde{\rho}(t) \cong 0 + \left(\int_0^t ds \mathcal{K}(t - s)\right) \tilde{\rho}(t). \tag{5.135}$$

It is called the *Redfield equation* and it is *not* Markovian, because the integrated superoperator still depends on time. Since the kernel is appreciable only at the origin it is reasonable to replace t in the upper integration limit by ∞.

These steps are summarized by the *Born–Markov approximation*:

$$\text{Assumption 2}: \quad \int_0^t ds \mathcal{K}(t-s)\tilde{\rho}(s) \cong \int_0^\infty ds \mathcal{K}(s)\tilde{\rho}(t)\,. \qquad (5.136)$$

It leads from (5.134) to a Markovian master equation provided $\langle \mathsf{H}_{\mathrm{int}} \rangle_{\rho_{\mathrm{E}}} = 0$.

However, by no means is such a master equation guaranteed to be completely positive. An example is the Caldeira–Leggett master equation discussed in Sect. 5.3.4. It can be derived by taking the environment to be a bath of bosonic field modes whose field amplitude is coupled linearly to the particle's position operator. A model assumption on the spectral density of the coupling then leads to the frictional behavior of (5.119) [17, 30].

A *completely positive* master equation can be obtained by a further simplification, the "secular" approximation, which is applicable if the system Hamiltonian H has a discrete, non-degenerate spectrum. The system operators A_k can then be decomposed in the system energy eigenbasis. Combining the contributions with equal energy differences

$$\mathsf{A}_k(\omega) = \sum_{E'-E=\hbar\omega} \langle E|\mathsf{A}_k|E'\rangle |E\rangle\langle E'| = \mathsf{A}_k^\dagger(\omega)\,, \qquad (5.137)$$

we have

$$\mathsf{A}_k = \sum_\omega \mathsf{A}_k(\omega)\,. \qquad (5.138)$$

The time dependence of the operators in the interaction picture is now particularly simple,

$$\widetilde{\mathsf{A}}_k(t) = \sum_\omega e^{-i\omega t}\mathsf{A}_k(\omega)\,. \qquad (5.139)$$

Inserting this decomposition we find

$$\partial_t \tilde{\rho}(t) = \sum_{k\ell} \sum_{\omega\omega'} e^{i(\omega-\omega')t} \Gamma_{k\ell}(\omega')\{\mathsf{A}_\ell(\omega')\tilde{\rho}(t)\mathsf{A}_k^\dagger(\omega) - \mathsf{A}_k^\dagger(\omega)\mathsf{A}_\ell(\omega')\tilde{\rho}(t)\} + \text{h.c.}$$

$$(5.140)$$

with

$$\Gamma_{k\ell}(\omega) = \frac{1}{\hbar^2}\int_0^\infty ds\, e^{i\omega s}\langle \widetilde{\mathsf{B}}_k(s)\mathsf{B}_\ell(0)\rangle_{\rho_{\mathrm{E}}}\,. \qquad (5.141)$$

For times t which are large compared to the time scale given by the smallest system energy spacings it is reasonable to expect that only equal pairs of frequencies ω, ω' contribute appreciably to the sum in (5.140), since all other contributions are averaged out by the wildly oscillating phase factor. This constitutes the *rotating wave approximation*, our third assumption

$$\text{Assumption 3}: \quad \sum_{\omega\omega'} e^{i(\omega-\omega')t} f(\omega,\omega') \simeq \sum_{\omega} f(\omega,\omega). \qquad (5.142)$$

It is now useful to rewrite

$$\Gamma_{k\ell}(\omega) = \frac{1}{2}\gamma_{k\ell}(\omega) + iS_{k\ell}(\omega) \qquad (5.143)$$

with $\gamma_{k\ell}(\omega)$ given by the full Fourier transform of the bath correlation function,

$$\gamma_{k\ell}(\omega) = \Gamma_{k\ell}(\omega) + \Gamma_{\ell k}^*(\omega) = \frac{1}{\hbar^2} \int_{-\infty}^{\infty} dt\, e^{i\omega t} \left\langle \tilde{B}_k(t) B_\ell(0) \right\rangle_{\rho_E}, \qquad (5.144)$$

and the hermitian matrix $S_{k\ell}(\omega)$ defined by

$$S_{k\ell}(\omega) = \frac{1}{2i}\left(\Gamma_{k\ell}(\omega) - \Gamma_{\ell k}^*(\omega)\right). \qquad (5.145)$$

The matrix $\gamma_{k\ell}(\omega)$ is *positive*[15] so that we end up with a master equation of the first Lindblad form (5.78),

$$\partial_t \tilde{\rho}(t) = \frac{1}{i\hbar}\left[H_{\text{Lamb}}, \tilde{\rho}(t)\right] + \sum_{k\ell\omega} \gamma_{k\ell}(\omega)\left(A_\ell(\omega)\tilde{\rho}(t)A_k^\dagger(\omega)\right.$$
$$\left. - \frac{1}{2}A_k^\dagger(\omega)A_\ell(\omega)\tilde{\rho}(t) - \frac{1}{2}\tilde{\rho}(t)A_k^\dagger(\omega)A_\ell(\omega)\right). \qquad (5.147)$$

The hermitian operator

$$H_{\text{Lamb}} = \hbar \sum_{k\ell\omega} S_{k\ell}(\omega)A_k^\dagger(\omega)A_\ell(\omega) \qquad (5.148)$$

---

[15] To see that the matrix $(\boldsymbol{\gamma}(\omega))_{k,\ell} \equiv \gamma_{k\ell}(\omega)$ is positive we write

$$(\boldsymbol{v}, \boldsymbol{\gamma v}) = \sum_{k\ell} v_k^* \,\gamma_{k\ell}(\omega) v_\ell$$
$$= \frac{1}{\hbar^2} \int dt\, e^{i\omega t} \sum_{k\ell} \left\langle e^{iH_E t/\hbar} B_k(0) v_k^* e^{-iH_E t/\hbar} B_\ell(0) v_\ell \right\rangle_{\rho_E}$$
$$= \int dt\, e^{i\omega t} \left\langle e^{iH_E t/\hbar} C^\dagger e^{-iH_E t/\hbar} C \right\rangle_{\rho_E} \qquad (5.146)$$

with $C := \hbar^{-1} \sum_\ell v_\ell B_\ell(0)$. One can now check that due to its particular form the correlation function

$$f(t) = \left\langle e^{iH_E t/\hbar} C^\dagger e^{-iH_E t/\hbar} C \right\rangle_{\rho_E}$$

appearing in (5.146) is *of positive type*, meaning that the $n \times n$ matrices $(f(t_i - t_j))_{ij}$ defined by an arbitrary choice of $t_1, \ldots, t_n$ and $n \in \mathbb{N}$ are positive. According to Bochner's theorem [33] the Fourier transform of a function which is of positive type is positive, which proves the positivity of (5.146).

describes a renormalization of the system energies due to the coupling with the environment, the *Lamb shift*. Indeed, one finds $[\mathsf{H}, \mathsf{H}_{\mathrm{Lamb}}] = 0$.

Reviewing the three approximations (5.129), (5.136), (5.142) in view of the decoherence problem one comes to the conclusion that they all seem to be well justified if the environment is generic and the coupling is sufficiently weak. Hence, the master equation should be alright for times beyond the short-time transient which is introduced due to the choice of a product state as initial state. Evidently, the problem of non-saturating decoherence rates encountered in Sect. 5.3.4 is rather due to the *linear* coupling assumption, corresponding to a "dipole approximation", which is clearly invalid once the system states are separated by a larger distance than the wavelength of the environmental field modes.

This shows the need to incorporate realistic, nonlinear environmental couplings with a finite range. A convenient way of deriving such master equations is discussed in the next section.

5.4.2 The Monitoring Approach

The following method to derive microscopic master equations differs considerably from the weak coupling treatment discussed above. It is not based on postulating an approximate "total" Hamiltonian of system plus environment, but on two operators, which can be characterized individually in an operational sense. This permits to describe the environmental coupling in a non-perturbative fashion and to incorporate the Markov assumption right from the beginning, rather than introducing it in the course of the calculation.

The approach may be motivated by the observation made in Sects. 5.1.3 and 5.3.3 that environmental decoherence can be understood as due to the information transfer from the system to the environment occurring in a sequence of indirect measurements. In accordance with this, we will picture the environment as monitoring the system continuously by sending probe particles which scatter off the system at random times. This notion will be applicable whenever the interaction with the environment can reasonably be described in terms of individual interaction events or "collisions", and it suggests a formulation in terms of scattering theory, like in Sect. 5.1.2. The Markov assumption is then easily incorporated by disregarding the change of the environmental state after each collision [34].

When setting up a differential equation, one would like to write the temporal change of the system as the rate of collisions multiplied by the state transformation due to an individual scattering. However, in general not only the transformed state will depend on the original system state but also the collision rate, so that such a naive ansatz would yield a nonlinear equation, violating the basic principles of quantum mechanics. To account for this state dependence of the collision rate in a proper way we will apply the concept of generalized measurements discussed in Sect. 5.1.3. Specifically, we shall assume that the system is surrounded by a hypothetical, minimally invasive

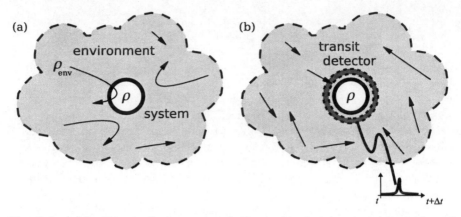

Fig. 5.1. (a) In the monitoring approach the system is taken to interact at most with one environmental (quasi-)particle at a time, so that three-body collisions are excluded. Moreover, in agreement with the Markov assumption, it is assumed that the environmental particles disperse their correlation with the system before scattering again. **(b)** In order to consistently incorporate the state dependence of the collision rate into the dynamic description of the scattering process, we imagine that the system is monitored continuously by a transit detector, which tells at a temporal resolution Δt whether a particle is going to scatter off the system, or not

detector, which tells at any instant whether a probe particle has passed by and is going to scatter off the system, see Fig. 5.1.

The rate of collisions is then described by a positive operator Γ acting in the system-probe Hilbert space. Given the uncorrelated state $\varrho_{\text{tot}} = \rho \otimes \rho_E$, it determines the probability of a collision to occur in a small time interval Δt,

$$\text{Prob}\left(C_{\Delta t}|\rho \otimes \rho_E\right) = \Delta t \, \text{tr}\left(\Gamma\left[\rho \otimes \rho_E\right]\right) . \tag{5.149}$$

Here, ρ_E is the stationary reduced single particle state of the environment. The microscopic definition of Γ will in general involve the current density operator of the relative motion and a total scattering cross section, see below.

The important point to note is that the information that a collision will take place changes our knowledge about the state, as described by the generalized measurement transformation (5.16). At the same time, we have to keep in mind that the measurement is not real, but is introduced here only for enabling us to account for the state dependence of the collision probability. It is therefore reasonable to take the detection process as *efficient*, see Sect. 5.1.3, and *minimally-invasive*, i.e., $U_\alpha = \mathbb{I}$ in (5.20), so that neither unnecessary uncertainty nor a reversible back-action is introduced. This implies that after a (hypothetical) detector click, but prior to scattering, the system-probe state will have the form

$$\mathcal{M}\left(\varrho_{\text{tot}}|C_{\Delta t}\right) = \frac{\Gamma^{1/2}\varrho_{\text{tot}}\Gamma^{1/2}}{\text{tr}\left(\Gamma\varrho_{\text{tot}}\right)} . \tag{5.150}$$

This measurement transformation reflects our improved knowledge about the incoming two-particle wave packet, and it may be viewed as enhancing those parts which are heading toward a collision. Similarly, the absence of a detection event during Δt changes the state, and this occurs with the probability $\text{Prob}\left(\overline{C}_{\Delta t}\right) = 1 - \text{Prob}\left(C_{\Delta t}\right)$.

Using the state transformation (5.150) we can now formulate the unconditioned system-probe state after a coarse-grained time Δt as the mixture of the colliding state transformed by the S-matrix and the untransformed non-colliding one, weighted with their respective probabilities,

$$\varrho'_{\text{tot}}(\Delta t) = \text{Prob}\left(C_{\Delta t}|\varrho_{\text{tot}}\right)\mathsf{S}\mathcal{M}\left(\varrho_{\text{tot}}|C_{\Delta t}\right)\mathsf{S}^{\dagger} + \text{Prob}\left(\overline{C}_{\Delta t}|\varrho_{\text{tot}}\right)\mathcal{M}\left(\varrho_{\text{tot}}|\overline{C}_{\Delta t}\right)$$
$$= \mathsf{S}\Gamma^{1/2}\varrho_{\text{tot}}\Gamma^{1/2}\mathsf{S}^{\dagger}\Delta t + \varrho_{\text{tot}} - \Gamma^{1/2}\varrho_{\text{tot}}\Gamma^{1/2}\Delta t . \tag{5.151}$$

Here, the complementary map $\mathcal{M}\left(\cdot|\overline{C}_{\Delta t}\right)$ is fixed by the requirement that the state ϱ_{tot} should remain unchanged both if the collision probability vanishes, $\Gamma = 0$, and if the scattering has no effect, $\mathsf{S} = \mathbb{I}$.

Focusing on the nontrivial part T of the two-particle S-matrix $\mathsf{S} = \mathbb{I} + i\mathsf{T}$ one finds that the unitarity of S implies that

$$\text{Im}(\mathsf{T}) \equiv \frac{1}{2i}\left(\mathsf{T} - \mathsf{T}^{\dagger}\right) = \frac{1}{2}\mathsf{T}^{\dagger}\mathsf{T} . \tag{5.152}$$

Using this relation we can write the differential quotient as

$$\frac{\varrho'_{\text{tot}}\left(\Delta t\right) - \varrho_{\text{tot}}}{\Delta t} = \mathsf{T}\Gamma^{1/2}\varrho_{\text{tot}}\Gamma^{1/2}\mathsf{T}^{\dagger} - \frac{1}{2}\mathsf{T}^{\dagger}\mathsf{T}\Gamma^{1/2}\varrho_{\text{tot}}\Gamma^{1/2} \tag{5.153}$$
$$- \frac{1}{2}\Gamma^{1/2}\varrho_{\text{tot}}\Gamma^{1/2}\mathsf{T}^{\dagger}\mathsf{T} + \frac{i}{2}\left[\mathsf{T} + \mathsf{T}^{\dagger}, \Gamma^{1/2}\varrho_{\text{tot}}\Gamma^{1/2}\right] .$$

It is now easy to arrive at a closed differential equation. We trace out the environment, assuming, in accordance with the Markov approximation, that the factorization $\varrho_{\text{tot}} = \rho \otimes \rho_E$ is valid prior to each monitoring interval Δt. Taking the limit of continuous monitoring $\Delta t \to 0$, approximating $\text{Tr}_E\left(\left[\text{Re}(\mathsf{T}), \Gamma^{1/2}\varrho_{\text{tot}}\Gamma^{1/2}\right]\right) \simeq \text{Tr}_E\left(\left[\Gamma^{1/2}\text{Re}(\mathsf{T})\Gamma^{1/2}, \varrho_{\text{tot}}\right]\right)$, and adding the generator H of the free system evolution we arrive at [34]

$$\frac{\text{d}}{\text{d}t}\rho = \frac{1}{i\hbar}[\mathsf{H}, \rho] + i\,\text{Tr}_E\left(\left[\Gamma^{1/2}\text{Re}(\mathsf{T})\Gamma^{1/2}, \rho \otimes \rho_E\right]\right)$$
$$+ \text{Tr}_E\left(\mathsf{T}\Gamma^{1/2}[\rho \otimes \rho_E]\Gamma^{1/2}\mathsf{T}^{\dagger}\right)$$
$$- \frac{1}{2}\text{Tr}_E\left(\Gamma^{1/2}\mathsf{T}^{\dagger}\mathsf{T}\Gamma^{1/2}[\rho \otimes \rho_E]\right)$$
$$- \frac{1}{2}\text{Tr}_E\left([\rho \otimes \rho_E]\Gamma^{1/2}\mathsf{T}^{\dagger}\mathsf{T}\Gamma^{1/2}\right) . \tag{5.154}$$

This general monitoring master equation, entirely specified by the rate operator Γ, the scattering operator $\mathsf{S} = \mathbb{I} + i\mathsf{T}$, and the environmental state ρ_E, is non-perturbative in the sense that the collisional interaction is nowhere assumed to be weak. It is manifestly Markovian even before the environmental trace is carried out, and one finds, by doing the trace in the eigenbasis of ρ_E, that is has the general Lindblad structure (5.79) of the generator of a quantum dynamical semigroup. The second term in (5.154), which involves a commutator, accounts for the renormalization of the system energies due to the coupling to the environment, just like (5.148), while the last three lines describe the incoherent effect of the coupling to the environment.

So far, the discussion was very general. To obtain concrete master equations one has to specify system and environment, along with the operators Γ and S describing their interaction. In the following applications, we will assume the environment to be an ideal Maxwell gas, whose single particle state

$$\rho_{\text{gas}} = \frac{\Lambda_{\text{th}}^3}{\Omega} \exp\left(-\beta \frac{\mathbf{p}^2}{2m}\right) \tag{5.155}$$

is characterized by the inverse temperature β, the normalization volume Ω, and the thermal de Broglie wave length Λ_{th} defined in (5.128).

5.4.3 Collisional Decoherence of a Brownian Particle

As a first application of the monitoring approach, let us consider the "localization" of a mesoscopic particle by a gaseous environment. Specifically, we will assume that the mass M of this Brownian particle is much greater than the mass m of the gas particles. In the limit $m/M \to 0$ the energy exchange during an elastic collision vanishes, so that the mesoscopic particle will not thermalize in our description, but we expect that the off-diagonal elements of its position representation will get reduced, as discussed in Sect. 5.3.4.

This can be seen by considering the effect of the S-matrix in the limit $m/M \to 0$. In general, a collision keeps the center-of-mass invariant, and only the relative coordinates are affected. Writing S_0 for the S-matrix in the center of mass frame and denoting the momentum eigenstates of the Brownian and the gas particle by $|\mathbf{P}\rangle$ and $|\mathbf{p}\rangle$, respectively, we have [35]

$$S|\mathbf{P}\rangle|\mathbf{p}\rangle = \int d^3Q |\mathbf{P}-\mathbf{Q}\rangle|\mathbf{p}+\mathbf{Q}\rangle\langle\frac{m_*}{m}\mathbf{p}-\frac{m_*}{M}\mathbf{P}+\mathbf{Q}|S_0|\frac{m_*}{m}\mathbf{p}-\frac{m_*}{M}\mathbf{P}\rangle , \tag{5.156}$$

where $m_* = Mm/(M+m)$ is the reduced mass and $\mathbf{Q}$ is the transfered momentum (and thus the change of the relative momentum). In the limit of a large Brownian mass we have $m_*/m \to 1$ and $m_*/M \to 0$, so that

$$S|\mathbf{P}\rangle|\mathbf{p}\rangle \to \int d^3Q |\mathbf{P}-\mathbf{Q}\rangle|\mathbf{p}+\mathbf{Q}\rangle\langle\mathbf{p}+\mathbf{Q}|S_0|\mathbf{p}\rangle \qquad [\text{for } M \gg m]. \tag{5.157}$$

It follows that a position eigenstate $|\mathbf{X}\rangle$ of the Brownian particle remains unaffected by a collision,

$$S|\boldsymbol{X}\rangle|\psi_{\text{in}}\rangle_{\text{E}} = |\boldsymbol{X}\rangle \underbrace{\left(e^{-i\mathbf{p}\cdot\boldsymbol{X}/\hbar}S_0 e^{i\mathbf{p}\cdot\boldsymbol{X}/\hbar} \right)|\psi_{\text{in}}\rangle_{\text{E}}}_{|\psi_{\text{out}}^{(\boldsymbol{X})}\rangle_{\text{E}}}, \qquad (5.158)$$

as can be seen by inserting identities in terms of the momentum eigenstates. Here, $|\psi_{\text{in}}\rangle_{\text{E}}$ denotes an arbitrary single-particle wave packet state of a gas atom. The exponentials in (5.158) effect a translation of S_0 from the origin to the position $\boldsymbol{X}$, so that the scattered state of the gas particle $|\psi_{\text{out}}^{(\boldsymbol{X})}\rangle_{\text{E}}$ depends on the location of the Brownian particle.

Just like in Sect. 5.1.1, a single collision will thus reduce the spatial coherences $\rho(\boldsymbol{X}, \boldsymbol{X}') = \langle\boldsymbol{X}|\rho|\boldsymbol{X}'\rangle$ by the overlap of the gas states scattered at positions $\boldsymbol{X}$ and $\boldsymbol{X}'$,

$$\rho'(\boldsymbol{X}, \boldsymbol{X}') = \rho(\boldsymbol{X}, \boldsymbol{X}')\langle\psi_{\text{out}}^{(\boldsymbol{X}')}|\psi_{\text{out}}^{(\boldsymbol{X})}\rangle_{\text{E}}. \qquad (5.159)$$

The reduction factor will be the smaller in magnitude the better the scattered state of the gas particle can "resolve" between the positions $\boldsymbol{X}$ and $\boldsymbol{X}'$.

In order to obtain the dynamic equation we need to specify the rate operator. Classically, the collision rate is determined by the product of the current density $j = n_{\text{gas}}v_{\text{rel}}$ and the total cross section $\sigma(p_{\text{rel}})$, and therefore Γ should be expressed in terms of the corresponding operators. This is particularly simple in the large mass limit $M \to \infty$, where $v_{\text{rel}} = |\mathbf{p}/m - \boldsymbol{P}/M| \to |\mathbf{p}|/m$, so that the current density and the cross section depend only on the momentum of the gas particle, leading to

$$\Gamma = n_{\text{gas}}\frac{|\mathbf{p}|}{m}\sigma(\mathbf{p}). \qquad (5.160)$$

If the gas particle moves in a normalized wave packet heading toward the origin then the expectation value of this operator will indeed determine the collision probability. However, this expression depends only on the modulus of the velocity so that it will yield a finite collision probability even if the particle is heading away form the origin. Hence, for (5.154) to make sense either the S-matrix should be modified to keep such a non-colliding state unaffected or Γ should contain in addition a projection to the subset of incoming states, see the discussion below.

In momentum representation, $\rho(\boldsymbol{P}, \boldsymbol{P}') = \langle\boldsymbol{P}|\rho|\boldsymbol{P}'\rangle$, (5.154) assumes the general structure[16]

$$\partial_t\rho(\boldsymbol{P}, \boldsymbol{P}') = \frac{1}{i\hbar}\frac{P^2 - (P')^2}{2M}\rho(\boldsymbol{P}, \boldsymbol{P}')$$
$$+ \int \mathrm{d}\boldsymbol{P}_0\mathrm{d}\boldsymbol{P}'_0\, \rho(\boldsymbol{P}_0, \boldsymbol{P}'_0)\,M(\boldsymbol{P}, \boldsymbol{P}'; \boldsymbol{P}_0, \boldsymbol{P}'_0)$$

---

[16] The second term in (5.154) describes forward scattering and vanishes for momentum diagonal ρ_E.

$$-\frac{1}{2}\int d\boldsymbol{P}_0\rho\left(\boldsymbol{P}_0,\boldsymbol{P}'\right)\int d\boldsymbol{P}_f\, M\left(\boldsymbol{P}_f,\boldsymbol{P}_f;\boldsymbol{P}_0,\boldsymbol{P}\right)$$

$$-\frac{1}{2}\int d\boldsymbol{P}'_0\rho\left(\boldsymbol{P},\boldsymbol{P}'_0\right)\int d\boldsymbol{P}_f\, M\left(\boldsymbol{P}_f,\boldsymbol{P}_f;\boldsymbol{P}',\boldsymbol{P}'_0\right).\quad(5.161)$$

The dynamics is therefore characterized by a single complex function

$$M\left(\boldsymbol{P},\boldsymbol{P}';\boldsymbol{P}_0,\boldsymbol{P}'_0\right)=\langle\boldsymbol{P}|\,\mathrm{tr}_{\mathrm{gas}}\left(\mathsf{T}\Gamma^{1/2}\left[|\boldsymbol{P}_0\rangle\langle\boldsymbol{P}'_0|\otimes\rho_{\mathrm{gas}}\right]\Gamma^{1/2}\mathsf{T}^\dagger\right)|\boldsymbol{P}'\rangle,$$
$$(5.162)$$

which has to be evaluated. Inserting the diagonal representation of the gas state (5.155)

$$\rho_{\mathrm{gas}}=\frac{(2\pi\hbar)^3}{\Omega}\int d\boldsymbol{p}_0\mu\left(\boldsymbol{p}_0\right)|\boldsymbol{p}_0\rangle\langle\boldsymbol{p}_0|\qquad(5.163)$$

it reads, with the choices (5.157) and (5.160) for S and Γ,

$$M\left(\boldsymbol{P},\boldsymbol{P}';\boldsymbol{P}-\boldsymbol{Q},\boldsymbol{P}'-\boldsymbol{Q}'\right)=\int d\boldsymbol{p}_1 d\boldsymbol{p}_0\mu\left(\boldsymbol{p}_0\right)\delta\left(\boldsymbol{Q}+\boldsymbol{p}_1-\boldsymbol{p}_0\right)\delta\left(\boldsymbol{Q}'+\boldsymbol{p}_0-\boldsymbol{p}_1\right)$$

$$\times\frac{n_{\mathrm{gas}}}{m}|\boldsymbol{p}_0|\,\sigma\left(\boldsymbol{p}_0\right)\frac{(2\pi\hbar)^3}{\Omega}|\langle\boldsymbol{p}_1|\mathsf{T}_0|\boldsymbol{p}_0\rangle|^2$$

$$=\delta\left(\boldsymbol{Q}-\boldsymbol{Q}'\right)\int d\boldsymbol{p}_0\,\mu\left(\boldsymbol{p}_0\right)\frac{n_{\mathrm{gas}}}{m}|\boldsymbol{p}_0|\,\sigma\left(\boldsymbol{p}_0\right)$$

$$\times\frac{(2\pi\hbar)^3}{\Omega}|\langle\boldsymbol{p}_0-\boldsymbol{Q}|\mathsf{T}_0|\boldsymbol{p}_0\rangle|^2$$

$$=:\delta\left(\boldsymbol{Q}-\boldsymbol{Q}'\right)M_{\mathrm{in}}\left(\boldsymbol{Q}\right).\qquad(5.164)$$

This shows that, apart from the unitary motion, the dynamics is simply characterized by momentum exchanges described in terms of gain and loss terms,

$$\partial_t\rho\left(\boldsymbol{P},\boldsymbol{P}'\right)=\frac{1}{i\hbar}\frac{P^2-(P')^2}{2M}\rho\left(\boldsymbol{P},\boldsymbol{P}'\right)+\int d\boldsymbol{Q}\,\rho\left(\boldsymbol{P}-\boldsymbol{Q},\boldsymbol{P}'-\boldsymbol{Q}\right)M_{\mathrm{in}}\left(\boldsymbol{Q}\right)$$

$$-\rho\left(\boldsymbol{P},\boldsymbol{P}'\right)\int d\boldsymbol{Q}M_{\mathrm{in}}\left(\boldsymbol{Q}\right).\qquad(5.165)$$

We still have to evaluate the function $M_{\mathrm{in}}\left(\boldsymbol{Q}\right)$, which can be clearly interpreted as the rate of collisions leading to a momentum gain $\boldsymbol{Q}$ of the Brownian particle,

$$M_{\mathrm{in}}\left(\boldsymbol{Q}\right)=\frac{n_{\mathrm{gas}}}{m}\int d\boldsymbol{p}_0\mu\left(\boldsymbol{p}_0\right)|\boldsymbol{p}_0|\,\sigma\left(\boldsymbol{p}_0\right)\frac{(2\pi\hbar)^3}{\Omega}|\langle\boldsymbol{p}_0-\boldsymbol{Q}|\mathsf{T}_0|\boldsymbol{p}_0\rangle|^2.$$
$$(5.166)$$

It involves the momentum matrix element of the on-shell T_0-matrix, $S_0 = 1 + iT_0$, which, according to elastic scattering theory [35], is proportional to the scattering amplitude f,

$$\langle \boldsymbol{p}_f | T_0 | \boldsymbol{p}_i \rangle = \frac{f(\boldsymbol{p}_f, \boldsymbol{p}_i)}{2\pi\hbar} \delta\left(\frac{p_f^2}{2} - \frac{p_i^2}{2}\right). \tag{5.167}$$

The delta function ensures the conservation of energy during the collision. At first sight, this leads to an ill-defined expression since the matrix element (5.167) appears as a squared modulus in (5.166), so that the three-dimensional integration is over a squared delta function.

The appearance of this problem can be traced back to our disregard of the projection to the subset of incoming states in the definition (5.160) of Γ. When evaluating M_{in} we used the diagonal representation (5.163) for ρ_{gas} in terms of (improper) momentum eigenstates, which comprise both incoming and outgoing characteristics if viewed as the limiting form of a wave packet. One way of implementing the missing projection to incoming states would be to use a different convex decomposition of ρ_{gas}, which admits a separation into incoming and outgoing contributions [36]. This way, M_{in} can indeed be calculated properly, albeit in a somewhat lengthy calculation. A shorter route to the same result sticks to the diagonal representation, but modifies the definition of S in a formal sense so that it keeps all outgoing states invariant.[17] The conservation of the probability current, which must still be guaranteed by any such modification, then implies a simple rule how to deal with the squared matrix element [36],

$$\frac{(2\pi\hbar)^3}{\Omega} |\langle \boldsymbol{p}_f | T_0 | \boldsymbol{p}_i \rangle|^2 \longrightarrow \frac{|f(\boldsymbol{p}_f, \boldsymbol{p}_i)|^2}{p_i \sigma(p_i)} \delta\left(\frac{p_f^2}{2} - \frac{p_i^2}{2}\right). \tag{5.168}$$

Here $\sigma(p) = \int d\Omega' |f(p\boldsymbol{n}', p\boldsymbol{n})|^2$ is the total elastic cross section. With this replacement we obtain immediately

$$M_{\text{in}}(\boldsymbol{Q}) = \frac{n_{\text{gas}}}{m} \int d\boldsymbol{p}_0 \, \mu(\boldsymbol{p}_0) \, |f(\boldsymbol{p}_0 - \boldsymbol{Q}, \boldsymbol{p}_0)|^2 \, \delta\left(\frac{p_0^2}{2} - \frac{(\boldsymbol{p}_0 - \boldsymbol{Q})^2}{2}\right). \tag{5.169}$$

As one would expect, the rate of momentum changing collisions is determined by a thermal average over the differential cross section $d\sigma/d\Omega = |f|^2$.

Also for finite mass ratios m/M a master equation can be obtained this way, although the calculation is more complicated [37, 38]. The resulting linear quantum Boltzmann equation then describes on equal footing the decoherence and dissipation effects of a gas on the quantum motion of a particle.

---

[17] In general, even a purely outgoing state gets transformed by S, since the definition of the S-matrix involves a backward time evolution [35].

The "localizing" effect of a gas on the Brownian particle can now be seen, after going into the interaction picture in order to remove the unitary part of the evolution, and by stating the master equation in position representation. From (5.165) and (5.169) one obtains

$$\partial_t \tilde{\rho}(\boldsymbol{X}, \boldsymbol{X}') = -F(\boldsymbol{X} - \boldsymbol{X}') \tilde{\rho}(\boldsymbol{X}, \boldsymbol{X}') \tag{5.170}$$

with *localization rate* [36]

$$F(\boldsymbol{x}) = \int_0^\infty \mathrm{d}v\, \nu(v)\, n_{\text{gas}}\, v \int \frac{\mathrm{d}\Omega_1 \mathrm{d}\Omega_2}{4\pi} \left(1 - e^{imv(\boldsymbol{n}_1 - \boldsymbol{n}_2) \cdot \boldsymbol{x}/\hbar} \right)$$
$$\times |f(mv\boldsymbol{n}_2, mv\boldsymbol{n}_1)|^2. \tag{5.171}$$

Here, the unit vectors $\boldsymbol{n}_1, \boldsymbol{n}_2$ are the directions of incoming and outgoing gas particles associated to the elements of solid angle $\mathrm{d}\Omega_1$ and $\mathrm{d}\Omega_2$ and $\nu(v)$ is the velocity distribution in the gas. Clearly, $F(\boldsymbol{x})$ determines how fast the spatial coherences corresponding to the distance $\boldsymbol{x}$ decay.

One angular integral in (5.171) can be performed in the case of isotropic scattering, $f(\boldsymbol{p}_f, \boldsymbol{p}_i) = f\left(\cos(\boldsymbol{p}_f, \boldsymbol{p}_i); E = p_i^2/2m\right)$. In this case,

$$F(\boldsymbol{x}) = \int_0^\infty \mathrm{d}v\, \nu(v)\, n_{\text{gas}} v \left\{ \sigma(mv) - 2\pi \int_{-1}^1 \mathrm{d}(\cos\theta) \left| f\left(\cos\theta; E = \frac{m}{2}v^2\right) \right|^2 \right.$$
$$\left. \times \operatorname{sinc}\left(2\sin\left(\frac{\theta}{2}\right) \frac{mv|\boldsymbol{x}|}{\hbar} \right) \right\}, \tag{5.172}$$

with $\operatorname{sinc}(x) = \sin(x)/x$ and θ the (polar) scattering angle.

The argument of the sinc function is equal to the momentum exchange during the collision times the distance in units of $\hbar$. As $|\boldsymbol{X} - \boldsymbol{X}'| \longrightarrow 0$ the sinc approaches unity and the angular integral yields the total cross section σ so that the localization rate vanishes, as required. At very small distances, a second order expansion in the distance $\boldsymbol{x}$ is permissible and one obtains a quadratic dependence [39], such as predicted by the Caldeira–Leggett model, see (5.127). However, once the distance $|\boldsymbol{X} - \boldsymbol{X}'|$ is sufficiently large so that the scattered state can resolve whether the collision took place at position $\boldsymbol{X}$ or $\boldsymbol{X}'$ the sinc function in (5.172) suppresses the integrand. It follows that in the limit of large distances the localization rate saturates, at a value given by the average collision rate $F(\infty) = \langle \sigma v n_{\text{gas}} \rangle$, see Fig. 5.2.

Decoherence in this saturated regime of large separations has been observed, in good agreement with this theory, in molecular interference experiments in the presence of various gases [40]. The intermediate regime between quadratic increase and saturation was also seen in such experiments on momentum-exchange mediated decoherence, by studying the influence of the heat radiation emitted by fullerene molecules on the visibility of their interference pattern [41].

As a conclusion of this section, we see that the scattering approach permits to incorporate realistic microscopic interactions transparently and without

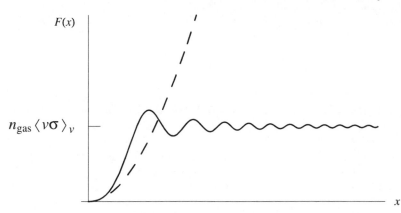

Fig. 5.2. The localization rate (5.172) describing the loss of wave-like behavior in a Brownian particle state saturates for large distances at the average collision rate. In contrast, the Caldeira–Leggett model predicts a quadratic increase beyond all bounds (*dashed line*), see (5.126). This indicates that linear coupling models should be taken with care if time scales are involved that differ strongly from the dissipation time scale

approximation in the interaction strength. The results show clearly that linear coupling models, which imply that decoherence rates grow above all bounds, have a limited range of validity. They cannot be judged by their success in describing dissipative phenomena. Frequent claims of "universality" in decoherence behavior, which are based on these linear coupling models, are therefore to be treated with care.

5.4.4 Decoherence of a Quantum Dot

As a second application of the monitoring approach, let us see how the dynamics of an immobile object with discrete internal structure, such as an implementation of a quantum dot, gets affected by an environment of ideal gas particles. For simplicity, we take the gas again in the Maxwell state (5.155), though different dispersion relations, e.g., in the case of phonon quasi-particles, could be easily incorporated. The interaction between system and gas will be described in terms of the in general inelastic scattering amplitudes determined by the interaction potential.

In the language of scattering theory the energy eigenstates of the non-motional degrees of freedom are called channels. In our case of a structureless gas they form a discrete basis of the system Hilbert space. In the following, the notation $|\alpha\rangle$, not to be confused with the coherent states of Sect. 5.3.4, will be used to indicate the system eigenstates of energy E_α. In this channel basis, $\rho_{\alpha\beta} = \langle\alpha|\rho|\beta\rangle$, the equation of motion (5.154) takes on the form of a general discrete master equation of Lindblad type,

$$\partial_t \rho_{\alpha\beta} = \frac{E_\alpha + \varepsilon_\alpha - E_\beta - \varepsilon_\beta}{i\hbar} \rho_{\alpha\beta} + \sum_{\alpha_0\beta_0} \rho_{\alpha_0\beta_0} M_{\alpha\beta}^{\alpha_0\beta_0}$$

$$-\frac{1}{2} \sum_{\alpha_0} \rho_{\alpha_0\beta} \sum_\gamma M_{\gamma\gamma}^{\alpha_0\alpha} - \frac{1}{2} \sum_{\beta_0} \rho_{\alpha\beta_0} \sum_\gamma M_{\gamma\gamma}^{\beta\beta_0} . \qquad (5.173)$$

The real energy shifts ε_α given below describe the coherent modification of the system energies due to the presence of the environment. They are due to the second term in (5.154) and are the analogue of the Lamb shift (5.148) encountered in the weak coupling calculation. The incoherent effect of the environment, on the other hand, is described by the set of complex rate coefficients

$$M_{\alpha\beta}^{\alpha_0\beta_0} = \langle\alpha| \operatorname{Tr}_E \left(\mathsf{T}\Gamma^{1/2} \left[|\alpha_0\rangle\langle\beta_0| \otimes \rho_{gas} \right] \Gamma^{1/2}\mathsf{T}^\dagger \right) |\beta\rangle . \qquad (5.174)$$

In order to calculate these quantities we need again to specify the rate operator Γ. In the present case, it is naturally given in terms of the current density operator $\mathsf{j} = n_{gas}\mathsf{p}/m$ of the impinging gas particles multiplied by the channel-specific total scattering cross sections $\sigma(\mathbf{p},\alpha)$,

$$\Gamma = \sum_\alpha |\alpha\rangle\langle\alpha| \otimes n_{gas} \frac{|\mathsf{p}|}{m} \sigma(\mathsf{p},\alpha) . \qquad (5.175)$$

Like in Sect. 5.4.3, this operator should in principle contain a projection to the subset of incoming states of the gas particle. Again, this can be accounted for in two different ways in the calculation of the rates (5.174). By using a non-diagonal decomposition of ρ_{gas}, which permits to disregard the outgoing states, one obtains[18]

$$M_{\alpha\beta}^{\alpha_0\beta_0} = \chi_{\alpha\beta}^{\alpha_0\beta_0} \frac{n_{gas}}{m^2} \int d\mathbf{p}\, d\mathbf{p}_0 \mu(\mathbf{p}_0) f_{\alpha\alpha_0}(\mathbf{p},\mathbf{p}_0)$$

$$\times f_{\beta\beta_0}^*(\mathbf{p},\mathbf{p}_0) \delta\left(\frac{\mathbf{p}^2 - \mathbf{p}_0^2}{2m} + E_\alpha - E_{\alpha_0} \right) , \qquad (5.176)$$

with the Kronecker-like factor

$$\chi_{\alpha\beta}^{\alpha_0\beta_0} := \begin{cases} 1 \text{ if } E_\alpha - E_{\alpha_0} = E_\beta - E_{\beta_0} \\ 0 \text{ otherwise} . \end{cases} \qquad (5.177)$$

The energy shifts are determined the real parts of the forward scattering amplitude,

$$\varepsilon_\alpha = -2\pi\hbar^2 \frac{n_{gas}}{m} \int d\mathbf{p}_0 \mu(\mathbf{p}_0) \operatorname{Re}[f_{\alpha\alpha}(\mathbf{p}_0,\mathbf{p}_0)] . \qquad (5.178)$$

---

[18] For the special case of factorizing interactions, $\mathsf{H}_{int} = \mathsf{A} \otimes \mathsf{B}_E$, and for times large compared to all system time scales this result can be obtained rigorously in a standard approach [42], by means of the "low-density limit" scaling method [2, 23].

Some details of this calculation can be found in [34]. Rather than repeating them here we note that the result (5.176) can be obtained directly by using the diagonal representation (5.163) of ρ_{gas} and the multichannel generalization of the replacement rule (5.168),

$$
\frac{(2\pi\hbar)^3}{\Omega} \langle \alpha \boldsymbol{p} | \mathsf{T} | \alpha_0 \boldsymbol{p}_0 \rangle \langle \beta_0 \boldsymbol{p}_0 | \mathsf{T}^\dagger | \beta \boldsymbol{p} \rangle \rightarrow \frac{\chi_{\alpha\beta}^{\alpha_0\beta_0}}{p_0 m} \frac{f_{\alpha\alpha_0}(\boldsymbol{p},\boldsymbol{p}_0) f_{\beta\beta_0}^*(\boldsymbol{p},\boldsymbol{p}_0)}{\sqrt{\sigma(p_0,\alpha_0)\,\sigma(p_0,\beta_0)}}
$$
$$
\times \delta\left(\frac{\boldsymbol{p}^2 - \boldsymbol{p}_0^2}{2m} + E_\alpha - E_{\alpha_0}\right) . \quad (5.179)
$$

The expression for the complex rates simplifies further if the scattering amplitudes are rotationally invariant, $f_{\alpha\alpha_0}\left(\cos(\boldsymbol{p},\boldsymbol{p}_0); E = p_0^2/2m\right)$. In this case we have

$$
M_{\alpha\beta}^{\alpha_0\beta_0} = \chi_{\alpha\beta}^{\alpha_0\beta_0} \int_0^\infty \mathrm{d}v\,\nu(v)\, n_{\text{gas}} v_{\text{out}}(v)\, 2\pi \int_{-1}^1 \mathrm{d}(\cos\theta)
$$
$$
\times f_{\alpha\alpha_0}\left(\cos\theta; E = \frac{m}{2}v^2\right) f_{\beta\beta_0}^*\left(\cos\theta; E = \frac{m}{2}v^2\right) \quad (5.180)
$$

with $\nu(v)$ the velocity distribution like in (5.172), and

$$
v_{\text{out}}(v) = \sqrt{v^2 - \frac{2}{m}(E_\alpha - E_{\alpha_0})} \quad (5.181)
$$

the velocity of a gas particle after a possibly inelastic collision.

This shows that limiting cases of (5.173) display the expected dynamics. For the populations $\rho_{\alpha\alpha}$ it reduces to a rate equation, where the cross sections $\sigma_{\alpha\alpha_0}(E) = 2\pi \int \mathrm{d}(\cos\theta)\,|f_{\alpha\alpha_0}(\cos\theta; E)|^2$ for scattering from channel α_0 to α determine the transition rates,

$$
M_{\alpha\alpha}^{\alpha_0\alpha_0} = \int \mathrm{d}v\,\nu(v)\, n_{\text{gas}} v_{\text{out}}(v)\, \sigma_{\alpha\alpha_0}\left(\frac{m}{2}v^2\right) . \quad (5.182)
$$

In the case of purely elastic scattering, on the other hand, i.e., for $M_{\alpha\beta}^{\alpha_0\beta_0} = M_{\alpha\beta}^{\alpha\beta} \delta_{\alpha\alpha_0} \delta_{\beta\beta_0}$, the coherences are found to decay exponentially,

$$
\partial_t |\rho_{\alpha\beta}| = -\gamma_{\alpha\beta}^{\text{elastic}} |\rho_{\alpha\beta}| . \quad (5.183)
$$

The corresponding pure dephasing rates are determined by the difference of the scattering amplitudes,

$$
\gamma_{\alpha\beta}^{\text{elastic}} = \pi \int \mathrm{d}v\,\nu(v)\, n_{\text{gas}} v_{\text{out}}(v) \int_{-1}^1 \mathrm{d}(\cos\theta)
$$
$$
\times \left| f_{\alpha\alpha}\left(\cos\theta; \frac{m}{2}v^2\right) - f_{\beta\beta}\left(\cos\theta; \frac{m}{2}v^2\right) \right|^2 . \quad (5.184)
$$

As one expects in this case, the better the scattering environment can distinguish between system states $|\alpha\rangle$ and $|\beta\rangle$ the more coherence is lost in this elastic process.

In the general case, the decay of off-diagonal elements will be due to a combination of elastic and inelastic processes. Although little can be said without specifying the interaction, it is clear that the integral over $|f_{\alpha\alpha} - f_{\beta\beta}|^2$ in (5.184), a "decoherence cross section" without classical interpretation, is not related to the inelastic cross sections characterizing the population transfer, and may be much larger. In this case, the resulting decoherence will be again much faster than the corresponding relaxation time scales.

5.5 Robust States and the Pointer Basis

We have seen that, even though the decoherence predictions of linear coupling models has to be taken with great care, the general observation remains valid that the loss of coherence may occur on a time scale $\gamma_{\mathrm{deco}}^{-1}$ that is much shorter the relaxation time γ^{-1}. Let us therefore return to the general description of open systems in terms of a semigroup generator $\mathcal{L}$, and ask what we can say about a general state after a time t which is still small compared to the relaxation time, but much larger than the decoherence time scale. From a classical point of view, which knows only about relaxation, the state has barely changed, but in the quantum description it may now be well approximated by a mixture determined by particular projectors P_ℓ,

$$\mathrm{e}^{\mathcal{L}t} : \rho \xrightarrow{\gamma_{\mathrm{deco}}^{-1} \ll t \ll \gamma^{-1}} \rho_t \simeq \rho' = \sum_\ell \mathrm{tr}(\rho \mathsf{P}_\ell)\, \mathsf{P}_\ell. \tag{5.185}$$

This set of projectors $\{\mathsf{P}_\ell\}$, which depend at most weakly on t, is called *pointer basis* [43] or set of *robust states* [44]. It is distinguished by the fact that a system prepared in such a state is hardly affected by the environment, while a superposition of two distinct pointer states decoheres so rapidly that it is never observed in practice.

We encountered this behavior with the damped harmonic oscillator discussed in Sect. 5.3.4. There the coherent oscillator states remained pure under Markovian dynamics, while superpositions between (macroscopically distinct) coherent states decayed rapidly. Hence, in this case the coherent states $\mathsf{P}_\alpha = |\alpha\rangle\langle\alpha|$ can be said to form an (over-complete) set of robust states, leading to the mixture

$$\rho' = \int \mathrm{d}\mu\,(\alpha)\, \mathrm{tr}(\rho \mathsf{P}_\alpha) \mathsf{P}_\alpha, \tag{5.186}$$

with appropriate measure μ.

The name *pointer basis* is well-fitting because the existence of such robust states is a prerequisite for the description of an ideal measurement device in a quantum framework. A macroscopic – and therefore decohering – apparatus implementing the measurement of an observable A is ideally constructed in such a way that macroscopically distinct positions of the "pointer" are obtained for the different eigenstates of A. Provided these pointer positions of the device are robust, the correct values are observed with certainty if the quantum system is in an eigenstate of the observable. Conversely, if the quantum system is not in an eigenstate of A, the apparatus will *not* end up in a superposition of pointer positions, but be found at a definite position, albeit probabilistically, with a probability given by the Born rule.

The main question regarding pointer states is, given the environmental coupling or the generator $\mathcal{L}$, what determines whether a state is robust or not, and how can we determine the set of pointer states without solving the master equation for all initial states. It is fair to say that this issue is not fully understood, except for very simple model environments, nor is it even clear how to quantify robustness.

An obvious ansatz, due to Zurek [6, 45], is to sort all pure states in the Hilbert space according to their (linear) entropy production rate, or rate of loss of purity,

$$\partial_t S_{\mathrm{lin}}[\rho] = -2\,\mathrm{tr}\left(\rho\mathcal{L}(\rho)\right) . \tag{5.187}$$

It has been called "predictability sieve" since the least entropy producing and therefore most predictable states are candidate pointer states [6].

In the following, a related approach will be described, following the presentation in [3, 46]. It is based on a time-evolution equation for robust states. Since such an equation must distinguish particular states from their linear superpositions it is necessarily nonlinear.

5.5.1 Nonlinear Equation for Robust States

We seek a nonlinear time-evolution equation for robust pure states P_t which, on the one hand, preserves their purity, and on the other, keeps them as close as possible to the evolved state following the master equation.

A simple nonlinear equation keeping a pure state pure is given by the following extension of the Heisenberg form for the infinitesimal time step,

$$P_{t+\delta t} = P_t + \delta t \left(\frac{1}{i}[A_t, P_t] + [P_t, [P_t, B_t]]\right) , \tag{5.188}$$

where A and B are hermitian operators. In fact, the unitary part can be absorbed into the nonlinear part by introducing the hermitian operator $X_t = -i[A_t, P_t] + B_t$. It "generates" the infinitesimal time translation of the projectors (and may be a function of P_t),

$$P_{t+\delta t} = P_t + \delta t[P_t, [P_t, X_t]] . \tag{5.189}$$

With this choice one confirms easily that the evolved operator has indeed the properties of a projector, to leading order in δt,

$$P^\dagger_{t+\delta t} = P_{t+\delta t} \tag{5.190}$$

and

$$(P_{t+\delta t})^2 = P_{t+\delta t} + O(\delta t^2). \tag{5.191}$$

The corresponding differential equation reads

$$\partial_t P_t = \frac{P_{t+\delta t} - P_t}{\delta t} = [P_t, [P_t, X_t]]. \tag{5.192}$$

To determine the operator X_t one minimizes the distance between the time derivatives of the truly evolved state and the projector. If we visualize the pure states as lying on the boundary of the convex set of mixed states, then a pure state will in general dive into the interior under the time evolution generated by $\mathcal{L}$. The minimization chooses the operator X_t in such a way that P_t sticks to the boundary, while remaining as close as possible to the truly evolved state.

The (Hilbert–Schmidt) distance between the time derivatives can be calculated as

$$\| \underbrace{\mathcal{L}(P_t)}_{\equiv Z} - \partial_t P_t \|^2_{\mathrm{HS}} = \mathrm{tr}\left[(Z - [P_t, [P_t, X_t]])^2\right]$$

$$= \mathrm{tr}\left(Z^2 - 2(Z^2 P_t - (Z P_t)^2)\right)$$

$$+ 2\,\mathrm{tr}\left((Z - X)^2 P_t - ((Z - X)P_t)^2\right). \tag{5.193}$$

We note that the first term is independent of X, whereas the second one is non-negative. With the obvious solution $X_t = Z \equiv \mathcal{L}(P_t)$ one gets a nonlinear evolution equation for robust states P_t, which is trace and purity preserving [46],

$$\partial_t P_t = [P_t, [P_t, \mathcal{L}(P_t)]]. \tag{5.194}$$

It is useful to write down the equation in terms of the vectors $|\xi\rangle$ which correspond to the pure state $P_t = |\xi\rangle\langle\xi|$,

$$\partial_t |\xi\rangle = [\mathcal{L}(|\xi\rangle\langle\xi|) - \underbrace{\langle\xi|\mathcal{L}(|\xi\rangle\langle\xi|)|\xi\rangle}_{\text{"decay rate"}}]|\xi\rangle. \tag{5.195}$$

If we take $\mathcal{L}$ to be of the Lindblad form (5.79) the equation reads

$$\partial_t |\xi\rangle = \frac{1}{i\hbar} H|\xi\rangle + \sum_k \gamma_k \left[\langle L_k^\dagger\rangle_\xi \left(L_k - \langle L_k\rangle_\xi\right) - \frac{1}{2}\left(L_k^\dagger L - \langle L_k^\dagger L_k\rangle\right)\right]|\xi\rangle$$

$$- \frac{1}{i\hbar}\langle H\rangle_\xi |\xi\rangle. \tag{5.196}$$

Its last term is usually disregarded because it gives rise only to an additional phase if $\langle H\rangle_\xi$ is constant. The meaning of the nonlinear equation (5.196) is best studied in terms of concrete examples.

5.5.2 Applications

Damped Harmonic Oscillator

Let us start with the damped harmonic oscillator discussed in Sect. 5.3.4. By setting $H = \hbar\omega a^\dagger a$ and $L = a$ (5.196) turns into

$$\partial_t|\xi\rangle = -i\omega a^\dagger a|\xi\rangle + \gamma\left(\langle a^\dagger\rangle_\xi(a - \langle a\rangle_\xi) - \frac{1}{2}\left(a^\dagger a - \langle a^\dagger a\rangle_\xi\right)\right)|\xi\rangle. \quad (5.197)$$

Note that the first term of the non-unitary part vanishes if $|\xi\rangle$ is a coherent state, i.e., an eigenstate of a. This suggests the ansatz $|\xi\rangle = |\alpha\rangle$ which leads to

$$\partial_t|\alpha\rangle = \left[\left(-i\omega - \frac{\gamma}{2}\right)\alpha a^\dagger + \frac{\gamma}{2}|\alpha|^2\right]|\alpha\rangle. \quad (5.198)$$

It is easy to convince oneself that this equation is solved by

$$|\alpha_t\rangle = |\alpha_0 e^{-i\omega t - \gamma t/2}\rangle = e^{-|\alpha_t|^2/2}e^{\alpha_t a^\dagger}|0\rangle \quad (5.199)$$

with $\alpha_t = \alpha_0 \exp\left(-i\omega t - \gamma t/2\right)$. It shows that the predicted robust states are indeed given by the slowly decaying coherent states encountered in Sect. 5.3.4.

Quantum Brownian Motion

A second example is given by the Brownian motion of a quantum particle. The choice

$$H = \frac{p^2}{2m} \quad \text{and} \quad L = \frac{\sqrt{8\pi}}{\Lambda_{\text{th}}}x \quad (5.200)$$

yields a master equation of the form (5.117) but without the dissipation term. Inserting these operators into (5.196) leads to

$$\partial_t|\xi\rangle = \frac{p^2}{2mi\hbar}|\xi\rangle - \gamma\frac{4\pi}{\Lambda_{\text{th}}^2}[\underbrace{(x - \langle x\rangle_\xi)^2 - \langle(x - \langle x\rangle_\xi)^2\rangle_\xi}_{\sigma_\xi^2(x)}]|\xi\rangle. \quad (5.201)$$

The action of the non-unitary term is apparent in the position representation, $\xi(x) = \langle x|\xi\rangle$. At positions x which are distant from mean position $\langle x\rangle_\xi$ as compared to the dispersion $\sigma_\xi(x) = \langle(x - \langle x\rangle_\xi)^2\rangle_\xi^{1/2}$ the term is negative and the value $\xi(x)$ gets suppressed. Conversely, the part of the wave function close to the mean position gets enhanced,

$$\langle x|\xi\rangle = \begin{cases} \text{suppressed if } |x - \langle x\rangle_\xi| > \sigma_\xi(x) \\ \text{enhanced} \quad \text{if } |x - \langle x\rangle_\xi| < \sigma_\xi(x). \end{cases} \quad (5.202)$$

This localizing effect is countered by the first term in (5.201) which causes the dispersive broadening of the wave function. Since both effects compete we expect stationary, soliton-like solutions of the equation.

Indeed, a Gaussian ansatz for $|\xi\rangle$ with ballistic motion, i.e., $\langle \mathsf{p}\rangle_\xi = p_0$, $\langle \mathsf{x}\rangle_\xi = x_0 + p_0 t/m$, and a *fixed* width $\sigma_\xi(\mathsf{x}) = \sigma_0$ solves (5.201) provided [44]

$$\sigma_0^2 = \frac{1}{4\pi}\sqrt{\frac{k_B T}{2\hbar\gamma}}\Lambda_{\text{th}}^2 = \left(\frac{\hbar^3}{8\gamma m^2 k_B T}\right)^{1/2}, \qquad (5.203)$$

see (5.128). As an example, let us consider a dust particle with a mass of $10\,\mu\mathrm{g}$ in the interstellar medium interacting only with the microwave background of $T = 2.7\,\mathrm{K}$. Even if we take a very small relaxation rate of $\gamma = 1/(13.7\times10^9\,\mathrm{y})$, corresponding to the inverse age of the universe, the width of the solitonic wave packet describing the center of mass is as small as $2\,\mathrm{pm}$. This subatomic value demonstrates again the remarkable efficiency of the decoherence mechanism to induce classical behavior in the quantum state of macroscopic objects.

Acknowledgments

Many thanks to Álvaro Tejero Cantero who provided me with his notes, typed with the lovely TEX_MACS program during the lecture. The present text is based on his valuable input. I am also grateful to Marc Busse and Bassano Vacchini for helpful comments on the manuscript. This work was supported by the Emmy Noether program of the DFG.

References

1. E. Joos, H. D. Zeh, C. Kiefer, D. Giulini, J. Kupsch, and I.-O. Stamatescu: *Decoherence and the Appearance of a Classical World in Quantum Theory*, 2nd edn. (Springer, Berlin 2003)
2. H.-P. Breuer and F. Petruccione: *The Theory of Open Quantum Systems* (Oxford University Press, Oxford 2002)
3. Strunz, W.T.: Decoherence in quantum physics. In: Buchleitner, A., Hornberger, K. (eds.) *Coherent Evolution in Noisy Environments*, Lect. Notes Phys. **611**, Springer, Berlin (2002)
4. G. Bacciagaluppi: The role of decoherence in quantum mechanics. In: Stansford *The Stanford Encyclopedia of Philosophy*, (Stanford University, Stanford 2005) http://plato.stanford.edu.
5. M. Schlosshauer: *Decoherence, the measurement problem, and interpretations of quantum mechanics*, Rev. Mod. Phys. **76**, 1267–1305 (2004)
6. W. H. Zurek: *Decoherence, einselection, and the quantum origins of the classical*, Rev. Mod. Phys. **75**, 715–775 (2003)

7. J. P. Paz and W. H. Zurek: Environment-induced decoherence and the transition from quantum to classical. In: *Les Houches Summer School Series*, vol. 72, ed. by R. Kaiser, C. Westbrook, and F. David (Springer-Verlag, Berlin 2001) p. 533

8. A. Bassi and G. Ghirardi: *Dynamical reduction models*, Phys. Rep. **379**, 257 (2003)

9. K. Kraus: *States, Effects and Operations: Fundamental notions of Quantum Theory* (Springer, Berlin 1983)

10. P. Busch, P. J. Lahti, and P. Mittelstaed: *The Quantum Theory of Measurement* (Springer-Verlag, Berlin 1991)

11. A. S. Holevo: *Statistical Structure of Quantum Theory* (Springer, Berlin 2001)

12. C. W. Helstrom: *Quantum Detection and Estimation Theory* (Academic Press, New York 1976)

13. A. Chefles: *Quantum state discrimination*, Contemp. Phys. **41**, 401–424 (2000)

14. G. M. Palma, K.-A. Suominen, and A. K. Ekert: *Quantum computers and dissipation*, Proc. R. Soc. Lond. A **452**, 567 (1996)

15. D. F. Walls and G. J. Milburn: *Quantum Optics* (Springer, Berlin 1994)

16. M. Hillery, R. F. O'Connell, M. O. Scully, and E. P. Wigner: *Distribution functions in physics: Fundamentals*, Phys. Rep. **106**, 121–167 (1984)

17. U. Weiss: *Quantum Dissipative Systems*, 2nd edn. (World Scientific, Singapore 1999)

18. P. Machnikowski: *Change of decoherence scenario and appearance of localization due to reservoir anharmonicity*, Phys. Rev. Lett. **96**, 140405 (2006)

19. Y. Imry: *Elementary explanation of the inexistence of decoherence at zero temperature for systems with purely elastic scattering*, Arxiv preprint cond-mat/0202044 (2002)

20. R. Doll, M. Wubs, P. Hänggi, and S. Kohler: *Limitation of entanglement due to spatial qubit separation*, Europhys. Lett. **76**, 547–553 (2006)

21. E. B. Davies: *Quantum Theory of Open Systems* (Academic Press, London 1976)

22. H. Spohn: *Kinetic equations from Hamiltonian dynamics: Markovian limits*, Rev. Mod. Phys. **52**, 569–615 (1980)

23. R. Alicki and K. Lendi: *Quantum Dynamical Semigroups and Applications* (Springer, Berlin 1987)

24. F. Petruccione and B. Vacchini: *Quantum description of Einstein's Brownian motion*, Phys. Rev. E **71**, 046134 (2005)

25. H. Carmichael: *An Open Systems Approach to Quantum Optics* (Springer, Berlin 1993)

26. K. Mølmer, Y. Castin, and J. Dalibard: *Monte Carlo wave-function method in quantum optics*, J. Opt. Soc. Am. B **10**, 524–538 (1993)

27. M. B. Plenio and P. L. Knight: *The quantum-jump approach to dissipative dynamics in quantum optics*, Rev. Mod. Phys. **70**, 101–144 (1998)

28. J. M. Raimond, M. Brune, and S. Haroche, *Colloquium: Manipulating quantum entanglement with atoms and photons in a cavity*, Rev. Mod. Phys. **73**, 565–582 (2001)

29. S. Haroche: Mesoscopic superpositions and decoherence in quantum optics. In: *Quantum entanglement and information processing*, Les Houches 2003, ed. by D. Estève, J.-M. Raimond, and J. Dalibard (Elsevier, Amsterdam 2004)

30. A. O. Caldeira and A. J. Leggett: *Path integral approach to quantum Brownian motion*, Physica A **121**, 587–616 (1983)
31. W. G. Unruh and W. H. Zurek: *Reduction of a wave packet in quantum brownian motion*, Phys. Rev. D **40**, 1071–1094 (1989)
32. L. Diósi: *On high temperature Markovian equation for quantum Brownian motion*, Europhys. Lett. **22**, 1-3 (1993)
33. L. Lukacs: *Characteristic Functions* (Griffin, London 1966)
34. K. Hornberger: *Monitoring approach to open quantum dynamics using scattering theory*, Europhys. Lett. **77**, 50007 (2007).
35. J. R. Taylor: *Scattering Theory* (John Wiley & Sons, New York 1972)
36. K. Hornberger and J. E. Sipe: *Collisional decoherence reexamined*, Phys. Rev. A **68**, 012105 (2003)
37. K. Hornberger: *Master equation for a quantum particle in a gas*, Phys. Rev. Lett. **97**, 060601 (2006)
38. K. Hornberger and B. Vacchini: *Monitoring derivation of the quantum linear Boltzmann equation*, Phys. Rev. A **77**, 022112 (2009)
39. E. Joos and H. D. Zeh: *The emergence of classical properties through interaction with the environment*, Z. Phys. B: Condens. Matter **59**, 223–243 (1985)
40. K. Hornberger, S. Uttenthaler, B. Brezger, L. Hackermüller, M. Arndt, and A. Zeilinger: *Collisional decoherence observed in matter wave interferometry*, Phys. Rev. Lett. **90**, 160401 (2003)
41. L. Hackermüller, K. Hornberger, B. Brezger, A. Zeilinger, and M. Arndt: *Decoherence of matter waves by thermal emission of radiation*, Nature **427**, 711–714 (2004)
42. R. Dümcke: *The low density limit for an N-level system interacting with a free bose or fermi gas*, Commun. Math. Phys. **97**, 331–359 (1985)
43. W. H. Zurek: *Pointer basis of quantum apparatus: Into what mixture does the wave packet collapse?*, Phys. Rev. D **24**, 1516–1525 (1981)
44. L. Diósi and C. Kiefer: *Robustness and diffusion of pointer states*, Phys. Rev. Lett. **85**, 3552–3555 (2000)
45. W. H. Zurek, S. Habib, and J. P. Paz: *Coherent states via decoherence*, Phys. Rev. Lett. **70**, 1187–1190 (1993)
46. N. Gisin and M. Rigo: *Relevant and irrelevant nonlinear Schrödinger equations*, J. Phys. A: Math. Gen. **28**, 7375–7390 (1995)

6 Diffusive Spin Transport

C.A. Müller

Physikalisches Institut, Universität Bayreuth, D-95440 Bayreuth, Germany

6.1 Introduction

Classical information processing uses charge encoding where the bit values
"0" and "1" are represented by a supplementary charge present or absent in
a register. In the quantum limit, one eventually is led to consider a single
elementary charge (electron or hole) in a quantum dot and hopes to realise
quantum superpositions of "qubit" states $|0\rangle$ and $|1\rangle$ and entanglement be-
tween distinct qubits. But since they interact via long-range Coulomb forces,
charge states suffer strongly from decoherence. Another discrete degree of
freedom is spin. Spin $\frac{1}{2}$ states, typically noted $|\pm 1\rangle$ or $|\uparrow, \downarrow\rangle$, are *the* natu-
ral realisation of a qubit. But spin, and more generally information as such,
needs a physical carrier. Candidates here are electrons or holes (massive particle of spin $s = \frac{1}{2}$) and photons (massless, spin $s = 1$). A new promising field
therefore is "spint(r)onics": spin-based information transport and processing
with elec*trons* and pho*tons*.

For these lectures, I chose to discuss the following model setting of diffusive spin transport (Fig. 6.1): spin-polarised particles are injected from the
left with probability $p_\uparrow(0) = 1$ into a disordered sample and move diffusively
towards the right, where a spin-sensitive detection reads out the final spin
polarisation $p_\uparrow(L)$ that we should calculate.

The leitmotiv of this lecture was as follows: Spin is a geometrical quantity,
and one should be able to use irreducible representations of the rotation group

$p_\uparrow(0) = 1$ $\qquad\qquad\qquad\qquad\qquad\qquad$ $p_\uparrow(L) = ?$

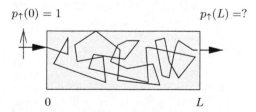

$0 \qquad\qquad\qquad\qquad\qquad\qquad\qquad L$

Fig. 6.1. Model setting of diffusive spin transport: spin-polarised particles are
injected from the left with $p_\uparrow(0) = 1$ into a disordered sample and move diffusively
towards the right, where a spin-sensitive detection reads out the spin-polarisation
$p_\uparrow(L)$

Müller, C.A.: *Diffusive Spin Transport*. Lect. Notes Phys. **768**, 277–314 (2009)
DOI 10.1007/978-3-540-88169-8_6 $\qquad\qquad$ © Springer-Verlag Berlin Heidelberg 2009

in order to take advantage of symmetries. Since the participants of the school were not required to have been educated in group theory, I decided to give a rather complete introduction into representations of the rotation group with the hope that the sometimes intimidating jargon of group-theoretical arguments may become more familiar to everybody. Attending myself other lectures, I adopted a master-equation approach parallelling Klaus Hornberger's lecture to which the present notes refer occasionally. During the lecture itself, only the first part on spin relaxation was delivered as presented in the following; in the second part on diffusion, I relied on arguments taken from diagrammatic perturbation theory that seemed to bewilder the audience more than anything else. The present notes remedy to this dissymmetry and treat also the diffusive part as momentum relaxation with a master equation. This parallel allows to combine both dynamics rather economically into a single, coherent picture of diffusive spin dynamics; it is my hope that this conceptual unity is appreciated by my readers.

These notes finish with a description of quantum corrections to diffusive spin transport including spin-flip effects. Other subjects covered during the school's lecture were of a more anecdotic type and, although hopefully enjoyed by the audience, did not seem to fit into the present format; readers interested in spintronics properly speaking are referred to the recent review by Žutić, Fabian, and Das Sarma [1]. I have to admit that I made no attempt to cover systematically the vast literature on the subject of irreversible spin dynamics and quantum transport, which would have been a hopeless task in any case; my apologies to many colleagues whose excellent contributions may not be duly cited in the following.

6.2 Spin Relaxation

"Spin" is internal angular momentum [2, 3]. This was recognised by Uhlenbeck and Goudsmit [4] following Pauli [5] who postulated the existence of a fourth quantum number to explain fine-structure features of atomic spectra. The clearest experimental manifestation of quantised spin is arguably the Stern–Gerlach experiment [6], where silver atoms are deviated by an inhomogeneous magnetic field into two distinct spots on a detector screen.

Dirac discovered that bispinors (vectors of four components, a spin $\frac{1}{2}$ spinor and an anti-spinor) appear naturally when one looks for a Schrödinger-type wave equation in the relativistic framework of the four-dimensional Minkowski space. Wigner [7] showed that spin is one of the fundamental quantum numbers that permits to identify an elementary particle in the first place: it characterises the particle's properties under rotations in its proper rest frame. Therefore, to understand spin is to understand rotations.

6.2.1 Spin – A Primer on Rotations

Rotation Group

Physical objects are described by coordinates $\boldsymbol{x} = (x_1, x_2, x_3) \in \mathbb{R}^3$ with respect to a reference frame in configuration space. Rotations (called "active" when the object is turned and "passive" when the reference frame is turned) are represented by 3×3 matrices: $\boldsymbol{x}' = R\boldsymbol{x}$. Proper rotations conserve the Euclidean scalar product $\boldsymbol{x} \cdot \boldsymbol{y} = \sum_i x_i y_i$ and the orientation of the frame. The rotation matrices are therefore members of SO(3), the set of orthogonal matrices $RR^{\mathrm{t}} = R^{\mathrm{t}}R = \mathbb{1}_3$ of unit determinant $\det R = +1$. With the usual matrix multiplication as an internal composition law, theses matrices form a *group*, satisfying the group axioms:

1. Internal composition: $\forall R_{1,2} \in \mathrm{SO}(3): R_{21} = R_2 R_1 \in \mathrm{SO}(3)$;
2. Existence of the identity: $\exists E: RE = ER = R\ \forall R \in \mathrm{SO}(3)$ with $E = \mathbb{1}_3$;
3. Existence of the inverse: $\forall R,\ \exists R^{-1}: RR^{-1} = R^{-1}R = E$.

The group is *non-Abelian* because the matrices do not commute: $R_2 R_1 \neq R_1 R_2$. An exception are rotations of the plane around one and the same axis, forming the Abelian group SO(2).

A possible parametrisation of a rotation is the polar description $(\hat{\boldsymbol{n}}, \theta) =: \boldsymbol{\theta}$ with $\hat{\boldsymbol{n}}$ the unit vector along the rotation axis and $\theta \in [0, \pi]$ the rotation angle. The following two rotations of configuration space are identical:

$$R(\hat{\boldsymbol{n}}, \theta = \pi) = R(-\hat{\boldsymbol{n}}, \theta = \pi). \tag{6.1}$$

These opposite points must be identified such that there are closed parameter curves that cannot be contracted into a single point, see Fig. 6.2. This means that SO(3) is a doubly connected manifold. Instead of studying this projective group, one may also turn to its *universal covering group* SU(2), the group of all unitary 2×2 matrices over $\mathbb{C}$ with unit determinant. SU(2) is simply

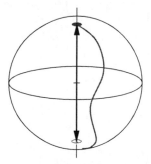

Fig. 6.2. The parameter space of the rotation group SO(3), a filled sphere of radius π. Two identical rotations (see (6.1)) can be connected by a closed curve that is not reducible to a single point

connected, and there is a *homomorphism* (a mapping preserving the group structure) linking every element $U \in \mathrm{SU}(2)$ to the rotation $R \in \mathrm{SO}(3)$: the rotation $\boldsymbol{x}' = R(\boldsymbol{\theta})\boldsymbol{x}$ is described by

$$\boldsymbol{x}' \cdot \boldsymbol{\sigma} = U(\boldsymbol{\theta})\boldsymbol{x} \cdot \boldsymbol{\sigma} U(\boldsymbol{\theta})^\dagger . \tag{6.2}$$

Here, $\boldsymbol{\sigma} = (\sigma_1, \sigma_2, \sigma_3)$ is a vector whose entries σ_i are the Pauli matrices such that

$$\boldsymbol{x} \cdot \boldsymbol{\sigma} = \begin{pmatrix} x_3 & x_1 - \mathrm{i}x_2 \\ x_1 + \mathrm{i}x_2 & -x_3 \end{pmatrix} \tag{6.3}$$

and the unitary rotation matrix acting from the left and from the right is given by

$$U(\boldsymbol{\theta}) = \mathbb{1}_2 \cos\frac{\theta}{2} - \mathrm{i}\hat{n} \cdot \boldsymbol{\sigma} \sin\frac{\theta}{2} = \exp(-\mathrm{i}\frac{\boldsymbol{\theta} \cdot \boldsymbol{\sigma}}{2}). \tag{6.4}$$

Note that this mapping is two-to-one because U and $-U$ yield the same rotation, and that it takes a rotation by an angle of 4π to recover the identity transformation: $U(\theta = 2\pi) = -\mathbb{1}_2$, but $U(\theta = 4\pi) = +\mathbb{1}_2$. This is not a mysterious quantum property as is sometimes stated, but reflects the double-connectedness of $\mathrm{SO}(3)$-rotations in our everyday reference frame.[1] Dirac's construction of a solid body connected by strings to a reference frame is supposed to convey an "experimental" idea of this property [3].

Representations

A group G can act in many different disguises that share the same abstract group structure, as defined by the multiplication law or group table. These different appearances are called (linear) *representations*. Mathematically, they are mappings $\mathcal{D}_i : G \to GL(V_i)$ from the group G to the general (linear) group of regular transformations $D : V_i \to V_i$ of a vector space V_i into itself.[2] Importantly, this mapping must be a *homomorphism* which means that the representation has the same group structure as G. Notably, for all elements $g_1, g_2 \in G$ (with the product $g_2 g_1 \in G$), the representing transformations verify $D(g_2 g_1) = D(g_2)D(g_1)$.

The dimension of the representation $\mathcal{D}_i$ is given by $\dim V_i$. Finite-dimensional linear representations are given in terms of quadratic matrices of size $\dim V_i$. A representation is called *reducible* if there is a basis of $V = V_1 \oplus V_2$ such that all transformations $D \in \mathcal{D}$ are written

$$D = \begin{pmatrix} D_1 & * \\ 0 & D_2 \end{pmatrix}. \tag{6.5}$$

---

[1] Cartan developed half-integer spin representations as early as 1913 within the theory of projective groups [8].

[2] $\mathcal{D}$ stands for the German word "Darstellung"; in anglo-saxon literature, often the symbol Γ is used.

In other words, the transformations $D_1 : V_1 \to V_1$ define already a representation $\mathcal{D}_1$ of its own. If the matrix is block-diagonal ($* = 0$), the representation is *completely reducible*. If it is not reducible, it is called *irreducible*, meaning that one has achieved to work in the smallest possible subspace.

The rotation group SU(2) is a manifold that depends on a continuous set of parameters $\boldsymbol{\theta} = (\theta_1, \theta_2, \theta_3)$ with respect to which it is infinitely differentiable. This structure is called a *Lie group*. The group multiplication law is completely determined by the commutation relation of its generators:

$$[J_j, J_k] := J_j J_k - J_k J_j = i\hbar \epsilon_{jkl} J_l \,. \tag{6.6}$$

Here, as in the following, the sum over repeated indices is understood. These generators are said to form a (representation of) the *Lie algebra* su(2). In the so-called natural representation of SU(2) by itself (see (6.4)), the generators are $\boldsymbol{J} = \hbar \frac{\boldsymbol{\sigma}}{2}$. Other representations will feature different generators, but all group representations share the same commutation relation! Finite rotations are generated by exponentiation: $U(\boldsymbol{\theta}) = \exp(-i\boldsymbol{\theta} \cdot \boldsymbol{J}/\hbar)$. Topologically speaking, SU(2) is compact. According to a general theorem, all representations of a compact Lie group are completely reducible to finite-dimensional irreducible representations.

Functional Representation of SO(3)

As an example for an infinite-dimensional representation of the rotation group SO(3) already useful in classical physics, consider the transformation of functions f describing the position of an object on the unit sphere. Saying that the object is rotated to $\boldsymbol{r}' = R\boldsymbol{r}$ implies that the function is transformed by

$$(Df)(\boldsymbol{r}) = f(R^{-1}\boldsymbol{r}) \,. \tag{6.7}$$

The action of D in the (infinite-dimensional) functional space can be written $D = \exp\{-i\boldsymbol{\theta} \cdot \boldsymbol{T}\}$, where $\boldsymbol{T} = -i\boldsymbol{r} \times \boldsymbol{\nabla}$ is a differential operator, as can be verified by considering an infinitesimal rotation around $\boldsymbol{\theta} = \theta\hat{\boldsymbol{n}}$: to first order in θ, we have $f(R^{-1}\boldsymbol{r}) = f(\boldsymbol{r} - \boldsymbol{\theta} \times \boldsymbol{r}) = f(\boldsymbol{r}) - (\boldsymbol{\theta} \times \boldsymbol{r}) \cdot \boldsymbol{\nabla} f(\boldsymbol{r}) = [1 - \boldsymbol{\theta} \cdot (\boldsymbol{r} \times \boldsymbol{\nabla})]f(\boldsymbol{r})$.

The finite-dimensional irreducible representations to which this infinite-dimensional one can be reduced are obtained by decomposing f into surface harmonics Y_{Lm}; each subspace $L = 0, 1, \ldots$ then admits an irreducible representation of dimension $2L+1$. Surface harmonics are a concept arising already with the multipole expansion of charge distributions in classical electrodynamics. But to cite Hermann Weyl [9]: "This reveals the true significance of surface harmonics; they are characterised by the fundamental symmetry properties here developed, and the solution of the potential equation in polar co-ordinates is merely an accidental approach to their theory."

What Is Quantum About Spin?

Spin, and especially half-integer spin, is much less mysterious than sometimes suggested by standard textbook wisdom. Half-integer spin does not require the kinematic framework of special relativity, it arises already in Galilean relativity. Also, as mentioned at the end of Sect. 6.2.1, reference frames for solid bodies already introduce half-integer spin. However, there are genuine quantum features to be aware of:

1. In classical mechanics, Noether's theorem assures that if the Lagrangian function is invariant under infinitesimal rotations, then the orbital angular momentum L is a conserved quantity. However, L a priori has nothing to do with the generator of rotations T introduced in the previous section. In quantum mechanics, thanks to the appearance of $\hbar$, the generators can be *identified* with *observables* $J = \hbar T$ with dimension of angular momentum. If the Hamiltonian is invariant under all rotations $U(\boldsymbol{\theta})$, their generators itself are conserved quantities, and Noether's theorem takes a very simple form:

$$H' = UHU^\dagger \Leftrightarrow [H, U] = 0 \Leftrightarrow [H, J] = 0 \Leftrightarrow \dot{J} = 0. \qquad (6.8)$$

Separating the orbital part L of angular momentum from the total angular momentum $J = L \otimes \mathbb{1}_S + \mathbb{1}_L \otimes S =: L + S$, one identifies the rest-frame angular momentum or spin S obeying the same fundamental commutation relations (6.6). In the remainder, we will only have to consider the spin part and let $L = 0$.

2. The observables S generate irreducible representations $\mathcal{D}^{(s)}$ of dimension $d_s = 2s + 1$ with $s = 0, \frac{1}{2}, \ldots$ and discrete magnetic quantum numbers $m = -s, -s+1, \ldots, s$. Pure states are noted $|sm\rangle$. The *Casimir operator* S^2 specifies the irreducible representation, $S^2|sm\rangle = \hbar^2 s(s+1)|sm\rangle$, whereas the magnetic quantum number gives the projection of the spin onto the quantisation axis (usually called the z-axis): $S_z|sm\rangle = \hbar m|sm\rangle$.

3. States can be classified regarding their transformation properties, but here with a more general importance than in classical mechanics due to the superposition principle. An atomic s-orbital, for instance, may be seen as a "coherent superposition of all possible Kepler orbits" and is invariant under all rotations. In Sect. 6.2.3, we will introduce the irreducible components of mixed quantum states, also known as *state multipoles*.

6.2.2 Master Equation Approach to Spin Relaxation

As a simple model for spin dynamics, we shall study the Hamiltonian

$$H = -\mu S \cdot B. \qquad (6.9)$$

It describes the coupling of the magnetic moment $\boldsymbol{\mu} = \mu S$ to a magnetic field B. For electrons, $\mu = -g\mu_B$ in terms of the Bohr magneton $\mu_B =$

$|e|\hbar/(2m_e c)$ and the gyromagnetic ratio $g = 2.003\ldots$ in vacuum. The spin operator has been chosen dimensionless such that its action in the irreducible representation $\mathcal{D}^{(s)}$ will be $S_z|sm\rangle = m|sm\rangle$ and $S^2|sm\rangle = s(s+1)|sm\rangle$ for the remainder of the lecture. The density matrix or statistical operator ρ of a spin S is a positive linear operator of trace unity on the state Hilbert space $\mathcal{H}_s = \mathbb{C}^{d_s}$ of dimension $d_s := \dim \mathcal{H}_s = 2s + 1$ that determines the expectation values of abitrary observables O as $\langle O \rangle = \text{tr}\{\rho O\}$.

Unitary Spin Dynamics

According to one of the fundamental axioms of quantum theory, any closed quantum system evolves unitarily according to the Liouville–von Neumann equation

$$i\hbar\partial_t\rho = [H, \rho]\,. \tag{6.10}$$

This equation of motion is formally solved as $\rho(t_2) = U(t_2, t_1)\rho(t_1)U(t_2, t_1)^\dagger$ by applying the time evolution operator for a time-dependent Hamiltonian,

$$U(t_2, t_1) = \text{T}\exp\left\{-\frac{i}{\hbar}\int_{t_1}^{t_2} H(t')\mathrm{d}t'\right\}\,, \tag{6.11}$$

where $\text{T}[H(t_1)H(t_2)\ldots H(t_n)] = H(t_i)H(t_j)\ldots H(t_k)$ for $t_i > t_j > \cdots > t_k$ is Dyson's time-ordering operation.

Non-unitary Spin Dynamics: A Classical Model Derivation

Phenomena like "relaxation, damping, dephasing, decoherence,..." have in common irreversible dynamics with an "arrow of time" [10] due to the irrevocable loss of energy and/or information into inobservable degrees of freedom, which are usually called "bath" or "environment".

As an introductory model, we consider the Hamiltonian $H = -\mu\boldsymbol{S}\cdot\boldsymbol{B}(t)$ with a randomly fluctuating magnetic field $\boldsymbol{B}(t)$. Predictions about the spin will involve an average over the field fluctuations, which we describe as a *classical* stochastic process [11]. This approach is typical for the physics of nuclear magnetic resonance; a regular driving magnetic field can of course be included in the treatment, but here we concentrate on the effect of random fluctuations.

To obtain the effective dynamics, we develop the time-propagated density matrix $\rho(t + \Delta t) = U(t + \Delta t, t)\rho(t)U(t + \Delta t, t)^\dagger$ to second order in the interaction Hamiltonian,

$$\rho(t + \Delta t) = \rho(t) - \frac{i}{\hbar}\int_t^{t+\Delta t} [H(t_1), \rho(t)]\mathrm{d}t_1 + \frac{1}{\hbar^2}\iint_t^{t+\Delta t} H(t_1)\rho(t)H(t_2)\mathrm{d}t_1\mathrm{d}t_2$$

$$- \frac{1}{2\hbar^2}\iint_t^{t+\Delta t} \text{T}(H(t_1)H(t_2)\rho(t) + \rho(t)H(t_1)H(t_2))\mathrm{d}t_1\mathrm{d}t_2\,, \tag{6.12}$$

and perform an average over all possible realisations of the fluctuating magnetic field $\boldsymbol{B}(t)$. As a stochastic process [11], it is completely specified by its correlation functions

$$C_{i_1 i_2 \ldots i_n}(t_1, t_2, \ldots, t_n) := \overline{B_{i_1}(t_1) B_{i_2}(t_2) \ldots B_{i_n}(t_n)}, \qquad (6.13)$$

the overline indicating an ensemble average over the field distribution. This distribution is taken to be centred Gaussian hence all correlation functions factorise into products of pair correlations. Therefore, only the first two moments need to be specified: $\overline{B_i(t)} = 0$ (zero mean) and

$$\overline{B_i(t_1) B_j(t_2)} =: B^2 c_{ij}(t_1, t_2). \qquad (6.14)$$

We assume a stationary process that depends only on the time difference $t_1 - t_2$ and has a very short internal correlation time τ_c such that $c_{ij}(t) = c_{ij} \tau_c \delta(t)$. This last assumption of "white noise" (the power spectrum $\tilde{c}_{ij}(\omega) = cst.$ contains all frequencies with equal weight) is valid if the noise correlation time τ_c is much shorter than the relevant timescale of the spin dynamics that is still to be determined. Lastly, we assume that the fluctuations are *isotropic*, $c_{ij} = \frac{1}{3} \delta_{ij}$.

Now we average the time-propagated density matrix (6.12) over the field fluctuations and use the *Born assumption* that there are no further correlations between the fields appearing explicitly and the average density matrix. Then, the ensemble average applies to the fields only. The term linear in H disappears because $\overline{B_i} = 0$. In the second-order terms, one of the time integrations is contracted by the $\delta(t_1 - t_2)$-distribution of the correlation function; the remaining integrand is time independent such that the integral gives just a factor Δt. The average time-evolved density matrix then becomes

$$\rho(t + \Delta t) = (1 - \gamma_s \Delta t) \rho(t) + \gamma_s \Delta t \sum_i \hat{S}_i \rho(t) \hat{S}_i + O((\gamma_s \Delta t)^2), \qquad (6.15)$$

where $\hat{S}_i := S_i / \sqrt{s(s+1)}$ is the "normalised spin operator" with $\sum_i \hat{S}_i^2 = \mathbb{1}$. The spin relaxation rate $\gamma_s := s(s+1) \omega_0^2 \tau_c$ is given in terms of the squared effective Larmor frequency $\omega_0^2 = \mu^2 B^2 / (3\hbar^2)$. The relevant timescale of evolution turns out to be $\tau_s := 1/\gamma_s$. The effective time evolution (6.15) is then valid for a small time step $\Delta t \ll \tau_s$ such that indeed $\gamma_s \Delta t \ll 1$.

Exercise 1 (Non-unitary spin dynamics: quantum derivation)

Consider the time-independent model Hamiltonian $H = -\hbar J \boldsymbol{S} \cdot \boldsymbol{\tau}$, where our spin $\boldsymbol{S}$ is coupled to a freely orientable magnetic impurity, here modelled as a spin $\frac{1}{2}$ with Pauli matrices $\boldsymbol{\tau}$. The effective spin dynamics of $\boldsymbol{S}$ is described by its reduced density matrix $\rho(t) = \mathrm{tr}_\tau \{\rho_{S\tau}(t)\}$ obtained by tracing out the uncontrolled impurity spin. Develop the time-evolved complete density matrix $\rho_{S\tau}(t)$ to second order in J as in Sect. 6.2.2 and take the trace over the impurity with initial statistical mixture $\rho_\tau(0) = \frac{1}{2} \mathbb{1}_2$ (it is helpful to use the identities $\text{tr}_\tau\{\tau_i\} = 0$ and $\text{tr}_\tau\{\tau_i\tau_j\} = 2\delta_{ij})$. Show that one finds exactly the evolution (6.15) with a relaxation rate $\gamma_s = s(s+1)J^2\Delta t$ up to higher-order terms that become negligible in the formal limit $\Delta t \to 0$ together with $J \to \infty$ taken at constant γ_s.[3]

The Quantum Channel and Its Operator Sum Representation

In the language of quantum information, the time evolution (6.15) up to order Δt defines a *quantum channel* $\rho = \rho(t) \mapsto \rho' = \rho(t + \Delta t)$ and is here given in the so-called *operator sum representation* (see [12] Sect. 8.2.3.)

$$\rho' = \sum_{i=0}^{3} W_i \rho W_i^\dagger \tag{6.16}$$

with

$$W_0 := \sqrt{1 - \gamma_s \Delta t}\,\mathbb{1}, \qquad W_i := \sqrt{\gamma_s \Delta t}\,\hat{S}_i, \quad i = 1, 2, 3. \tag{6.17}$$

It is easy to verify that $\sum_{i=0}^{3} W_i^\dagger W_i = \mathbb{1}$, which guarantees the trace conservation $\text{tr}\rho' = \text{tr}\rho$. Kraus has proved that a channel of this form assures that the final density matrix is again completely positive. Therefore, it is also known as the *Kraus representation*, and the W_i are commonly referred to as *Kraus operators*, cf. Sect. 5.3.1. In contrast to the unitary evolution $\rho' = U\rho U^\dagger$ of (6.10), the appearance of several independent terms in the sum (6.16) signals non-unitary dynamics.

For a spin $\frac{1}{2}$ with $\hat{S}_i = \sigma_i/\sqrt{3}$, this quantum channel is the *qubit depolarising channel* (see [12] Sect. 8.3.4.):

$$\rho' = (1 - p_1)\rho + \frac{p_1}{3}\sum_i \sigma_i\rho\sigma_i, \qquad p_1 = \gamma_s\Delta t. \tag{6.18}$$

With equal probability $p_1/3$, the qubit is affected by the action of one of the Pauli matrices σ_i, and with probability $1 - p_1$, it remains untouched. Since for spin $\frac{1}{2}$ one may write $\sum_i \sigma_i\rho\sigma_i = 2\mathbb{1}_2 - \rho$, the channel also takes the suggestive form

$$\rho' = (1 - p_2)\rho + \frac{p_2}{2}\mathbb{1}_2. \tag{6.19}$$

This means that with probability $p_2 = 4p_1/3$ (remember $p_1 \ll 1$ such that also $p_2 \ll 1$), the density matrix is taken to a complete statistical mixture and remains identical with probability $(1 - p_2)$.

---

[3] But attention: at finite coupling J, the dynamics of our spin shows *recurrence* on a timescale given by the so-called Poincaré time $t_{\text{rec}} \propto 1/J$. One could obtain a truly irreversible dynamics only by supposing that the single impurity spin is reset rapidly enough in order to dispose the coherence. Alternatively, one may imagine the setting treated in Sect. 6.4.1: our spin is moving and encounters different impurity spins such that in the thermodynamic limit, the Poincaré recurrence time goes to infinity and true irreversibility sets in.

By convention, the "depolarising channel" for higher spin $S \geq 1$ (i.e., $d_s \times d_s$ density matrices with $d_s = 2s + 1 \geq 3$) is still defined via the relation (6.19) with the statistical mixture $\frac{1}{d_s}\mathbb{1}_{d_s}$ on the right-hand side. This channel is also called "SU(n) channel" with $n = d_s$ because the corresponding Kraus operators are the $n^2 - 1$ generators of the Lie algebra $\mathsf{su}(n)$ [13]. Note that our physical model of an arbitrary spin S coupled to a fluctuating magnetic field does *not* lead to this specific Lie algebra channel: obviously, the operator-sum representation (6.16) contains only the three generators of $\mathsf{su}(2)$, albeit in a representation of dimension $d_s = 2s + 1$. We will see in the following that this makes the spin dynamics richer and its description more involved. Group-theoretical methods will be introduced that are well adapted to cope with this complexity.

The Liouvillian

The linear operators on the state Hilbert space $\mathcal{H}_s$ are themselves elements of a linear vector space (we can add operators and multiply them by complex numbers). This vector space is called *Liouville space* $L(\mathcal{H}_s)$ and is spanned, for example, by the basis of *dyadics* induced by basis vectors $|n\rangle$ of $\mathcal{H}_s$:

$$|m\rangle\langle n| =: |mn), \quad n, m = 1, \ldots, d_s. \tag{6.20}$$

The Liouville–von Neumann equation of motion $i\partial_t \rho = \mathcal{L}\rho$ for a closed quantum system defines the *Liouvillian*

$$\mathcal{L} = \frac{1}{\hbar}[H, \cdot], \tag{6.21}$$

whose matrix elements in the dyadic basis are

$$(mn|\mathcal{L}|m'n') = \mathcal{L}_{mn,m'n'} = H_{mm'}\delta_{nn'} - H_{n'n}\delta_{mm'}. \tag{6.22}$$

Exercise 2 (Liouvillian eigenvalues)

Show that the eigenvalues of the Liouvillian in the basis $\{|mn)\}$ induced by the energy basis $H|m\rangle = \varepsilon_m|m\rangle$ are the possible transition frequencies $\omega_{mn} = (\varepsilon_m - \varepsilon_n)/\hbar$. These are experimentally accessible quantities, in contrast to the absolute energy eigenvalues of the Hamiltonian H.

As an operator between operators, the Liouvillian $\mathcal{L} : L(\mathcal{H}_s) \to L(\mathcal{H}_s)$ is called a *superoperator* [14–17]. The superoperator formalism is a convenient starting point for projection operator techniques in statistical dynamics [15], effective dynamics of open quantum systems [18] and time-dependent perturbation theory [19].

The Lindbladian

The effective dynamics of our spin coupled to a randomly fluctuating field can also be formulated in terms of a superoperator. Taking the formal limit $\lim_{\Delta t \to 0} \frac{\rho(t + \Delta t) - \rho(t)}{\Delta t} =: \partial_t \rho(t)$ in (6.15) leads to the *master equation*

$$\partial_t \rho(t) = \overline{\mathcal{L}}\rho(t), \tag{6.23}$$

a linear equation of motion for the effective density operator, whose effective generator of time evolution is the *Lindbladian* [17]

$$\overline{\mathcal{L}}\rho(t) = -\frac{\gamma_s}{2} \sum_i [\hat{S}_i, [\hat{S}_i, \rho(t)]] = \frac{\gamma_s}{2} \sum_i \left(2\hat{S}_i \rho(t)\hat{S}_i - \hat{S}_i \hat{S}_i \rho(t) - \rho(t)\hat{S}_i \hat{S}_i \right).$$
$$\tag{6.24}$$

This is a pure relaxation superoperator in the Lindblad form, which assures the complete positivity of the time-evolved density matrix (see Sect. 5.3.2). The symbolic limit $\Delta t \to 0$ really means $\Delta t \gamma_s \ll 1$ but is still assumed to be "coarse-grained" compared to the field correlations, $\Delta t \gg \tau_c$. In this limit, the master equation (6.23) is a linear equation for the density matrix $\rho(t)$ local in time t and thus describes Markovian dynamics without any memory effects.

Exercise 3 (Lindbladian matrix elements)
Show that the superoperator matrix elements of the Lindbladian (6.24) are given by

$$\overline{\mathcal{L}}_{mn,m'n'} = \gamma_s \left(\hat{\boldsymbol{S}}_{mm'} \cdot \hat{\boldsymbol{S}}_{n'n} - \delta_{mm'}\delta_{nn'} \right), \tag{6.25}$$

and verify the trace-preserving property $\sum_m \overline{\mathcal{L}}_{mm,m'n'} = 0$ from this expression.

6.2.3 Irreducible Scalar Spin Relaxation Rates

Formally, the solution of the master equation (6.23) is very simple:

$$\rho(t) = \exp[\overline{\mathcal{L}}t]\rho(0). \tag{6.26}$$

The dynamics induced by the Lindbladian is called a "quantum dynamic semigroup", cf. Sect. 5.3.1. Indeed, the time evolution superoperator satisfies $\exp[\overline{\mathcal{L}}(t_2 + t_1)] = \exp[\overline{\mathcal{L}}t_2] \exp[\overline{\mathcal{L}}t_1]$ and $\exp[\overline{\mathcal{L}}0] = \mathbb{1}$, which indicates a group structure. "Semigroup" means that the inverse to each element does not need to exist, and indeed here it does not since the non-unitary dynamics obtained by tracing out the environment has induced an arrow of time. The Kraus representation (6.16) and the Lindblad form (6.24) guarantee the complete positivity of the final density matrix if the initial one is completely positive, but the inverse is not true: the Lindbladian is not invertible (see [17] Sect. 3.4.1). We will indeed see in Sect. 6.2.3 that $\overline{\mathcal{L}}$ has one vanishing eigenvalue.

In Liouville space, the master equation is a matrix equation $\partial_t |\rho) = \overline{\mathcal{L}}|\rho)$ of dimension $d_s^2 \times d_s^2$. As always when dealing with matrix equations, we have to diagonalise the Lindbladian in Liouville space in order to be able to use the formal solution (6.26). For spin $\frac{1}{2}$ and $d_s = 2$, diagonalising a 4×4 matix is elementary and can be done by hand, but already for spin 1 with $d_s^2 = 9$, this becomes cumbersome. For higher spin, one definitely needs to resort to an efficient strategy to find the eigenstructure of the Lindbladian. In the following, we show how to take maximal advantage of rotational symmetries by using group-theoretical methods that lead to a very simple and physically transparent description for the effective spin dynamics.

Scalar Relaxation Process: What Results We Should Expect

The Lindbladian was obtained by an isotropic average and is thus a *scalar* object, i.e., invariant under rotations. A rather high-brow way of expressing this simple property is to say "it transforms under the trivial representation $\mathcal{D}^{(0)}$". We may anticipate that the statistical operator can be decomposed into parts that transform under the irreducible representations $\mathcal{D}^{(K)}$ of the rotation group, $\rho = \sum_K \rho^{(K)}$. The Lindbladian as a scalar object can only connect subspaces of equal rank K. Furthermore, inside each subspace, it cannot distinguish between different orientations. Thus, in an adapted basis of Liouville space, it can be written as a purely diagonal matrix

$$\overline{\mathcal{L}} = \begin{pmatrix} \lambda_0 & 0 & \cdots & 0 \\ 0 & \lambda_1 & \cdots & 0 \\ \vdots & \vdots & \ddots & \vdots \\ 0 & 0 & \cdots & \lambda_{d_s^2} \end{pmatrix}. \tag{6.27}$$

How many *different* eigenvalues may we expect? The total number is the dimension of the Liouville space, d_s^2. Each subspace of rank K will have dimension $d_K = 2K + 1$. Therefore, we will have to find only $2s + 1$ different eigenvalues λ_K, each of which has degeneracy $2K + 1$:

particle	s	d_s	d_s^2	eigenvalues	degeneracy
electron	$\frac{1}{2}$	2	4	λ_0, λ_1	1, 3
photon	1	3	9	$\lambda_0, \lambda_1, \lambda_2$	1, 3, 5

Can some of the eigenvalues be identical which would imply an even larger degeneracy? A trivial example for this would be any operator proportional to the identity. But we will see below that this is not the case for the Lindbladian (6.24): the eigenvalues pertaining to different subspaces $K \neq K'$ are indeed different, $\lambda_K \neq \lambda_{K'}$. We are therefore sure to have reduced the problem to the simplest possible formulation.

Irreducible Tensor Operators

Before we can define irreducible superoperators, we first had better understand the simpler concept of ordinary irreducible operators. An irreducible tensor operator of rank K is, by definition, a set of $2K + 1$ components T_Q^K,

$Q = -K, -K+1, \ldots, K$, that transform under irreducible representations of rank K (i.e., whose transformation does not mix different K):

$$(T_Q^K)' = U T_Q^K U^\dagger = \sum_{Q=-K}^{K} D_{QQ'}^{(K)} T_{Q'}^K . \tag{6.28}$$

Equivalently, one specifies the infinitesimal rotation properties by requiring

$$[J_\pm, T_Q^K] = \hbar\sqrt{K(K+1) - Q(Q\pm 1)} T_{Q\pm 1}^{(K)}, \tag{6.29}$$

$$[J_0, T_Q^K] = \hbar Q \, T_Q^K . \tag{6.30}$$

Here, the angular momentum raising and lowering operators are $J_\pm = J_x \pm i J_y$. The simplest examples for irreducible tensor operators we need to know for the following are

- $K = 0$ or scalar operator T_0^0, a single operator that commutes with all components of the total angular momentum $\boldsymbol{J}$. For instance S^2, the Casimir operator indexing the irreducible representations $\mathcal{D}^{(s)}$.
- $K = 1$ or vector operator with Euclidean components $\boldsymbol{A} = (A_1, A_1, A_3)$ satisfying

$$[J_j, A_k] = i\hbar \epsilon_{jkl} A_l , \tag{6.31}$$

and spherical components $A_0 = A_3$ and $A_{\pm 1} = \mp \frac{1}{\sqrt{2}}(A_1 \pm i A_2)$.

Exercise 4 (Irreducible or Not?)

Is the Hamiltonian $H = \frac{p^2}{2m} - \mu \boldsymbol{S} \cdot \boldsymbol{B}$ of a free massive particle coupled to an external magnetic field $\boldsymbol{B}$ via the magnetic moment $\boldsymbol{\mu} = \mu \boldsymbol{S}$ of its spin $\boldsymbol{S}$ a scalar? An irreducible tensor operator?

State Multipoles

In the usual Hilbert space basis, the statistical operator reads

$$\rho = \sum_{m,m'} \rho_{mm'} |sm\rangle\langle sm'| . \tag{6.32}$$

Here, the ket $|sm\rangle$ transforms under the irreducible representation $\mathcal{D}^{(s)}$, whereas the bra $\langle sm'|$, as its complex conjugate, transforms under $(\mathcal{D}^{(s)})^*$, the contragredient representation. The ket-bra $|sm\rangle\langle sm'|$ transforms under the direct product $\mathcal{D}^{(s)} \otimes (\mathcal{D}^{(s)})^*$, which is reducible. One introduces therefore an ensemble of elements that do transform under the irreducible representation $\mathcal{D}^{(K)}$,

$$T_Q^K := T_Q^K(s,s) := \sum_{m,m'} (-)^{s-m} \langle ssm' -m|KQ\rangle |sm'\rangle\langle sm| \tag{6.33}$$

with $K = 0, 1, \ldots, 2s$ and $Q = -K, -K + 1, \ldots, K$. The Clebsch–Gordan coefficients $\langle s_1 s_2 m_1 m_2 | KQ \rangle$ are the coefficients of the unitary basis change from the direct product $\mathcal{H}_{s_1} \otimes \mathcal{H}_{s_1}$ towards the Hilbert subspace $\mathcal{H}_K$ of spin K that should be familiar from the addition of two spins. The CG coefficients are non-zero only if two selection rules are satisfied: (i) the two magnetic quantum numbers on the left add up to the one on the right, $m_1 + m_2 = Q$; (ii) the angular momentum on the right satisfies the triangle inequality $|s_1 - s_2| \leq K \leq s_1 + s_2$.

In our case, we do not couple two spins, but a spin and its complex conjugate. Since there is a contragredient representation in the game, the CG coefficients in (6.33) feature a characteristic minus sign in front of $-m$. The triangle selection rule implies for us $0 \leq K \leq 2s$.

The irreducible tensor operators $T_Q^K =: |KQ)$ form a basis of Liouville space that is properly orthonormal with respect to the trace scalar product of matrices:

$$(KQ|K'Q') := \mathrm{tr}\{(T_Q^K)^\dagger T_{Q'}^{K'}\} = \delta_{KK'}\delta_{QQ'} . \tag{6.34}$$

The Hermitian conjugate is $(T_Q^K)^\dagger = (-)^Q T_{-Q}^K$. Any linear operator A can be decomposed in this basis,

$$A = \sum_{KQ} A_{KQ} T_Q^K \quad \text{with} \quad A_{KQ} := (KQ|A) = \mathrm{tr}\{(T_Q^K)^\dagger A\} . \tag{6.35}$$

The irreducible components

$$\rho_{KQ} = \mathrm{tr}\{(T_Q^K)^\dagger \rho\} = \langle (T_Q^K)^\dagger \rangle \tag{6.36}$$

of the density matrix are called *state multipoles* or statistical tensors, and have been introduced already in the 1950s by Fano and Racah [20–22].

Exercise 5 (Irreducible Tensor Operators)
(0) Show that all T_Q^K except T_0^0 have zero trace and calculate the state monopole moment ρ_{00} (use $\sqrt{2s+1}\langle ssm'-m|00\rangle = (-)^{s-m}\delta_{mm'}$).
(1) Show that the irreducible vector operator is proportional to the spin operator, $T_Q^1 = c_s^{-1/2} S_Q$. Hint: consider the action of T_0^1 ($\sqrt{c_s}\langle ssm-m|10\rangle = (-)^{s-m} m$), argue with rotational invariance and fix the proportionality constant $c_s = s(s+1)d_s/3$ by computing $\mathrm{tr}[(T^1)^2]$.

Irreducible Spin Superoperators

By inserting the decomposition (6.35) on the left and right side of an arbitrary superoperator acting like $\mathcal{L}A$, one obtains

$$\mathcal{L}A = \sum_{KQK'Q'} |K'Q')\,(K'Q'|\mathcal{L}|KQ)\,(KQ|A) , \tag{6.37}$$

such that with the notation $\mathcal{L}_{K'Q',KQ} := (K'Q'|\mathcal{L}|KQ)$ the superoperator reads

$$\mathcal{L} = \sum_{KQK'Q'} \mathcal{L}_{K'Q',KQ} |K'Q')(KQ|, \tag{6.38}$$

where the Liouville-space dyadics on the right-hand side transform under $\mathcal{D}^{(K')} \otimes (\mathcal{D}^{(K)})^*$, in an analogous manner to (6.32). Following the same strategy as previously, we recouple the elements again using the appropriate CG-coefficients to get irreducible superoperators [23]

$$T_M^L(K, K') := \sum_{Q,Q'} (-)^{K-Q} \langle K'KQ'{-}Q|LM \rangle |K'Q')(KQ|. \tag{6.39}$$

Finally, any superoperator in completely decomposed form reads

$$\mathcal{L} = \sum_{LM} \sum_{K,K'} \mathcal{L}_{LM}(K, K') T_M^L(K, K') \tag{6.40}$$

with coefficients $\mathcal{L}_{LM}(K, K') = \sum_{Q,Q'} (-)^{K-Q} \langle K'KQ'{-}Q|LM \rangle \mathcal{L}_{K'Q',KQ}$. This decomposition is completely general and applies to arbitrary superoperators. It is only worth the effort, however, if the superoperator has rotational symmetries. The greatest gain in computational speed and conceptual clarity is obtained if the superoperator is a scalar such that its only non-zero component is $L = 0, M = 0$. In that case, which applies to our Lindbladian (6.24), one finds by virtue of the triangle rule that $K = K'$: as promised, $\mathcal{L}$ indeed connects subspaces of equal rank. We can choose the decomposition

$$\mathcal{L} = \sum_{K=0}^{2s} \lambda_K T^{(K)}, \tag{6.41}$$

where the $T^{(K)}$ are orthogonal projectors onto the subspaces $L(\mathcal{H}_s)^{(K)}$ of irreducible tensor operators of rank K:

$$T^{(K)} = \sqrt{2K+1}\, T_0^0(K, K) = \sum_Q |KQ)(KQ|. \tag{6.42}$$

They are indeed orthogonal, $T^{(K)}T^{(K')} = \delta_{KK'} T^{(K)}$, by virtue of (6.34), and resolve the identity in Liouville space, $\sum_K T^{(K)} = 1$, by virtue of a completeness relation of CG coefficients.

Scalar Relaxation Rates

Once the invariant subspaces are known, the eigenvalues are obtained by projecting the superoperator onto an arbitrary basis element:

$$\lambda_K = (KQ|\overline{\mathcal{L}}|KQ) = \text{tr}\left\{ (T_Q^K)^\dagger \overline{\mathcal{L}} T_Q^K \right\}. \tag{6.43}$$

For the spin relaxation Lindbladian (6.24), one must calculate commutators of the form $[S_i, [S_i, T_Q^K]]$. In order to use the defining commutation relations (6.29) and (6.30) for irreducible tensor operators, one writes the scalar product of spin operators in terms of the spherical components $S_0 = S_z$, $S_{\pm 1} = \mp \frac{1}{\sqrt{2}}(S_x \pm iS_y)$,

$$\sum_i S_i S_i = \sum_{p=0,\pm 1} (S_p)^\dagger S_p = \sum_{p=0,\pm 1} (-)^p S_p S_{-p} , \qquad (6.44)$$

such that

$$\lambda_K = -\frac{\gamma_s}{2s(s+1)} \sum_{p=0,\pm 1} (-)^p \text{tr} \left\{ (T_Q^K)^\dagger [S_{-p}, [S_p, T_Q^K]] \right\} . \qquad (6.45)$$

It is now a simple exercise to show with the help of (6.29) and (6.30) (by paying attention to the supplementary factor $\sqrt{2}\hbar$ in the definition of the raising/lowering components $J_\pm = \sqrt{2}\hbar S_{\pm 1}$) that the double commutator gives back the tensor operator itself, $[S_{-p}, [S_p, T_Q^K]] = c(p, Q, K) T_Q^K$. Summing all three terms gives remarkably simple eigenvalues,

$$\lambda_K = -\gamma_s \frac{K(K+1)}{2s(s+1)}, \quad K = 0, 1, \ldots, 2s . \qquad (6.46)$$

These real, negative eigenvalues describe a pure relaxation process as expected from the definition of the Lindbladian (6.24). They are scalar objects, also known as *rotational invariants*, and can be expressed in terms of $6j$-coefficients that are constructed out of the irreducible representations $\mathcal{D}^{(K)}$ for the state multipole, $\mathcal{D}^{(s)}$ for the spin S itself and $\mathcal{D}^{(1)}$ for its coupling to the vector field B [3, 23]. This type of consideration is of considerable importance in many different fields of physics involving angular momentum or spin; for example, relaxation coefficients very similar to (6.46) characterise spatial correlations in certain ground states of quantum spin chains [24].

Isotropic Spin Relaxation

The master equation $\partial_t \rho(t) = \overline{\mathcal{L}} \rho(t)$ separates into uncoupled equations for each invariant subspace:

$$\partial_t \rho(t) = \sum_{KQ} \partial_t \rho_{KQ}(t) T_Q^K = \sum_{KQ} \rho_{KQ}(t) \overline{\mathcal{L}} T_Q^K = \sum_{KQ} \rho_{KQ}(t) \lambda_K T_Q^K . \qquad (6.47)$$

The resulting differential equation $\partial_t \rho_{KQ}(t) = \lambda_K \rho_{KQ}(t)$ for the state multipoles is easily solved to yield a simple exponential decay

$$\rho_{KQ}(t) = e^{-\gamma_K t} \rho_{KQ}(0) \qquad (6.48)$$

with relaxation rates $\gamma_K = |\lambda_K|$. This is a particular example for a state multipole relaxation as described by Blum in Chap. 8 of [20]. The first two values deserve a special discussion. The scalar mode relaxation rate $\gamma_0 = 0$ assures the trace preservation of $\rho(t)$. At the same time, this vanishing eigenvalue is responsible for the fact that the Lindbladian is not invertible. The vector mode relaxation rate $\gamma_1 = \gamma_s/[s(s+1)] =: 1/\tau_1$ describes the relaxation of the *orientation* or average spin vector:

$$\langle \boldsymbol{S}(t) \rangle = \text{tr}\{\rho(t)\boldsymbol{S}\} = \text{e}^{-t/\tau_1} \langle \boldsymbol{S}(0) \rangle \tag{6.49}$$

since $\text{tr}\{T_Q^K S_{Q'}\}$ projects onto $K = 1$ (remember exercise 5(1) and the orthogonality relation (6.34)).

For a qubit spin $\frac{1}{2}$, this is all that needs to be calculated since any 2×2-density matrix can be parameterised as $\rho_2(t) = \frac{1}{2}\mathbb{1}_2 + \langle \boldsymbol{S}(t) \rangle \cdot \boldsymbol{\sigma}$ and thus

$$\rho_2(t) = \text{e}^{-t/\tau_1}\rho(0) + (1 - \text{e}^{-t/\tau_1})\frac{1}{2}\mathbb{1}_2 . \tag{6.50}$$

This isotropic spin $\frac{1}{2}$ relaxation therefore is for all times given by the depolarising channel (6.19) with $p_2(t) = 1 - \text{e}^{-t/\tau_1}$. The Kraus operators for the operator sum representation (6.16), valid for *all* times, follow by using $p_1(t) = \frac{3}{4}p_2(t)$:

$$W_0(t) = \frac{\sqrt{1 + 3\text{e}^{-t/\tau_1}}}{2}\,\mathbb{1}_2, \qquad W_i(t) = \frac{\sqrt{1 - \text{e}^{-t/\tau_1}}}{2}\,\sigma_i, \quad i = 1, 2, 3 . \tag{6.51}$$

Naturally, by developing these expressions to first order in $t/\tau_1 = \frac{4}{3}\gamma_s t$, one finds the Kraus operators for an infinitesimal time step derived in Sect. 6.2.2. In general, it is easy to show by derivation that to each quantum dynamical semigroup described by an exponential superoperator $\exp[\mathcal{L}t]$ corresponds a Lindblad-type master equation [25], cf. Sect. 5.3.2. However, as always, the inverse operation of integrating the infinitesimal time evolution to finite times is much harder. Deriving a set of Kraus operators for an arbitrary quantum channel in general requires the complete diagonalisation of the microscopic Hamiltonian. Luckily, for spin $\frac{1}{2}$ everything is so simple that the complete calculation is possible.

Naturally, one thus wonders whether the full-fledged angular momentum formalism is necessary at all to describe isotropic spin relaxation. A simple calculation shows that in the "depolarising channel" defined by (6.19), all non-scalar eigenvalues $\lambda_1, \lambda_2, \ldots$ are identical, such that again the Kraus operators are the generators of the $\mathsf{su}(n)$ Lie algebra. For that channel, one does not need to employ angular momentum theory, and the Kraus operators are the generators of the $\mathsf{su}(n)$ Lie algebra [13]. But please be aware that this is *not* the case for our arbitrary spin coupled to a freely fluctuating impurity spin where all higher state multipoles $K \geq 1$ come with their own different decay rates (6.46). For this channel, the author has not been able to determine the Kraus operators for finite times (but would certainly be happy to receive any valuable information on that point by his readers). In this case, there seems to be no easier way to describe the spin relaxation than in terms of the irreducible components:

$$\rho(t) = \sum_{KQ} e^{-\gamma_K t} \rho_{KQ}(0) T_Q^K .$$

(6.52)

This implies the simple exponential decay

$$A_{KQ}(t) = e^{-\gamma_K t} A_{KQ}(0)$$

(6.53)

for the irreducible components (6.35) of any observable A.

6.3 Diffusion

6.3.1 Transport

We may call "transport" a movement from a point r to a point r' that is induced by an external cause. In free space, propagation is *ballistic*: the average square of the distance covered after a time t scales like $\langle r^2 \rangle \sim v^2 t^2$, where v is the particle's velocity. A disordered medium contains *impurities* that interrupt the ballistic movement. So-called "quenched" disorder is fixed for each realisation of an experiment, but varies from experiment to experiment when samples are changed. Predictions about observables will involve an average $\overline{(\ldots)}$ by integrating over a classical disorder distribution or by tracing out uncontrolled quantum degrees of freedom. Generically, the averaged expectation value behaves as a diffusive quantity: $\overline{\langle r^2 \rangle} \sim 2Dt$ with D the *diffusion constant*.

In a hydrodynamic description, diffusion is a direct consequence of two very basic hypotheses:

(i) a local conservation law $\partial_t n + \boldsymbol{\nabla} \cdot \boldsymbol{j} = 0$, also known as the continuity equation, linking the local density $n(\boldsymbol{r}, t)$ and the local current density $\boldsymbol{j}(\boldsymbol{r}, t)$, and

(ii) a linear response relation $\boldsymbol{j} = -D\boldsymbol{\nabla} n$, known as Fourier's law in the context of heat transport and Fick's law in the context of particle diffusion.

By inserting the second relation into the first, one finds immediately the diffusion equation $(\partial_t - D\nabla^2)n = 0$. The hydrodynamic description is only valid for times and distances large compared to the scales on which microscopic scattering takes place. The linear response coefficient D has to be determined microscopically. In essence, the simplest physical process leading to diffusion is a random walk or repeated elastic scattering. A kinetic description where point particles collide with obstacles permits to derive the diffusion constant D associated with this process as function of microscopic scattering parameters. In this section, we will derive the appropriate kinetic equation for elastic momentum scattering from first principles using a master equation approach and determine the relevant diffusion constant. A largely equivalent presentation with more details can be found in Chap. 4 of the recommendable book "Quantum Transport Theory" by J. Rammer [26].

6.3.2 Momentum – a Primer on Translations

If spin is to be understood by considering rotations, then momentum is understood by considering translations.

Let a particle be prepared at a point $r_0 = 0$ with a spreading described by a function $f(r)$. After translation of the entire preparation apparatus by a vector $a \in \mathbb{R}^d$, the new position is described by the function

$$[T_a f](r) := f(r - a) \tag{6.54}$$

such that the particle is indeed centred around $r_0' = a$. In this so-called "active" formulation of translations, relation (6.54) defines the action of the translation operator T_a in a functional space of, say, probability distributions, in exact analogy to the case of rotations treated in Sect. 6.2.1. Here, it is an infinite-dimensional representation of the group $(\mathbb{R}, +)$ of real numbers with the addition "$+$" as a group law; to be precise, in d dimensions it is the d-fold direct product of such representations. Translations have the identity element $E = T_0$ and inverse $T_a^{-1} = T_{-a}$. This group is Abelian because different translations commute: $T_a T_b = T_{a+b} = T_{b+a} = T_b T_a$. Furthermore, this group is a simply connected Lie group, and all translations can be generated by exponentiation $T_a = \exp\{-ia \cdot T\}$ of $i = 1, \ldots, d$ generators T_i that form a Lie algebra.[4]

In the functional representation (6.54), all translations $T_a = \exp\{-ia \cdot T\}$ are generated by $T = -i\nabla$, the derivative with respect to the position coordinate, which becomes apparent through a Taylor series expansion

$$f(r - a) = f(r) - a \cdot \nabla f(r) + \cdots = \exp\{-a \cdot \nabla\} f(r). \tag{6.55}$$

In quantum physics, the fundamental commutation relation between position and momentum, $[\hat{r}_i, \hat{p}_j] = i\hbar \delta_{ij}$, implies that the momentum observable $\hat{p} = \hbar T$ is the translation generator $\hat{p} = -i\hbar\nabla$ in position representation. Translations are implemented by the unitary operator

---

[4] If, however, the possible positions lie on a lattice, only discrete translations by lattice vectors are allowed, and translations are a representation of $(\mathbb{Z}, +)$, the additive group of integer numbers. In an infinite volume, the translation group is not compact because it is not bounded such that its representation theory is quite different from the compact rotation groups $SU(n)$. Notably, it has no finite-dimensional irreducible representations. An exception is the discrete translations on a lattice with periodic boundary conditions, often used for classification of crystalline lattices. This group is the cyclic group, and its irreducible representations are labelled by the admissible wave vectors of the reciprocal lattice [27]. Physicists are taught to know this under the epithet of Bloch's theorem.

$$U(\boldsymbol{a}) = \exp\{-i\boldsymbol{a} \cdot \hat{\boldsymbol{p}}/\hbar\}, \tag{6.56}$$

such that the particle's position operator $\hat{\boldsymbol{r}}$ transforms as

$$\hat{\boldsymbol{r}}' = U(\boldsymbol{a})\hat{\boldsymbol{r}}U(\boldsymbol{a})^\dagger = \hat{\boldsymbol{r}} - \boldsymbol{a}, \tag{6.57}$$

quite similarly to the corresponding identity (6.2) for rotations. In contrast to the angular momentum commutation relations (6.6), the simpler version $[\hat{p}_i, \hat{p}_j] = 0$ for the translation generators reflects their Abelian structure.

The momentum operator is diagonal in the basis of momentum eigenstates, $\hat{\boldsymbol{p}}|\boldsymbol{p}\rangle = \boldsymbol{p}|\boldsymbol{p}\rangle$. Any operator O is translation invariant if and only if it commutes with the momentum operator, $[O, \hat{\boldsymbol{p}}] = 0$. Therefore, it is diagonal in the momentum representation, $\langle\boldsymbol{p}'|O|\boldsymbol{p}\rangle = \delta_{\boldsymbol{p}\boldsymbol{p}'}O_{\boldsymbol{p}}$. Notably, in absence of any external perturbation, the Hamiltonian H_0 should be translation invariant, $H_0|\boldsymbol{p}\rangle = \varepsilon_{\boldsymbol{p}}|\boldsymbol{p}\rangle$, a property sometimes referred to by the expression "the p_is are good quantum numbers". Henceforth, we will choose units such that $\hbar = 1$ and thus drop the distinction between momentum and wave vectors: $\boldsymbol{p} = \hbar\boldsymbol{k} = \boldsymbol{k}$.

The spatial form of the wave functions $\psi_{\boldsymbol{p}}(\boldsymbol{r}) = \langle\boldsymbol{r}|\boldsymbol{p}\rangle$ is fixed by their translational properties: using (6.54) together with (6.56) yields $\langle\boldsymbol{r} - \boldsymbol{a}|\boldsymbol{p}\rangle = \langle\boldsymbol{r}|[U(\boldsymbol{a})|\boldsymbol{p}\rangle] = \exp\{-i\boldsymbol{a} \cdot \boldsymbol{p}\}\langle\boldsymbol{r}|\boldsymbol{p}\rangle$ such that $\langle\boldsymbol{r}|\boldsymbol{p}\rangle = C\exp\{i\boldsymbol{r} \cdot \boldsymbol{p}\}$ up to a normalisation factor. In a finite volume $\Omega = L^d$, these plane waves are square integrable and can be properly normalised. We choose to work in the limit $\Omega \to \infty$ and fix $C = 1$. The identity is resolved by

$$\mathbb{1} = \int \mathrm{d}^d r |\boldsymbol{r}\rangle\langle\boldsymbol{r}| = \int \frac{\mathrm{d}^d p}{(2\pi)^d}|\boldsymbol{p}\rangle\langle\boldsymbol{p}|. \tag{6.58}$$

Using the plane-wave expansion, Fourier transformation is written using the convention

$$f(\boldsymbol{x}) = \int \frac{\mathrm{d}^d p}{(2\pi)^d} e^{i\boldsymbol{x}\cdot\boldsymbol{p}} f_{\boldsymbol{p}} \qquad \text{with} \qquad f_{\boldsymbol{p}} = \int \mathrm{d}^d x\, e^{-i\boldsymbol{x}\cdot\boldsymbol{p}} f(\boldsymbol{x}). \tag{6.59}$$

This choice is convenient because factors of 2π are always associated with p-integrals which, if required, can be easily converted back to finite-volume sums, $\int \frac{\mathrm{d}^d p}{(2\pi)^d} F(\boldsymbol{p}) = L^{-d} \sum_{\boldsymbol{p}} F(\boldsymbol{p})$.

6.3.3 Master Equation Approach to Diffusion

Hamiltonian

We will show that a microscopic quantum derivation of diffusive behaviour is possible starting from the single-particle Hamiltonian

$$H = H_0 + V. \tag{6.60}$$

Here the translation-invariant part H_0 describes free propagation in momentum eigenstates $|p\rangle$ with eigenenergies $\varepsilon_p = p^2/2m$; the generalisation to more general dispersion relations is straightforward. The random impurity potential

$$V = \sum_i v(\hat{r} - x_i) = \sum_i e^{-i\hat{p}\cdot x_i} v(\hat{r}) e^{i\hat{p}\cdot x_i} \qquad (6.61)$$

describes momentum scattering by an arbitrary potential $v(r)$, typically quite short-ranged, centred on random classical positions $\{x_i\}$. Deriving observable quantities will involve averages over all possible realisations of the disorder. The ensemble average of any quantity $O(\{x_i\})$ is the integral

$$\overline{O} = \int (\prod_i d^d x_i) P(\{x_i\}) O(\{x_i\}) \qquad (6.62)$$

over all inpurity positions weighted by their distribution $P(\{x_i\})$. The simplest distribution $P(\{x_i\}) = \prod_i P(x_i) = \prod_{i=1}^{N} \Omega^{-1}$ describes N uncorrelated impurities with average uniform density $n = N/\Omega$ in any finite volume Ω. This distribution will be used in the following, and we will go to the thermodynamic limit $N, \Omega \to \infty$ with fixed density n.

This description is valid if the mass M of scattering impurities is much larger than the mass m of scattered particles. This is the case for electrons scattered by lattice defects in solid-state devices or for photons scattered by cold atoms when recoil can be neglected. The impurities then have no internal dynamics and simply realise an external potential $v(r)$; this description can be obtained in the limit $m/M \to 0$ from the more general model, where also the dynamics of impurities is taken into account. Note that this limit is just the opposite of the usual picture used for quantum Brownian motion, cf. Sect. 5.4.3, where one tracks the movement of a large test particle bombarded frequently by smaller ones.

Derivation of the Master Equation

Here, we follow the standard derivation of a master equation for open quantum systems [18, 28] by adapting the Born–Markov or weak-coupling recipe to our case, cf. Sect. 5.4.1. The ensemble average (6.62) now plays the role of a trace over bath variables.

Starting from the Liouville–von Neumann equation $\partial_t \tilde{\rho}(t) = -i[\tilde{V}(t), \tilde{\rho}(t)]$ in the interaction representation $\tilde{A}(t) = U_0^\dagger(t) A U_0(t)$ and developing to second order in the interaction leads to the pre-master equation for the averaged density matrix $\overline{\tilde{\rho}(\{x_i\}, t)} =: \overline{\tilde{\rho}(t)}$:

$$\partial_t \overline{\tilde{\rho}(t)} = -i\overline{[\tilde{V}(t), \rho(0)]} - \int_0^t \overline{\left[\tilde{V}(t), [\tilde{V}(t-t'), \tilde{\rho}(\{x_i\}, t-t')]\right]} dt'. \qquad (6.63)$$

We may assume that the initial density matrix $\rho(0)$ does not depend on the disorder configuration – it may, for instance, represent an initial wave-packet prepared far from impurities whose temporal evolution we wish to follow. Then, the first term $[\tilde{V}(t), \rho(0)]$ vanishes because

$$\overline{\tilde{V}(t)} = \overline{\tilde{V}}(t) = \overline{V} \tag{6.64}$$

is a constant real number that shifts all energy levels ε_p of H_0 by a uniform amount and thus gives no contribution under the commutator.

The resulting equation is still exact, but not useful: it is not a closed equation for the ensemble-averaged $\bar{\tilde{\rho}}(t)$ since the density matrix inside the integral still depends on the disorder configuration. In order to cope with the integrand, one typically proceeds with the so-called *Born approximation*, replacing the exact density matrix inside the integral by its average:

$$\partial_t \bar{\tilde{\rho}}(t) = - \int_0^t \overline{\left[\tilde{V}(t), [\tilde{V}(t - t'), \bar{\tilde{\rho}}(t - t')]\right]} dt' . \tag{6.65}$$

Now, we are left with an effective Gaussian model of disorder since everything depends on the pair correlations $\overline{VV}$. Concerning the time dependence, we still face a difficult integro-differential equation for $\bar{\tilde{\rho}}(t)$. If the timescale of scattering is much smaller than the timescale of evolution we are interested in, we can perform the *Markov approximation* by replacing $\bar{\tilde{\rho}}(t - t') \mapsto \bar{\tilde{\rho}}(t)$ inside the integral and by letting the upper limit t of integration go to ∞ such that now we have a closed differential equation for $\bar{\tilde{\rho}}(t)$. Reverting to the Schrödinger representation we find the following master equation for scattering by fixed impurities:

$$\partial_t \bar{\rho}(t) = -\mathrm{i}[H_0, \bar{\rho}(t)] + \mathcal{D}\bar{\rho}(t) \tag{6.66}$$

with a scattering superoperator $\mathcal{D}$ defined by

$$\mathcal{D}\bar{\rho}(t) = \overline{V\bar{\rho}(t)W} + \overline{W\bar{\rho}(t)V} - \overline{VW}\bar{\rho}(t) - \bar{\rho}(t)\overline{WV} , \tag{6.67}$$

where

$$W := \int_0^\infty \tilde{V}(-t')\mathrm{d}t' = \int_{-\infty}^0 U_0^\dagger(t')VU_0(t')\mathrm{d}t' =: \sum_j \mathrm{e}^{-\mathrm{i}\hat{\boldsymbol{p}}\cdot\boldsymbol{x}_j} w(\hat{\boldsymbol{r}})\mathrm{e}^{\mathrm{i}\hat{\boldsymbol{p}}\cdot\boldsymbol{x}_j} . \tag{6.68}$$

It will become apparent in Sect. 6.3.3 that the weak coupling or Born approximation discards genuine quantum corrections and entails purely classical dynamics. Instead of "approximation", we had better speak of "simplification" because at this stage we have no means of knowing whether the resulting description is truly an approximation or perhaps qualitatively wrong. And really, in Sect. 6.4.2 we will see that quantum corrections need to be considered in phase-coherent samples.

Momentum Representation

In order to see what kind of evolution the master equation (6.66) describes, we evaluate it in the momentum representation in which the free Hamiltonian H_0 is diagonal. In the short-hand notation $|1\rangle = |\boldsymbol{p}_1\rangle$ and $\varepsilon_1 = \varepsilon_{\boldsymbol{p}_1}$, the first scattering contribution reads

$$\langle 1|\overline{V\bar{\rho}(t)W}|4\rangle = \sum_{2,3}\sum_{i,j}\overline{\langle 1|e^{-i\hat{\boldsymbol{p}}\cdot\boldsymbol{x}_i}v(\hat{\boldsymbol{r}})e^{i\hat{\boldsymbol{p}}\cdot\boldsymbol{x}_i}|2\rangle\langle 3|e^{-i\hat{\boldsymbol{p}}\cdot\boldsymbol{x}_j}w(\hat{\boldsymbol{r}})e^{i\hat{\boldsymbol{p}}\cdot\boldsymbol{x}_j}|4\rangle}\langle 2|\bar{\rho}(t)|3\rangle .$$

$$(6.69)$$

The terms with $i \neq j$ are proportional to $\overline{V}^2$ and cancel with an equivalent contribution in (6.67) that comes with a minus sign (as before, the average $\overline{V}$ gives no contribution thanks to the commutator structure of the equation of motion). In the terms $i = j$, we can take the translation operators outside the matrix elements and perform the ensemble average:

$$\sum_i \overline{e^{-i(\boldsymbol{p}_1-\boldsymbol{p}_2+\boldsymbol{p}_3-\boldsymbol{p}_4)\cdot\boldsymbol{x}_i}} = \frac{N}{L^d}\int d^dx\, e^{-i(\boldsymbol{p}_1-\boldsymbol{p}_2+\boldsymbol{p}_3-\boldsymbol{p}_4)\cdot\boldsymbol{x}}$$

$$= n(2\pi)^d\delta(\boldsymbol{p}_1+\boldsymbol{p}_3-\boldsymbol{p}_2-\boldsymbol{p}_4) .$$

$$(6.70)$$

As expected, the average over a uniform distribution restores translational invariance which is equivalent to the conservation of total momentum expressed by $(2\pi)^d\delta(\boldsymbol{p}_1+\boldsymbol{p}_3-\boldsymbol{p}_2-\boldsymbol{p}_4) =: \delta_{1+3,2+4}$.

Proceeding in the evaluation of (6.69), the first matrix element

$$\langle 1|v(\hat{\boldsymbol{r}})|2\rangle = \int d^dr\, e^{-i(\boldsymbol{p}_1-\boldsymbol{p}_2)\cdot\boldsymbol{x}}v(\boldsymbol{r}) = v_{\boldsymbol{p}_1-\boldsymbol{p}_2} =: v_{12} \qquad (6.71)$$

is the Fourier transform of the scattering potential. In the second matrix element $\langle 3|w(\hat{\boldsymbol{r}})|4\rangle = \int_{-\infty}^0\langle 3|U_0^\dagger(t')v(\hat{\boldsymbol{r}})U_0(t')|4\rangle dt'$, we can pull out the time integration

$$\int_{-\infty}^0 e^{i(\varepsilon_3-\varepsilon_4)t'}dt' = \frac{i}{\varepsilon_4-\varepsilon_3+i0} =: \Gamma_{34} . \qquad (6.72)$$

Readers with a background in perturbation theory will recognise this as the matrix element $\Gamma_{34} = i\langle\boldsymbol{p}_3|G_0^R(\varepsilon_4)|\boldsymbol{p}_3\rangle$ of the free retarded resolvent operator $G_0^R(\omega) = (\omega - H_0 + i0)^{-1}$. This gives us a hint on the applicability of the Markov approximation: the rapid timescale here is the inverse energy difference $\varepsilon_3 - \varepsilon_4$ of incident and scattered state. The effective evolution on much longer timescales into a new state $|\boldsymbol{p}_3\rangle$ is constrained by (6.72) to the energy shell ε_4 of the incident state $|\boldsymbol{p}_4\rangle$. The imaginary contribution of Γ produces the *Lamb shift* that renormalises the original energy levels, cf. Sect. 5.4.1, whereas the real part yields the relaxation rates that render the dynamics irreversible.

Altogether, this first contribution to the collision functional reads

$$\langle 1|\overline{V\bar{\rho}(t)W}|4\rangle = n\sum_{2,3}\delta_{1+3,2+4}v_{12}v_{34}\Gamma_{34}\langle 2|\bar{\rho}(t)|3\rangle . \qquad (6.73)$$

Collecting all four contributions gives

$$
\begin{aligned}
\langle 1|\mathcal{D}\bar{\rho}(t)|4\rangle = \quad & n\sum_{2,3}\Big[\delta_{1+3,2+4}v_{12}v_{34}(\Gamma_{12}+\Gamma_{34})\langle 2|\bar{\rho}(t)|3\rangle \\
& -\delta_{2,3}(|v_{12}|^2\Gamma_{21}+|v_{34}|^2\Gamma_{43})\langle 1|\bar{\rho}(t)|4\rangle\Big].
\end{aligned}
\tag{6.74}
$$

We choose to use the parametrisation

$$
\begin{aligned}
\boldsymbol{p}_1 &= \boldsymbol{p}+\boldsymbol{q}/2, & \boldsymbol{p}_2 &= \boldsymbol{p}'+\boldsymbol{q}/2,\\
\boldsymbol{p}_4 &= \boldsymbol{p}-\boldsymbol{q}/2, & \boldsymbol{p}_3 &= \boldsymbol{p}'-\boldsymbol{q}/2,
\end{aligned}
\tag{6.75}
$$

that complies with the conservation of total momentum (6.70). With this parametrisation, $v_{12}=v_{\boldsymbol{p}-\boldsymbol{p}'}=v_{34}^*$. Furthermore, energy differences become

$$
\varepsilon_1-\varepsilon_2 =: \varepsilon_{\boldsymbol{p}}-\varepsilon_{\boldsymbol{p}'}+\frac{(\boldsymbol{p}-\boldsymbol{p}')\cdot\boldsymbol{q}}{2m}
\tag{6.76}
$$

and $\varepsilon_1-\varepsilon_4=\boldsymbol{p}\cdot\boldsymbol{q}/m$, and we write the sum of matrix elements (6.72) in the form

$$
\Gamma_{12}+\Gamma_{34} =: \Gamma_{\boldsymbol{p}\boldsymbol{p}'}(\boldsymbol{q}).
\tag{6.77}
$$

Finally, disposing of the overbar $\bar{\rho}(t)\mapsto\rho(t)$, the density matrix elements $\rho(\boldsymbol{p},\boldsymbol{q},t):=\langle\boldsymbol{p}+\frac{\boldsymbol{q}}{2}|\rho(t)|\boldsymbol{p}-\frac{\boldsymbol{q}}{2}\rangle$ obey the master equation

$$
\partial_t\rho(\boldsymbol{p},\boldsymbol{q},t)=-\mathrm{i}\frac{\boldsymbol{q}\cdot\boldsymbol{p}}{m}\rho(\boldsymbol{p},\boldsymbol{q},t)+\mathcal{D}[\rho(\boldsymbol{p},\boldsymbol{q},t)]
\tag{6.78}
$$

with the scattering functional

$$
\mathcal{D}[\rho(\boldsymbol{p},\boldsymbol{q},t)]=n\int\frac{\mathrm{d}^d p'}{(2\pi)^d}|v_{\boldsymbol{p}-\boldsymbol{p}'}|^2\left[\Gamma_{\boldsymbol{p}\boldsymbol{p}'}(\boldsymbol{q})\rho(\boldsymbol{p}',\boldsymbol{q},t)-\Gamma_{\boldsymbol{p}'\boldsymbol{p}}(\boldsymbol{q})\rho(\boldsymbol{p},\boldsymbol{q},t)\right].
\tag{6.79}
$$

Trace Conservation and Continuity Equation

Clearly, the master equation (6.78) has the form of a kinetic balance equation where the scattering functional (6.79) contains transitions $\boldsymbol{p}'\to\boldsymbol{p}$ that increase the magnitude of $\rho(\boldsymbol{p},\boldsymbol{q},t)$ and also negative contributions of depleting transitions $\boldsymbol{p}\to\boldsymbol{p}'$. Since we have neither sinks nor external sources, the net effect must be zero which should be apparent in a conservation of the local probability density. And really, the master equation first of all preserves the trace, $\partial_t\mathrm{tr}\{\rho(t)\}=\int\frac{\mathrm{d}^d p}{(2\pi)^d}\partial_t\rho(\boldsymbol{p},0,t)=0$, since the scattering functional is antisymmetric under the exchange $\boldsymbol{p}\leftrightarrow\boldsymbol{p}'$ and thus

$$
\int\frac{\mathrm{d}^d p}{(2\pi)^d}\mathcal{D}[\rho(\boldsymbol{p},\boldsymbol{q},t)]=0.
\tag{6.80}
$$

This antisymmetry is inherited from the double-commutator structure of (6.65) and holds for all spatial Fourier momenta $\boldsymbol{q}$. Summing the $\boldsymbol{q}$-dependent master equation over $\boldsymbol{p}$ thus leads to the *continuity equation*

$$\partial_t n_{\boldsymbol{q}}(t) + \mathrm{i}\boldsymbol{q} \cdot \boldsymbol{j}_{\boldsymbol{q}}(t) = 0 \qquad (6.81)$$

that links the Fourier transforms of the local density $n(\boldsymbol{r}, t)$ and local current density $\boldsymbol{j}(\boldsymbol{r}, t)$, given as the first two $\boldsymbol{p}$-moments of the density distribution:

$$n_{\boldsymbol{q}}(t) = \int \mathrm{d}^d r \, e^{-\mathrm{i}\boldsymbol{q}\cdot\boldsymbol{r}} n(\boldsymbol{r}, t) = \int \frac{\mathrm{d}^d p}{(2\pi)^d} \rho(\boldsymbol{p}, \boldsymbol{q}, t) \,, \qquad (6.82a)$$

$$\boldsymbol{j}_{\boldsymbol{q}}(t) = \int \mathrm{d}^d r \, e^{-\mathrm{i}\boldsymbol{q}\cdot\boldsymbol{r}} \boldsymbol{j}(\boldsymbol{r}, t) = \int \frac{\mathrm{d}^d p}{(2\pi)^d} \frac{\boldsymbol{p}}{m} \rho(\boldsymbol{p}, \boldsymbol{q}, t) \,. \qquad (6.82b)$$

The current vanishes by parity for isotropic distributions $\rho(p, \boldsymbol{q}, t)$.

Momentum Isotropisation

What kind of dynamics does the master equation describe? A first, simple answer is possible by considering the limit $\boldsymbol{q} = 0$ that describes spatially averaged quantities. The definitions (6.72), (6.76) and (6.77) imply $\Gamma_{\boldsymbol{p}\boldsymbol{p}'}(0) = 2\pi\delta(\varepsilon_p - \varepsilon_{p'})$, which assures the conservation of energy during elastic scattering. Since the isotropic energy ε_p fixes the modulus of $\boldsymbol{p}' = p\hat{\boldsymbol{n}}'$, the angular probability distribution $\rho(p\hat{\boldsymbol{n}}, 0, t) =: f_\varepsilon(\hat{\boldsymbol{n}}, t)$ at fixed energy satisfies

$$\partial_t f_\varepsilon(\hat{\boldsymbol{n}}, t) = 2\pi n \int \frac{\mathrm{d}^d p'}{(2\pi)^d} \delta(\varepsilon - \varepsilon_{p'}) |v_{\boldsymbol{p}-\boldsymbol{p}'}|^2 [f_\varepsilon(\hat{\boldsymbol{n}}', t) - f_\varepsilon(\hat{\boldsymbol{n}}, t)] \,. \qquad (6.83)$$

For an isotropic point-scatterer potential $v(\boldsymbol{r}) = v_0\delta(\boldsymbol{r})$ that has no dependence on momentum, the equation of motion takes the simple form

$$\partial_t f_\varepsilon(\hat{\boldsymbol{n}}, t) = -\gamma_{\mathrm{el}}(\varepsilon) [f_\varepsilon(\hat{\boldsymbol{n}}, t) - \langle f_\varepsilon(t) \rangle] \,, \qquad (6.84)$$

where $\langle f_\varepsilon(t) \rangle := \int \mathrm{d}\Omega_d' f_\varepsilon(\hat{\boldsymbol{n}}', t)$ is the angular average of the distribution, properly normalised: $\mathrm{d}\Omega_d'$ is $\mathrm{d}\theta'/2\pi$ in $d = 2$ dimensions and $\mathrm{d}\phi' \mathrm{d}(\cos\theta')/4\pi$ in $d = 3$. The elastic scattering rate

$$\gamma_{\mathrm{el}}(\varepsilon) = 2\pi\nu(\varepsilon) n v_0^2 \qquad (6.85)$$

is defined in terms of the density of states $\nu(\varepsilon) = \int \frac{\mathrm{d}^d p'}{(2\pi)^d} \delta(\varepsilon - \varepsilon_{p'})$ and can equally well be obtained by Fermi's golden rule. Clearly, equation (6.83) describes a simple exponential decay of the initial angular distribution $f_\varepsilon(\hat{\boldsymbol{n}}, t = 0)$ towards a completely isotropic distribution $\langle f_\varepsilon(t) \rangle$. Thus, our model of elastic momentum scattering by fixed impurities leaves the kinetic energy conserved, but describes the isotropisation of the momentum distribution with a rate γ_{el}.

By the same token, a global net current $\boldsymbol{j}_0(t)$ initially different from zero decreases to zero exponentially fast. In other words, an initial wave packet launched with a definite velocity loses the memory of its initial direction on a timescale $\tau_{\mathrm{el}} = 1/\gamma_{\mathrm{el}}$.

Boltzmann–Lorentz Equation

In order to know on what spatial scale the momentum isotropisation occurs and how the average position of the wave packet evolves in time, we have to consider the master equation (6.78) at finite $\boldsymbol{q}$. We expect diffusive behaviour to appear on large spatial scales and thus make a Taylor expansion around $\boldsymbol{q} = 0$. Retaining only lowest-order terms in $q/p \ll 1$ (which corresponds to spatial scales much larger than the particle's wavelength), the equation of motion becomes

$$\partial_t \rho(\boldsymbol{p}, \boldsymbol{q}, t) + \mathrm{i}\frac{\boldsymbol{q} \cdot \boldsymbol{p}}{m}\rho(\boldsymbol{p}, \boldsymbol{q}, t) = \mathcal{C}[\rho(\boldsymbol{p}, \boldsymbol{q}, t)] \tag{6.86}$$

with the purely *elastic* collision integral

$$\mathcal{C}[\rho(\boldsymbol{p}, \boldsymbol{q}, t)] = 2\pi n \int \frac{\mathrm{d}^d p'}{(2\pi)^d}\delta(\varepsilon - \varepsilon_{p'})|v_{\boldsymbol{p}-\boldsymbol{p}'}|^2 [\rho(\boldsymbol{p}', \boldsymbol{q}, t) - \rho(\boldsymbol{p}, \boldsymbol{q}, t)] \; . \tag{6.87}$$

The only explicit occurrence of $\boldsymbol{q}$ in (6.86) is now in the ballistic term on the left-hand side that originates from the free evolution with H_0.

The parametrisation (6.75) has the additional advantage that the density matrix elements $\rho(\boldsymbol{p}, \boldsymbol{q}, t)$ are the spatial Fourier transform of the Wigner distribution [29]

$$\begin{aligned} W(\boldsymbol{p}, \boldsymbol{r}, t) &= \frac{1}{(2\pi\hbar)^d} \int \mathrm{d}^d r' \langle \boldsymbol{r} - \frac{\boldsymbol{r}'}{2}|\rho(t)|\boldsymbol{r} + \frac{\boldsymbol{r}'}{2}\rangle \mathrm{e}^{\mathrm{i}\boldsymbol{p}\cdot\boldsymbol{r}'/\hbar} \\ &= \frac{1}{(2\pi\hbar)^d} \int \frac{\mathrm{d}^d q}{(2\pi)^d}\mathrm{e}^{\mathrm{i}\boldsymbol{q}\cdot\boldsymbol{r}}\rho(\boldsymbol{p}, \boldsymbol{q}, t) \end{aligned} \tag{6.88}$$

that represents the quantum density operator in a classical phase space, see also Sect. 4.6. Here, we have momentarily restored $\hbar$'s visibility. With this standard normalisation, the probability density $n(\boldsymbol{r}) = \langle \boldsymbol{r}|\rho|\boldsymbol{r}\rangle$ with normalisation $\int \mathrm{d}^d r n(\boldsymbol{r}) = 1$ is given as the marginal $n(\boldsymbol{r}) = \int \mathrm{d}^d p W(\boldsymbol{p}, \boldsymbol{r})$. Conversely, $w(\boldsymbol{p}) = \int \mathrm{d}^d r W(\boldsymbol{p}, \boldsymbol{r}) = (2\pi\hbar)^{-d}\rho(\boldsymbol{p}, 0)$ is the momentum distribution, and $\int \mathrm{d}^d r \mathrm{d}^d p W(\boldsymbol{p}, \boldsymbol{r}) = 1$.

A Fourier transform with respect to $\boldsymbol{q}$ now yields the kinetic equation

$$\partial_t W(\boldsymbol{p}, \boldsymbol{r}, t) + \frac{1}{m}\boldsymbol{p} \cdot \boldsymbol{\nabla}_{\boldsymbol{r}} W(\boldsymbol{p}, \boldsymbol{r}, t) = \mathcal{C}[W(\boldsymbol{p}, \boldsymbol{r}, t)] \tag{6.89}$$

with the same elastic scattering integral (6.87). Remarkably, we have obtained precisely the linear Boltzmann–Lorentz equation for the classical phase-space density $W(\boldsymbol{p}, \boldsymbol{r}, t)$ under elastic scattering from fixed impurities [30]. For photons, this equation is known as the radiative transfer equation. Ex post, we can therefore conclude that the Born approximation (6.65) was the crucial step that discarded quantum corrections to propagation amplitudes and left us with a classical phase-space distribution. This interpretation transpires also by analysing Feynman diagrams in perturbation theory and path-integral approaches (see Chap. 4 of [26] for details).

The *weak disorder limit* in which the Born approximation and Boltzmann transport theory are expected to be valid corresponds to the regime where disorder corrections to the free energy of the particle are small, $\gamma_{\rm el}(\varepsilon) \ll \varepsilon$. The scattering time defines a typical length scale, the elastic scattering mean free path $\ell_{\rm el} = v_0\tau_{\rm el}$ or average distance between successive scattering events. The weak disorder limit can also be stated as $1/(k\ell_{\rm el}) \ll 1$ which requires that successive scatterers are placed in the scattering far field: the average distance $\ell_{\rm el}$ must be larger than the wavelength $\lambda = 2\pi/k$. This is a low-density argument because $\ell_{\rm el} = 1/(n\sigma_{\rm el})$ in terms of the density n of scatterers and their total elastic scattering cross section $\sigma_{\rm el}$.

It is allowed to neglect the explicit $\boldsymbol{q}$-dependence inside the collision integral (6.87), which contributes already a factor $\gamma_{\rm el}$, on hydrodynamic scales $q\ell_{\rm el} \ll 1$. Consistently, it is precisely in the hydrodynamic regime that we wish to determine the diffusion constant. We have already derived the continuity equation (6.81). Making step (ii) of the general argument presented in Sect. 6.3.1, we now turn to the calculation of the diffusion coefficient in the linear response regime.

6.3.4 Linear Response and Diffusion Constant

Any phase-space distribution $W_{\rm eq}(p)$ that is homogeneous, independent of time and rotation invariant, i.e., depends only on the modulus of $\boldsymbol{p}$, is a solution of the Boltzmann–Lorentz equation (6.89) since each term vanishes separately. The corresponding density matrix elements are of the form $\rho_{\rm eq}(\boldsymbol{p},\boldsymbol{q},t) = (2\pi)^d\delta(\boldsymbol{q})\rho_{\rm eq}(p)$. This type of solution is called a *global equilibrium*. The underlying statistics could be a Fermi–Dirac or Bose distribution, or their classical limit, the Boltzmann distribution.

By creating a small gradient of concentration, one can then induce a linear-response current that is proportional to the driving gradient; the coefficient of proportionality is the diffusion constant. Kinetic theory permits to calculate linear response coefficients. We will follow the linearisation method developed by Chapman and Enskog [30] in order to derive the diffusion constant, but in terms of the density matrix components $\rho(\boldsymbol{p},\boldsymbol{q},t)$ instead of the space-dependent Wigner distribution because it will prove useful to work with Fourier-transformed quantities.

Linearisation à la Chapman–Enskog

Suppose that initially, the distribution function $\rho(p,\boldsymbol{q},0)$ is a *local equilibrium* solution (i.e., isotropic in momentum) established by local scattering on a rapid timescale $\tau_{\rm el} = 1/\gamma_{\rm el}$. However, we assume a non-delta-like dependence on $\boldsymbol{q}$, i.e., a finite gradient in real space. It is therefore no longer a global equilibrium solution of the Boltzmann–Lorentz equation. The time τ_{eq} it takes to reach global equilibrium is much longer than the local scattering time such that we have a small parameter $\tau_{el}/\tau_{eq} \ll 1$ (sometimes called the "Knudsen number"). The linearisation method of Chapman and Enskog works by expanding the distribution function formally in powers of this small parameter,

$$\rho(\boldsymbol{p},\boldsymbol{q},t) = \rho_0(\boldsymbol{p},\boldsymbol{q},t) + \tau_{el}\rho_1(\boldsymbol{p},\boldsymbol{q},t) + O(\tau_{el}^2). \tag{6.90}$$

For the linear response calculation, these first two terms will suffice. The collision integral effectively multiplies the distribution by $\gamma_{el} = 1/\tau_{el}$ such that $\mathcal{C}[\tau_{el}^n\rho_n] = O(\tau_{el}^{n-1})$. Identifying equal orders on both sides of the kinetic equation, we find to order τ_{el}^{-1}:

$$0 = \mathcal{C}[\rho_0], \tag{6.91}$$

which is satisfied if $\rho_0(p,\boldsymbol{q},t)$ is *locally* isotropic. By parity, the current density (6.82b) then is entirely generated by the correction,

$$\boldsymbol{j_q}(t) = \int \frac{d^dp}{(2\pi)^d} \frac{\boldsymbol{p}}{m} \tau_{el}(\varepsilon_p)\rho_1(\boldsymbol{p},\boldsymbol{r},t). \tag{6.92}$$

The continuity equation (6.81) then implies that to lowest order, the local density remains time independent: $\partial_t\rho_0(p,\boldsymbol{q},t) = 0$. To this order $\tau_{el}^0 = 1$, the kinetic equation reduces to

$$i\frac{\boldsymbol{q}\cdot\boldsymbol{p}}{m}\rho_0(p,\boldsymbol{q}) = \mathcal{C}[\tau_{el}\rho_1(\boldsymbol{p},\boldsymbol{q})]. \tag{6.93}$$

In order to calculate the current (6.92), we need to solve this equation for ρ_1 as function of ρ_0. We content ourselves with the simple case of an isotropic point-scattering potential $v_p = v_0$ treated in Sect. 6.3.3 (other cases can be treated by an expansion in angular eigenfunctions, see [31] for the case of a potential with scattering anisotropy). Then, using the right-hand side of (6.84), we find

$$\mathcal{C}[\tau_{el}\rho_1] = -\rho_1(\boldsymbol{p},\boldsymbol{q}) \tag{6.94}$$

plus an isotropic term $\langle\rho_1\rangle$ that does not contribute to the current (6.92) anyway and can be dropped such that altogether, the distribution takes the stationary form

$$\rho_1(\boldsymbol{p},\boldsymbol{q}) = -i\frac{\boldsymbol{q}\cdot\boldsymbol{p}}{m}\rho_0(p,\boldsymbol{q}). \tag{6.95}$$

Diffusion Coefficient

Inserting (6.95) into (6.92), the resulting current reads

$$\boldsymbol{j_q} = -\frac{i}{m^2}\int \frac{d^dp}{(2\pi)^d}\boldsymbol{p}(\boldsymbol{q}\cdot\boldsymbol{p})\tau_{el}(\varepsilon_p)\rho_0(p,\boldsymbol{q}). \tag{6.96}$$

By isotropy, $\int d^d p\, p_i p_j f(p) = (\delta_{ij}/d) \int d^d p\, p^2 f(p)$, and the current is collinear with q. To lowest order in q we find

$$\boldsymbol{j_q} = -iqD_0 n_q \,, \tag{6.97}$$

the Fourier-transformed version of the linear response relation $\boldsymbol{j} = -D_0 \boldsymbol{\nabla} n$. Kinetic theory has allowed us to calculate the diffusion coefficient

$$D_0 = \frac{\overline{v_0^2 \tau}}{d} \tag{6.98}$$

as the product of an effective velocity and scattering time averaged over the momentum distribution,

$$\overline{v_0^2 \tau} = \frac{1}{m^2} \int \frac{d^d p}{(2\pi)^d} p^2 \tau_{\mathrm{el}}(\varepsilon_p) \rho_0(p) \,. \tag{6.99}$$

Often, the distribution $\rho_0(p)$ is a sharply peaked function around a certain momentum p_0 (for instance, the Fermi momentum p_{F} for electrons), whereas $p^2 \tau_{\mathrm{el}}(\varepsilon_p)$, according to (6.85), varies smoothly with the density of states such that $\overline{v_0^2 \tau} = p_0^2 \tau_{\mathrm{el}}(\varepsilon_{p_0})/m^2$. In terms of the scattering mean-free path ℓ_{el}, the diffusion constant for isotropic point scatteres can also be written

$$D_0 = \frac{\overline{v}_0 \ell_{\mathrm{el}}}{d} \,. \tag{6.100}$$

For anisotropic scattering, one has to replace the scattering mean-free path ℓ_{el} by the transport mean-free path ℓ_{tr} [31].

Diffusion

Inserting the linear response current (6.97) in the continuity equation leaves us with a simple differential equation $\partial_t n_q(t) = -D_0 q^2 n_q(t)$ that is immediately integrated to give an exponential decay

$$n_q(t) = e^{-D_0 q^2 t} n_q(0) \tag{6.101}$$

for the Fourier components of the initial density fluctuation. Long-range fluctuations ($q \to 0$) take a very long characteristic time $\tau_q = 1/(D_0 q^2) \to \infty$ to relax because of the constraint imposed by local conservation. This diagonal decomposition into Fourier modes with their continuous momentum index q is the analogue of the spin relaxation (6.53), where the discrete index K separates high-K irreducible modes with rather large relaxation rates γ_K from the isotropic component $K = 0$ or trace that is conserved.[5]

---

[5] This analogy becomes even clearer if positions in a finite volume are restricted to a lattice of L^d sites, because the irreducible representations of the discrete translation group (which under periodic boundary conditions is the d-fold direct product of the cyclic group $\mathbb{Z}_L$) are precisely labelled by the different allowed q-vectors of the reciprocal lattice [27].

The expectation value of the average radius squared,

$$\overline{\langle r^2 \rangle}(t) = \text{tr}\{r^2 \bar{\rho}(t)\} = -\nabla_q^2 n_q(t)|_{q=0} \,, \tag{6.102}$$

for the diffuse density (6.101) reads:

$$\overline{\langle r^2 \rangle}(t) = \langle r^2 \rangle_0 + 2dD_0 t \,. \tag{6.103}$$

As expected, the long-time behaviour of the particle's displacement is indeed governed by the Boltzmann diffusion constant (6.100).

6.4 Diffusive Spin Transport

Having described in Sect. 6.2 the relaxation of a single motionless spin in a fluctuating field, and in Sect. 6.3 the diffusion of spin-less massive particles, we now combine these two pictures and consider a spin on a massive carrier particle that moves and encounters impurities. The Hamiltonian is still of the form $H = H_0 + V$, where H_0 describes ballistic propagation in momentum and spin eigenstates $|p\sigma\rangle := |p\rangle \otimes |s\sigma\rangle$ with spin-independent eigenenergies $\varepsilon_p = p^2/2m$ (spin quantum numbers will from now be called σ in order to avoid confusion with the particle's mass). The impurity potential V could describe momentum scattering, spin-flip scattering, spin-orbit coupling, and the like, by randomly distributed scatterers.

In the following, we will consider in detail the case of freely orientable magnetic impurities that induce *spin-flips*. Other mechanisms such as spin-orbit scattering can be treated along the same lines. Actual laboratory realisations include electronic spin-flip scattering, quite relevant even for very low impurity concentrations, and the randomisation of photon polarisation under the influence of scattering by atoms with degenerate dipole transitions in cold atomic clouds.

6.4.1 Master Equation Approach to Diffusive Spin Transport

Deriving the Master Equation

In addition to the elastic momentum scattering potential (6.61) that for clarity we now call V_{el}, consider then a spin-flip interaction potential

$$V_{\text{sf}} = \sum_j v_{\text{sf}}(\hat{r} - x_j) S \cdot \tau_j \tag{6.104}$$

between the spin S and a collection of freely orientable magnetic impurities modelled as spin $\frac{1}{2}$ with Pauli matrices τ_j centred at sites x_j. The magnitude of spin–spin coupling has been included in the short-ranged potential $v_{\text{sf}}(r)$

whose spatial dependence induces momentum scattering. The ensemble average now contains the usual average (6.62) over random sites as well as a trace $\mathrm{tr}_{\{\tau\}}(\rho_{\{\tau\}}\cdot)$ over impurity spins. We will assume that these impurities are distributed independently and isotropically, $\rho_{\{\tau\}} = \bigotimes_j \rho_j$ with $\rho_j = \frac{1}{2}\mathbb{1}_2$.

It is now a simple task to derive a spin-diffusion master equation for the density matrix $\rho(t)$ that operates on the combined Hilbert space $\mathcal{H} = \mathcal{H}_s \otimes \mathcal{H}_p$ of spin and momentum by retracing exactly the steps of Sect. 6.3.3 with the new potential $V = V_{\mathrm{el}} + V_{\mathrm{sf}}$ instead of just V_{el}. Within the Born approximation second order in V, mixed terms $\overline{V_{\mathrm{sf}}V_{\mathrm{el}}} = \overline{V_{\mathrm{sf}}}\,\overline{V_{\mathrm{el}}}$ give no contribution (just as the product of averages before), and we can consider the impact of V_{sf} separately. A typical term arising in the new spin-flip part is the counterpart of (6.69), $\langle 1|V_{\mathrm{sf}}\rho(t)W_{\mathrm{sf}}|4\rangle$. Together with the momentum-scattering factors appearing already in (6.73), we find now an additional sum over spin indices that defines the action of a spin-flip superoperator

$$
\begin{aligned}
\langle \sigma_1|\mathcal{V}\rho|\sigma_4\rangle &= \sum_{\sigma_2,\sigma_3} \overline{\langle \sigma_1|\boldsymbol{S}\cdot\boldsymbol{\tau}|\sigma_2\rangle\langle \sigma_3|\boldsymbol{S}\cdot\boldsymbol{\tau}|\sigma_4\rangle}\langle \sigma_2|\rho|\sigma_3\rangle \\
&= \sum_{\sigma_2,\sigma_3} \boldsymbol{S}_{\sigma_1\sigma_2}\cdot\boldsymbol{S}_{\sigma_3\sigma_4}\langle \sigma_2|\rho|\sigma_3\rangle,
\end{aligned}
\tag{6.105}
$$

where the isotropic average over the impurity spin leads to the scalar contraction $\frac{1}{3}\mathrm{tr}_\tau\{\tau_i\tau_j\} = \delta_{ij}$ (cf. Exercise 1).

Collecting all terms then gives the master equation that describes elastic and spin-flip scattering in the small-$\boldsymbol{q}$ limit:

$$
\partial_t\rho(\boldsymbol{p},\boldsymbol{q},t) + \mathrm{i}\frac{\boldsymbol{q}\cdot\boldsymbol{p}}{m}\rho(\boldsymbol{p},\boldsymbol{q},t) = \mathcal{C}[\rho] + \mathcal{L}[\rho].
\tag{6.106}
$$

Here, the elastic collision integral

$$
\mathcal{C}[\rho] = \frac{\gamma}{\nu(\varepsilon)}\int\frac{\mathrm{d}^d p'}{(2\pi)^d}\delta(\varepsilon - \varepsilon_{p'})\left[\rho(\boldsymbol{p}',\boldsymbol{q},t) - \rho(\boldsymbol{p},\boldsymbol{q},t)\right]
\tag{6.107}
$$

describes momentum isotropisation with a total scattering rate

$$
\gamma = \gamma_{\mathrm{el}} + \gamma_{\mathrm{sf}}
\tag{6.108}
$$

that includes the spin-flip contribution $\gamma_{\mathrm{sf}} = s(s+1)2\pi\nu(\varepsilon)v_{\mathrm{sf}}^2 n_{\mathrm{sf}}$. Summing the rates, a prescription known as "Mathiessen's rule" [32], is permitted if the scattering mechanisms do not interfere with each other, which is the case in the low-density Born approximation considered here. Note that the collision integral (6.107) acts solely on the momentum degrees of freedom, but is the identity in spin space. Genuine spin-flips are generated by

$$
\mathcal{L}[\rho] = \frac{1}{\nu(\varepsilon)}\int\frac{\mathrm{d}^d p'}{(2\pi)^d}\delta(\varepsilon - \varepsilon_{p'})\overline{\mathcal{L}}\rho(\boldsymbol{p}',\boldsymbol{q},t),
\tag{6.109}
$$

where $\overline{\mathcal{L}} = \gamma_{\mathrm{sf}}\hat{\mathcal{L}}$ is precisely the spin relaxation Lindbladian (6.24) derived in Sect. 6.2.2, now with the spin-flip relaxation rate γ_{sf}.

Diffusive Spin Relaxation

Solving the spin-flip master equation (6.106) in the linear response regime is now exceedingly simple by projecting it onto irreducible spin components $\rho^{(K)} = \sum_Q \rho_{KQ} T_Q^K$ since the spin relaxation Lindbladian is diagonal in that basis, $\overline{\mathcal{L}}\rho^{(K)} = \lambda_K \rho^{(K)}$, with eigenvalues λ_K given by (6.46).

Summing the master equation for the Kth spin sector over p gives the continuity equation

$$\partial_t n_q^{(K)}(t) + \mathrm{i}q \cdot j_q^{(K)}(t) = -\gamma_K n_q^{(K)}(t) \tag{6.110}$$

for the density $n_q^{(K)} = \int \frac{\mathrm{d}^d p}{(2\pi)^d} \rho^{(K)}(p,q)$ and associated current density $j_q^{(K)}$. The spin-flip Lindbladian is responsible for the appearance of a source term on the right-hand side, or rather a sink with spin relaxation rate

$$\gamma_K = |\lambda_K| = \gamma_{\mathrm{sf}} \frac{K(K+1)}{2s(s+1)} . \tag{6.111}$$

In the limit $q \to 0$, we recover exactly the global spin relaxation (6.48) of Sect. 6.2.3. In particular, the total trace is conserved since the spin trace is the $K = 0$ sector with vanishing eigenvalue $\lambda_0 = 0$.

In the linear response regime, the Chapman–Enskog method of Sect. 6.3.4 carries through in each spin sector K. A difference occurs for the time derivative of the locally isotropic component $\rho_0^{(K)}(p,q,t)$. The master equation (6.106) implies

$$\partial_t \rho_0^{(K)}(p,q,t) = -\gamma_K \rho_0^{(K)}(p,q,t) , \tag{6.112}$$

which is solved by $\rho_0^{(K)}(p,q,t) = \mathrm{e}^{-\gamma_K t} \rho_0^{(K)}(p,q,0)$ instead of just being constant as in Sect. 6.3.4. Finally, the spin-sector density components show diffusive as well as spin-relaxation dynamics

$$n_q^{(K)}(t) = \mathrm{e}^{-D_0 q^2 t} \mathrm{e}^{-\gamma_K t} n_q^{(K)}(0) , \tag{6.113}$$

where the Boltzmann diffusion constant D_0 is evaluated with the total momentum relaxation time $\tau = 1/\gamma$ from (6.108).

Now we can answer the question that was the starting point of the lecture (cf. Fig. 6.1): Imagine that we can inject spin-polarised particles $|\uparrow\rangle := |s, \sigma = +s\rangle$ with probability $p_\uparrow(0) = \langle\uparrow|\rho_0|\uparrow\rangle = 1$ on one end of a diffusive medium of length L. What is the probability $p_\uparrow(L)$ of retaining the spin polarisation at the other end, assuming that we have spin-sensitive detection?

By taking the matrix elements $\langle\uparrow|\rho^{(K)}(t)|\uparrow\rangle$ of each irreducible spin component, we find that the probability relaxes during the transmission time $t = L^2/2D_0$ as

$$\begin{aligned} p_\uparrow(t) &= \frac{1}{2s+1} + \frac{3s^2}{s(s+1)(2s+1)} \mathrm{e}^{-t/\tau_1} + \dots \\ &= \frac{1}{2}(1 + \mathrm{e}^{-t/\tau_1}) \end{aligned} \tag{6.114}$$

with the last line valid for electrons for which $1/\tau_1 = 4\gamma_{sf}/3$. Equivalently, the *degree of spin polarisation*

$$\pi(t) = \frac{p_\uparrow(t) - p_\downarrow(t)}{p_\uparrow(t) + p_\downarrow(t)} = e^{-t/\tau_1} \qquad (6.115)$$

simply relaxes from unity to zero. Naturally, for long enough times, the distribution relaxes towards its equilibrium value $p_{eq} = \frac{1}{2s+1}$ to have any magnetic quantum number $\sigma = -s, \ldots, s$.

In terms of length scales, the spin relaxation time permits to define the spin relaxation length $\lambda_{sf} = \sqrt{2D_0\tau_{sf}}$. In solid-state devices, the density of magnetic impurities can be controlled such that $\gamma_{sf} \ll \gamma_{el}$ which means that spin coherence can be maintained quite efficiently, even on scales $L \gg \ell_{el}$ where the momentum dynamics is no longer ballistic, but already diffusive.

6.4.2 Quantum Corrections

In the Boltzmann transport theory developed in Sect. 6.3, one propagates effectively classical probabilities. However, in an environment that preserves the phase coherence of the wave, one must propagate probability *amplitudes* that, by the superposition principle, allow for interference phenomena. Elastic impurity scattering does preserve the phase coherence of the propagating wave which means that we have to expect quantum corrections to the Boltzmann transport theory whenever external phase-breaking mechanism are so rare that the corresponding dephasing time τ_ϕ is much longer than the elastic scattering time τ_{el}.

Weak Localisation

A prominent example for such a quantum correction is weak localisation (WL): the resistance of weakly disordered metallic samples shows a negative magnetoresistance $\partial\rho/\partial B > 0$ at small fields if spin-orbit scattering is absent [33, 34]. This contradicts the classical Boltzmann–Drude picture that predicts that the resistance should increase with a magnetic field. The reason can be understood by considering the quantum return probability to a point which includes the constructive interference of counter-propagating amplitudes and is therefore larger than the classical probability, see Fig. 6.3. The interference effect is masked by a large enough magnetic field since loops of different sizes pick up different Aharonov–Bohm phases.

Without any dephasing mechanism nor an external magnetic field, the interference correction $D = D_0 + \Delta D$ to the classical diffusion constant D_0 is given by summing the contributions of all closed diffusive paths,

$$\frac{\Delta D}{D_0} = -\frac{1}{\pi\hbar\nu(\epsilon)} \int \frac{d^d q}{(2\pi)^d} \int_0^\infty dt e^{-D_0 q^2 t} \propto \int \frac{d^d q}{(2\pi)^d} \frac{1}{D_0 q^2}. \qquad (6.116)$$

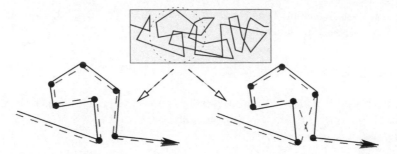

Fig. 6.3. Diffusive path (*above*) with the classical Boltzmann contribution of co-propagating amplitudes (*left*) and the corresponding counter-propagating amplitudes (*right*) that lead to weak localisation corrections

In low dimensions $d = 1, 2$, this q-integral diverges for small q which indicates that WL corrections become very important whenever the interfering amplitudes can explore large length scales. The formal divergence is cured by a cutoff that can be either the system size itself or a finite phase-coherence length $L_\phi = \sqrt{D_0 \tau_\phi}$ due to dephasing mechanisms on a timescale τ_ϕ.[6]

Dephasing of Weak Localisation by Spin-Flip Scattering

Including spin-flip scattering into the weak-localisation picture can be done by diagonalising the spin-flip vertex appropriately [23]. The irreducible subspaces turn out to be the usual singlet and triplet state subspaces spanned by the Hilbert-space vectors $|KQ\rangle$.[7] To sketch the result, the WL correction is written

$$\frac{\Delta D}{D_0} = -\frac{1}{\pi \hbar \nu(\epsilon)} \int \frac{\mathrm{d}^d q}{(2\pi)^d} \sum_{K=0}^{2s} \frac{w_K}{D_0 q^2 + \tau_c(K)^{-1}}, \qquad (6.117)$$

where each spin channel comes with a weight $w_K = (-)^{2s+K}(2K+1)/(2s+1)$. For electrons, $w_0 = -\frac{1}{2}$ and $w_1 = \frac{3}{2}$. More importantly, each spin channel is damped with its coherence time $\tau_c(K)$. They are given in terms of the spin relaxation rates (6.46) by a recoupling relation that reduces to

$$\frac{1}{\tau_c(K)} = \frac{2}{\tau_{\mathrm{sf}}} + \lambda_K = \frac{2}{\tau_{\mathrm{sf}}}\left(1 - \frac{K(K+1)}{4s(s+1)}\right) \qquad (6.118)$$

such that for electrons $\tau_c(0) = \tau_{\mathrm{sf}}/2$ and $\tau_c(1) = 3\tau_{\mathrm{sf}}/2$.

---

[6] The ultraviolet divergence for large q is cut off by the shortest scale of scattering, ℓ_{el}, and irrelevant for the present considerations.

[7] Not operators $|KQ)$ in Liouville space as for the propagated intensity: the difference is due to the fact that the two amplitudes in weak localisation loops propagate in opposite directions such that two states have to be recoupled instead of a state and its complex conjugate.

Clearly, the interference in all spin channels is dephased rather efficiently with a rate given essentially by the spin-flip rate, the numerical prefactor being entirely fixed by geometry. This effect is visible even in quite pure samples as admirably shown by F. Pierre and coworkers [35], see Fig. 6.4. The complete theory needed for linking the experimental data to impurity concentrations requires to take into account the interaction between electronic excitations near the magnetic impurity (Kondo physics) as well as the temperature-dependent dynamics of impurity spins (Korringa physics) that have both been neglected within the present lecture; for details, see [35].

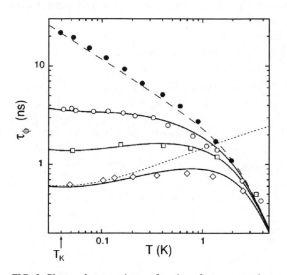

FIG. 5. Phase coherence time as function of temperature in several silver wires. Sample Ag(6N)c ($\bullet$) is made of the purest silver source. Samples Ag(5N)b ($\bigcirc$), Ag(5N)$c_{Mn0.3}$ ($\square$), and Ag(5N)d_{Mn1} ($\diamond$) were evaporated simultaneously using our 5N silver source. Afterward, 0.3 ppm and 1 ppm of manganese were added by ion implantation respectively in samples Ag(5N)$c_{Mn0.3}$ and Ag(5N)d_{Mn1}. The presence of very dilute manganese atoms, a magnetic impurity of Kondo temperature $T_K = 40$ mK, reduces τ_ϕ leading to an apparent "saturation" at low temperature. Continuous lines are fits of $\tau_\phi(T)$ taking into account the contributions of electron–electron and electron–phonon interactions (dashed line) and spin–flip collisions using the concentration c_{mag} of magnetic impurity as a fit parameter (dotted line is τ_{sf} for $c_{mag} = 1$ ppm). Best fits are obtained using $c_{mag} = 0.13$, 0.39, and 0.96 ppm, respectively, for samples Ag(5N)b, Ag(5N)$c_{Mn0.3}$, and Ag(5N)d_{Mn1}, in close agreement with the concentrations implanted and consistent with the source material purity used.

Fig. 6.4. This "set of experiments suggests that the frequently observed 'saturation' of τ_ϕ in weakly disordered metallic thin films can be attributed to spin-flip scattering from extremely dilute magnetic impurities, at a level undetectable by other means"; reprinted with permission from F. Pierre et al. [35]. Copyright (2003) by the American Physical Society

Multiple coherent scattering of photons in clouds of laser-cooled atoms is also subject to spin-flip physics: the so-called coherent backscattering effect (an interference enhancement of backscattered intensity) is severely reduced when photons are scattered by dipole transitions with a Zeeman degeneracy [36]. The analytical theory for multiple coherent scattering of polarised photons by degenerate dipole transitions employs the concepts of irreducible decompositions exposed in Sect. 6.2. Compared to electrons, the theory appears much simpler because at low light intensity, photons do not interact. However, the treatment of photon propagation is more involved because the field transversality adds another source of polarisation relaxation that needs to be taken into account [37].

Thus, magnetic impurities are a very efficient source of dephasing for interference of spin-carrying particles: the large ground-state degeneracy implied by the random orientations of freely orientable impurity spins permits dephasing even at zero temperature – when other decoherence processes like electron–phonon or electron–electron scattering are suppressed – because no energy exchange is involved and stocking which-path information comes for free. In return, whenever the impurity degrees of freedom can be constrained by other means, then perfect coherence is restored. This has been shown in Aharonov–Bohm interference experiments with electronic samples subject to a strong external magnetic field that aligns the impurity spins [38, 39]. Similarly, in atomic clouds, an external magnetic field that lifts the internal atomic Zeeman degeneracy can be used to enhance the effective phase coherence length of diffusing photons [40].

Acknowledgements

It is a pleasure to thank the organisers and participants of the summer school for providing a stimulating atmosphere of scientific exchange in which numerous questions and remarks from students and colleagues permitted me to refine my arguments. Sincere thanks to Klaus Hornberger for accompanying my first steps into Kraus representations, to R. F. Werner for bringing some of the mathematical background and reference [24] to my attention, to Michael Wolf for further clarifying discussions and to Hugues Pothier for helpful remarks concerning weak localisation measurements in mesoscopic electronic devices. Finally, I am grateful to the École normale supérieure (Paris) for its hospitality that helped me finalise these lecture notes and to Christian Wickles for a careful reading of the manuscript.

References

1. I. Žutić, J. Fabian, and S. Das Sarma: *Spintronics: Fundamentals and applications*, Rev. Mod. Phys. **76**, 323 (2004)
2. M. Chaichian and R. Hagedorn: *Symmetries in Quantum Mechanics* (IOP Publishing, Bristol 1998)

3. L. C. Biedenharn and J. D. Louck: Angular Momentum in Quantum Mechanics. In:*Encyclopedia of Mathematics*, vol. 8 (Addison Wesley, New York 1981)
4. G. E. Uhlenbeck and S. Goudsmit: *Ersetzung der Hypothese vom unmechanischen Zwang durch eine Forderung bezüglich des inneren Verhaltens jedes einzelnen Elektrons*, Naturwiss. **13**, 953 (1925)
5. W. Pauli: *Über den Zusammenhang des Abschlusses der Elektronengruppen im Atom mit der Komplexstruktur der Spektren*, Z. Phys. **31**, 765 (1925)
6. W. Gerlach and O. Stern: *Der experimentelle Nachweis der Richtungsquantelung im Magnetfeld*, Z. Phys. **9**, 349 (1922)
7. D. J. Gross: *Symmetry in Physics: Wigner's legacy*, Physics Today (December 1995) p. 46.
8. E. Cartan: *Les groupes projectifs qui ne laissent invariante aucune multiplicité plane*, Bull. Soc. Math. France **41**, 53 (1913)
9. H. Weyl: *The Theory of Groups and Quantum Mechanics* (Dover Publications, New York 1950), first published as *Gruppentheorie und Quantenmechanik* in 1928
10. H.-D. Zeh: *The Physical Basis of the Direction of Time* (Springer, New York 2001) Chap. 3
11. N. van Kampen, *Stochastic Processes in Physics and Chemistry* (North Holland, Amsterdam 1990)
12. M. A. Nielsen and I. L. Chuang: *Quantum Computation and Quantum Information* (Cambridge University Press, Cambridge 2002)
13. W. G. Ritter: *Quantum channels and representation theory*, J. Math. Phys. **46**, 082103 (2005)
14. E. Fick and G. Sauermann: *The Quantum Statistics of Dynamic Processes*, Series in Solid-State Science vol. 86 (Spinger, New York 1990)
15. R. Zwanzig: *On the identity of three generalized master equations*, Physica **30**, 1109 (1964)
16. H. Gabriel: *Theory of the influence of environment on the angular distribution of nuclear radiation*, Phys. Rev. **181**, 506 (1969), especially the appendix
17. Preskill, J.: *Quantum Information and Computing*. Lect. Notes Phys. **229**, Chap. 3 (1998), http://www.theory.caltech.edu/people/preskill/ph229/
18. C. W. Gardiner and P. Zoller: *Quantum Noise* (Springer, New York 2004)
19. S. Mukamel: *Superoperator representation of nonlinear response: Unifying quantum field and mode coupling theories*, Phys. Rev. E **68**, 021111 (2004)
20. K. Blum, *Density Matrix Theory and Applications* (Plenum Press, New York 1996)
21. U. Fano and G. Racah: *Irreducible Tensorial Sets* (Academic Press, New York 1959)
22. U. Fano: *Description of states in quantum mechanics by density matrix and Operator techniques*, Rev. Mod. Phys. **29**, 74 (1957)
23. C. A. Müller, C. Miniatura, E. Akkermans and G. Montambaux: *Mesoscopic scattering of spin s particles*, J. Phys. A: Math. Gen. **38**, 7807 (2005)
24. M. Fannes, B. Nachtergaele, and R. F. Werner: *Exact antiferromagnetic ground states of quantum spin chains*, Europhys. Lett. **10**, 633–637 (1989)
25. T. F. Havel: *Robust procedures for converting among Lindblad, Kraus and matrix representations of quantum dynamical semigroups*, J. Math. Phys. **44**, 534 (2003)
26. J. Rammer, *Quantum Transport Theory* (Perseus Books, Reading, 1998)

27. J. F. Cornwell: *Group Theory in Physics. An Introduction* (Academic Press, New York 1997)
28. H.-P. Breuer and F. Petruccione: *The Theory of Open Quantum Systems* (Oxford University Press, Oxford 2002)
29. M. Hillery, R. F. O'Connell, M. O. Scully, and E. P. Wigner: *Distribution function in physics: fundamentals*, Phys. Rep. **106**, 121–167 (1984)
30. R. Balian: *From Microphysics to Macrophysics*, vol. II (Springer, New York 1992)
31. M. Le Bellac, F. Mortessagne, and G. Batrouni: *Equilibrium and Nonequilibrium Statistical Thermodynamics* (Oxford University Press, Oxford, 2002)
32. N. W. Ashcroft and N. D. Mermin: *Solid State Physics* (Harcourt Brace, New York 1976)
33. G. Bergmann: *Weak localization in this films: a time-of-flight experiment with conduction electrons*, Phys. Rep. **107**, 1 (1984)
34. W. Zwerger: Theory of coherent transport. In: *Quantum Transport and Dissipation*, ed. T. Dittrich et al., (Wiley–VCH, New York 1998)
35. F. Pierre, A. B. Gougam, A. Anthore, H. Pothier, D. Estève, and N. Birge, *Dephasing of electrons in mesoscopic metal wires*, Phys. Rev. B **68**, 085413 (2003)
36. G. Labeyrie, D. Delande, C. A. Müller, C. Miniatura, and R. Kaiser: *Coherent backscattering of light by cold atoms: theory meets experiment*, Europhys. Lett. **61**, 327 (2003)
37. C. A. Müller and C. Miniatura: *Multiple scattering of light by atoms with internal degeneracy*, J. Phys. A: Math. Gen. **35**, 10163 (2002)
38. S. Washburn and R. Webb: *Aharonov–Bohm effect in normal metal: quantum coherence and transport*, Adv. Phys. **35**, 375 (1986)
39. F. Pierre and N. Birge: *Dephasing by extremely dilute magnetic impurities revealed by Aharonov–Bohm oscillations*, Phys. Rev. Lett. **89**, 206804 (2002)
40. O. Sigwarth, G. Labeyrie, T. Jonckheere, D. Delande, R. Kaiser, and C. Miniatura: *Magnetic field enhanced coherence length in cold atomic gases*, Phys. Rev. Lett. **93**, 143906 (2004)

Index